Advances in Simulation

Volume 6

Series Editors:

Paul A. Luker
Bernd Schmidt

Advances in Simulation

R.P. van Wijk van Brievingh   D.P.F. Möller
Editors

Biomedical Modeling and Simulation on a PC

A Workbench for Physiology and Biomedical Engineering

With 170 Figures and Six 5¹/₄″ Diskettes

Springer-Verlag
New York Berlin Heidelberg London Paris
Tokyo Hong Kong Barcelona Budapest

Editors:

Rogier P. van Wijk van Brievingh
Department of Biophysics
Faculty of Medicine and Health
  Sciences
Al Ain, United Arab Emirates

Dietmar P.F. Möller
Institut für Informatik
Technische Universität Clausthal
D-3392 Clausthal-Zellerfeld
Germany

Series Editors:

Paul A. Luker
Department of Computer Science
California State University, Chico
Chico, CA 95929-0410
USA

Bernd Schmidt
Institut für Informatik
Universität Erlangen-Nürnberg
Erlangen, Germany

Library of Congress Cataloging-in-Publication Data
Biomedical Modeling and Simulation on a PC
    Rogier P. van Wijk van Brievingh, Dietmar Möller, editors.
      p. cm. — (Advances in simulation; v. 6)
    Includes bibliographical references and index.
      1. Physiology—Computer simulation.   2. Biotechnology—computer
    simulation.   3. Microcomputers.   I. Wijk van Brievingh, Rogier P. van.
    II. Möller, Dietmar.   III. Series.
    QP33.6.D38B68   1991
    612'.001'13—dc20   91-20615

Printed on acid-free paper.

© 1993 Springer-Verlag New York, Inc.
Softcover reprint of the hardcover 1st edition 1993

Additional material to this book can be downloaded from http://extras.springer.com.

Production managed by Francine Sikorski; manufacturing supervised by Vincent Scelta.
Camera-ready copy prepared by the editors.

9 8 7 6 5 4 3 2 1

ISBN-13:978-1-4613-9165-4   e-ISBN-13:978-1-4613-9163-0
DOI: 10.1007/978-1-4613-9163-0

Series Preface

I have long had an interest in the life sciences, but have had few opportunities to indulge that interest in my professional activities. It has only been through simulation that those opportunities have arisen. Some of my most enjoyable classes were those I taught to students in the life sciences, where I attempted to show them the value of simulation to their discipline. That there is such a value cannot be questioned. Whether you are interested in population ecology, pharmacokinetics, the cardiovascular system, or cell interaction, simulation can play a vital role in explaining the underlying processes and in enhancing our understanding of these processes. This book comprises an excellent collection of contributions, and clearly demonstrates the value of simulation in the particular areas of physiology and bioengineering.

My main frustration when teaching these classes to people with little or no computer background was the lack of suitable simulation software. This directly inspired my own attempts at producing software usable by the computer novice. It is especially nice that software is available that enables readers to experience the examples in this book for themselves.

I would like to congratulate and thank the editors, Rogier P. van Wijk van Brievingh and Dietmar P.F. Möller, for all of their excellent efforts. They should be proud of their achievement.

This is the sixth volume in the Advances in Simulation series, and other volumes are in preparation. Clearly, the series is now well established, and we are constantly seeking to expand it. We would like to cover all aspects of advances in simulation, whether theoretical, methodological, or hardware or software related. An important part of publishing material that constitutes an advance in some discipline is to make the material available while it is still of considerable use. The editorial and production staff at Springer-Verlag see to it that this is the case. I urge anybody who is eager to share their advances in simulation to contact Bernd Schmidt or myself. We would love to hear from you, whether you propose a single-author monograph or, as in this volume, a compilation of contributions by many authors.

Chico, California   Paul A. Luker

Preface

Simulation of physiological processes and bioengineering concepts is a subject in most conferences on simulation (with sessions on biomedical engineering) or on bioengineering (with sessions on simulation). Generally, however, educational aspects are not explicitly presented. Thus, the editors have sought colleagues interested in the presentation of their results for educational purposes and are happy to have found so many renowned researchers willing to contribute to this book.

At preliminary meetings with the authors, it was concluded that

— simulation on a personal computer offers a good educational setting for "investigative learning" in physiology and biomedical engineering courses;
— a sufficient variety of published and validated material is available;
— simulation exercises with the models available could best be designed by the model-makers themselves;
— corresponding chapters should be written, with reference to standard textbooks for background information;
— chapters on general aspects should be added by experts;
— a user interface should be designed, from which all simulations could be run, and should include context-sensitive "help"; and
— a suitable simulation language should be chosen.

The editors are happy to have met with such great enthusiasm from the authors. We gratefully acknowledge the support received from the foundation "Meducation"; from Ms. Erika van Verseveld, who prepared the camera-ready copy; and from the software house "Boza" for permission to include the simulation tool BIOPSI.

Al Ain, United Arab Emirates   Rogier P. van Wijk van Brievingh
Clausthal-Zellerfeld, Germany   Dietmar P.F. Möller

Contents

Contributors

Jos E.C.M. Aarts, M.S., Chairperson, Department of Medical Informatics, Hogeschool Midden Nederland, Sektor Ontwikkeling, Larikslaan 10, NL 3833 AM Leusden, The Netherlands.

Matheus G.J. Arts, Ph.D., Associate Professor of Biophysics, Department of Biophysics, University of Limburg, P.O. Box 616, NL 6200 MD Maastricht, The Netherlands.

Jan E.W. Beneken, Ph.D., M.S.E.E., Professor of Biomedical Electrical Engineering, Department of Electrical Engineering, Eindhoven University of Technology, P.O. Box 513, NL 5600 MB Eindhoven, The Netherlands.

Aad Berkenbosch, Ph.D., B.S.C.E., Lecturer and Researcher, Department of Physiology, Faculty of Medicine, Leiden University, P.O. Box 9604, NL 2300 RC Leiden, The Netherlands.

Jacob de Goede, Ph.D. M.S.C.E., Associate Professor of Physiology, Department of Physiology, Faculty of Medicine, Leiden University, P.O. Box 9604, NL 2300 RC Leiden, The Netherlands.

Dirk A. de Jong, B.S.E.E., Research Engineer, Department of Neurology, Faculty of Medicine, Erasmus University, P.O. Box 1738, NL 3000 DR Rotterdam, The Netherlands.

Mark J. de Leeuw van Weenen, M.S.A.P., Research Assistant, Laboratory for Biomedical Electrical Engineering, Department of Electrical Engineering, Delft University of Technology, P.O. Box 5031, NL 2600 GA Delft, The Netherlands.

Antoine J.C. de Reus, Research Assistant, Laboratory for Biomedical Electrical Engineering, Department of Electrical Engineering, Delft University of Technology, P.O. Box 5031, NL 2600 GA Delft, The Netherlands.

Richard J.M.G. de Zwart, M.S.Biol., Research Assistant, Department of Physiology, Faculty of Medicine, Leiden University, P.O. Box 9604, NL 2300 RC Leiden, The Netherlands.

Ignacio A. García Alves, M.S.E.E., Research Assistant, Laboratory for Biomedical Electrical Engineering, Department of Electrical Engineering, Delft University of Technology, P.O. Box 5031, NL 2600 GA Delft, The Netherlands.

Ben J. Jansen, Ph.D., Assistant Professor of Physics, Institute for Educational Physics, Faculty of Physics and Astronomy, University of Utrecht, Transitorium I, Leuvenlaan 21, 3584 CE Utrecht, The Netherlands.

Eugene J.H. Kerckhoffs, Ph.D., M.S.A.P., Associate Professor of Informatics, Department of Technical Mathematics and Informatics, Delft University of Technology, Julianalaan 132, NL 2628 BL Delft, The Netherlands.

Henk Koppelaar, Ph.D., Professor of Informatics, Department of Technical Mathematics and Informatics, Delft University of Technology, Julianalaan 132, NL 2628 BL Delft, The Netherlands.

Erik W. Kruyt, M.S.E.E., Lecturer, Head, Information Processing Group, Department of Physiology, Faculty of Medicine, Leiden University, P.O. Box 9604, NL 2300 RC Leiden, The Netherlands.

Fredericus B.M. Min, Ph.D., M.S.E.E., Associate Professor of Instrumentation Technology, Department of Applied Education Science, Twente University, P.O. Box 217, NL 7500 AE Enschede, The Netherlands.

Patrick Min, Research Assistant, Department of Computer Science, Leiden University, Niels Bohrweg 1, NL 2333 CA Leiden, The Netherlands.

Dietmar P.F. Möller, Ph.D., M.S.E.E., Professor of Informatics, Institut für Informatik, Technische Universität Clausthal, Clausthal-Zellerfeld, Germany; address for correspondence: Product Group Anaesthesia, Drägerwerk AG, Moislinger Allee 53-56, D 2400 Lübeck 1, Germany.

Frank J. Pasveer, M.S.E.E., Scientific Advisor, Admiraal Trompstraat 79, 3333 TQ Zwijndrecht, The Netherlands.

Jiří Potůcěk, Ph.D., M.S.E.E., Scientific Director, MediSoft Computer Servis Centrum, Hajceka 10/1651, CS 100 000 Prague 10, Czechoslovakia.

Lourens J. van Briemen, B.S.A.P., Research Engineer, Department of Neurology, Faculty of Medicine, Erasmus University, P.O. Box 1738, NL 3000 DR Rotterdam, The Netherlands.

Paul P.J. van den Bosch, Ph.D., M.S.E.E., Professor of Control Engineering, Control Laboratory, Department of Electrical Engineering, Delft University of Technology, P.O. Box 5031, NL 2600 GA Delft, The Netherlands.

Bert van Duijn, M.S.Biol., Research Assistant, Department of Physiology, Faculty of Medicine, Leiden University, P.O. Box 9604, NL 2300 RC Leiden, The Netherlands.

Willem A. van Duyl, Ph.D., M.S.E.E., Associate Professor of Medical Technology, Laboratory for Biological and Medical Physics and Technology, Faculty of Medicine, Erasmus University, P.O. Box 1738, NL 3000 DR Rotterdam, The Netherlands.

Gerard L. van Eendenburg, M.S.Airospace.E., Research Assistant, Laboratory for Biomedical Electrical Engineering, Department of Electrical Engineering, Delft University of Technology, P.O. Box 5031, NL 2600 GA Delft, The Netherlands.

John H.M. van Eijndhoven, Ph.D., M.S.M.E., Research Engineer, Central Department of Automation Informatics, Academic Hospital Dijkzigt, dr. Molewaterplein 40, NL 3015 GD Rotterdam, The Netherlands.

Rogier P. van Wijk van Brievingh, Ph.D., M.S.E.E., Professor of Biomedical Engineering, Department of Biophysics, Faculty of Medicine and Health Sciences, Al Ain, United Arab Emirates; address for correspondence: Department of Electrical Engineering, Delft University of Technology, P.O. Box 5031, NL 2600 GA Delft, The Netherlands.

Alettus A. Verveen, Ph.D., Professor of Physiology, Department of Physiology, Faculty of Medicine, Leiden University, P.O. Box 9604, NL 2300 RC Leiden, The Netherlands.

Karel H. Wesseling, M.S.E.E., Head, Biomedical Instrumentation Group, Organization for Applied Research TNO, c/o Academical Medical Centre, Meibergdreef 15, NL 1105 AZ Amsterdam, The Netherlands.

Zdenek Wünsch, M.D., Associate Professor of Cybernetics, Department for Biocybernetics, Institute of Physiology, Charles University, Albertov 5, CS 128 00 Prague 2 Nove Mesto, Czechoslovakia.

Dirk L. Ypey, Ph.D., Professor of Physiology (Twente University), Department of Physiology, Faculty of Medicine, Leiden University, P.O. Box 9604, NL 2300 RC Leiden, The Netherlands.

1
State of the Art and Future Aspects of Modeling and Simulation in Physiology and Biomedical Engineering

Dietmar P.F. Möller

1.1 Introduction

During the past 20 years sophisticated models have been developed for a large variety of physiological systems. In the meantime model-building and simulation of physiological systems have widely been accepted as an interdisciplinary method by biomedical scientists, requiring direct cooperation between engineers, computer science experts, applied mathematicians, physiologists and physicians. Hence mathematical, biochemical or physical models of physiological systems, e.g. the circulatory system, the renal system, the respiratory system, the endocrine system, and the thermoregulatory system, have been derived and their dynamical behaviour has been studied by simulation. Moreover these models have also been successfully applied to parameter estimation techniques either to optimize the model behaviour by optimizing the parameter sets used, or to estimate even those parameters, not directly measurable, that are important for closed loop monitoring concepts in health care units.

These brief comments show why it is helpful to model and simulate physiological systems. The common problems arising from modeling and simulation in engineering and in biomedical or life science indicate the possibility of applying the same methods to a wide rang of disciplines, while at the same time increasing the cooperation between these different fields of study. However, modeling and simulation of physiological systems is generally the reverse of the simulation process in engineering science and systems theory, since the physiological systems have already been created and optimized by evolution.

Under normal circumstances, the biomedical scientist is not solely interested in the mathematical model of a physiological system under healthy or orthological conditions. He might prefer the developed mathematical model that adequately describes pathological behaviour in disease. Hence a disease can roughly be seen as a system of behaviour outside normality.

In contrast, the main goals in engineering applications are systems synthesis and optimization, and the systems engineer is primarily interested in the mathematical model of a process under normal operating conditions. His aim is to use the model in order to optimally control the process or at least to keep it within a range of conditions that ensures safe operation or at least to keep it in a range of conditions that avoids the possibility of drifting near the boundaries of conditions [Ingram, 1985, Möller, 1981].

The simulation process of physiological systems is an iterative one, consisting of model-building and computer-assisted simulation by changing the structure of the model and its parameters in an effort to match the real biological system. In fact, the derived model has served its purpose when an optimal match is obtained between the simulation results and the data obtained from the real physiological system under test. This investigation technique is the basis for the development of models of physiological systems.

1.2   Derivation of Model Equations of Physiological Systems

1.2.1   General Aspects

In general the model-building process of a physiological system entails the utilization of two types of information:

- a-priori physiological knowledge of the system being modeled (the physiological system must be observable)
- experimental data consisting of measurements of the system inputs and outputs.

With respect to the spectrum of available models, many levels of conceptual and mathematical representation are evident, depending on the purpose for which the model was intended and the extent of the a-priori physiological knowledge available. These different levels of representation are shown in the following chapters, outlining the models in detail. Two major facts are of importance when modeling a physiological system:

- a model always is a simplification of reality, but it should never be so simple that its answers are not true
- a model has to be simple and easy to use.

These are the two relevant boundary conditions for model-building, because it is a compromise between model goodness (i.e. the exactness of the results obtained from the model) and expenditure of modeling (i.e. the cost for developing the model, its implementation on the computer and its simulation, which is shown in Fig. 1.1).

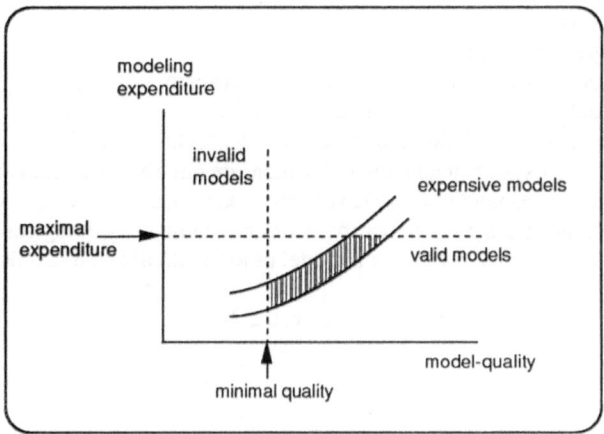

Figure 1.1. Dependence of the modeling expenditure (costs) versus the degree of
accuracy (model quality)

From Fig. 1.1 it can be concluded that there is no reason to develop expensive
models, because the increment of goodness is less than the increase of costs. This
point is of importance since a mathematical model is a very compact way to
describe a physiological system: a complex model describes not only the
relationships between the system inputs and outputs, but also gives a detailed
insight into the system structure and internal relationships.

This is because the main relationships between the physical variables of the
physiological system to be modelled are mapped into appropriate mathematical
expressions. For instance, the relation between input- and output-variables of a
system will be described – depending on the system's complexity – by an ordinary
second-order differential equation or a set of first-order differential equations that
represent the mathematical model of a physiological system. In principle, there
exist two different approaches in obtaining a system model: a pure theoretical one
based on the derivation of the essential physical (including the chemical and
biochemical) relationships within the system, and a pure empirical one based on
experiments on the system itself. Practical approaches use a combination of both,
which might be the most advantageous. Concerning the theoretical derivation of
model equations for a physiological system, as for example the balance equations
of the system, we must take into account, the law of conservation of mass, energy
and momentum, as well as some further phenomenological laws typical of the
system under test, and the existing boundary conditions.

Normally there are different ways of describing a physiological system mathematically.

The most common are:
- state space description
- input-output description in the frequency domain
- input-output description in the time domain
- the operator formulation associated with Rosenbrock and Wolvowich.

For describing a system mathematically, there are three basic notations: input, state and output. Correspondingly we have a state space X, a set U of input values and a set Y of output values. Systems can be static or dynamic. Physiological systems are normally dynamic. A system (or a model of the system) is a dynamic one if

$$S = (X, Y, U, v, t, a, b)$$

with
- state space X
- set of output values Y
- set of input values U
- set of admissible controls v
- time domain t
- state transition map a
- read out map b,

or direct expressed in the state-space description:

$$\dot{\underline{X}} = \underline{f}(\underline{X}, \underline{U}, Z, \underline{\Theta}_s)$$

with the state vector $\underline{X}$, which are the system variables such as systemic pressure, pulmonary pressure, central venous pressure, intracranial pressure, blood volume, and extracellulary fluid volume. $\underline{U}$ is the control vector of the system – tonus, heart rate, muscle strength – and $\underline{\Theta}_s$ is the system parameter vector, based on passive elements such as the compliance, the resistance, and the inertia of blood. Z is the system excitation, e.g. stress, workload, pharmacokinetic stimulation, etc.

The block-diagram (Fig. 1.2) in state-space notation of the cardiovascular system illustrates the general aspects outlined above.

In Fig. 1.2 the state vector

$$\underline{X} = [PAS, PVS, PAP, PVP]^T$$

describes the transient behavior of the arterial systemic pressure (PAS), the central venous pressure (PVS), the arteriopulmonary pressure (PAP), and the venopulmonary pressure (PVP). The control vector $\underline{U}$ is based upon the heart frequency (HF) and the peripheral resistance (RA) of the systemic vascular pathway. The system parameter vector $\underline{\Theta}_s$ describes the capacitive and resistive elements of the cardiovascular system. The sensing of PAS by the baroreceptor at the afferent pathway to the centers of the medulla can be modeled by the linear output equation

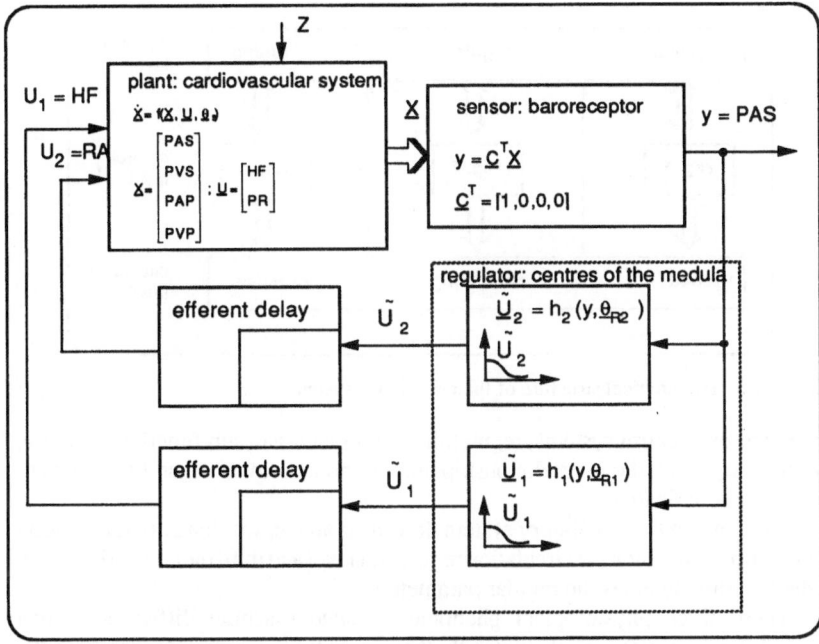

Figure 1.2. Block-diagram of the cardiovascular system

$$y = \underline{c}^T \underline{x}$$

$$\underline{c}^T = [1,0,0,0].$$

The centers of the medulla are interpreted as the system-controller with PAS as controlled variable modeled by the two non-linear controller-equations

$$\tilde{U}_1 = h_1 (y, \underline{\Theta}_{R1})$$

$$\tilde{U}_2 = h_2 (y, \underline{\Theta}_{R2})$$

with $\underline{\Theta}_{R1}$ and $\underline{\Theta}_{R2}$ as the respective controller parameter vectors. The two system control variables $\tilde{U}_1$ and $\tilde{U}_2$ are given by the controller outputs $\tilde{U}_1$ and $\tilde{U}_2$ respectively but delayed by the efferent pathway [Möller, 1981].

It is easy to conclude from this example that modeling is not a simple task since it requires a detailed knowledge of physiology and general physical laws. A further, inherent difficulty of modeling a physiological structure is the consequence of the hierarchical system structure. Generally speaking, each cell of a living organism is a small part of the complete system and is under the supervision of many local control systems for regulation of individual physical, chemical and biochemical processes. At the next hierarchical level, each organ of the organism is supervised

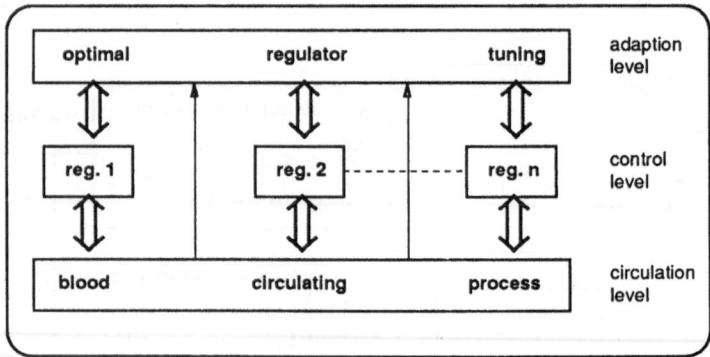

Figure 1.3. Hierarchical structure of the circulatory system

by numerous control systems, regulating individual organ sub-functions. Finally, at the body level, the interrelationship among the individual organ functions are centrally coordinated.

Using again the circulatory system described above, we find at first the blood circulation level, then the circulation regulation level and third the adaptation level, which optimally tunes the regular parameters.

These three physiological phenomena create essential difficulties when describing the systems mathematically that are, at least in principle, related to the problem of dimensionality of the system model. The complexity of a living system, such as the circulatory system, needs a nonlinear model of extremely high order to describe it precisely; the capillary system itself requires the model dimension of 1000 or more.

Besides the mathematical difficulties already discussed here, there are some experimental difficulties due to the limitations in excitation of living systems and data collection in vivo. This shows the importance of using simulation to investigate physiological systems, since a biological simulator can be excited as desired, but never a patient.

1.2.2   Some Modeling Concepts

In this section we are introducing some useful systems concepts to model physiological processes. From a more general point of view, there are three types of concepts, which depend on a-priori knowledge based on

- knowledge of the inputs;
- knowledge of the outputs; or
- knowledge of the system state

for the decision of the unknowns, as shown in Fig. 1.4.

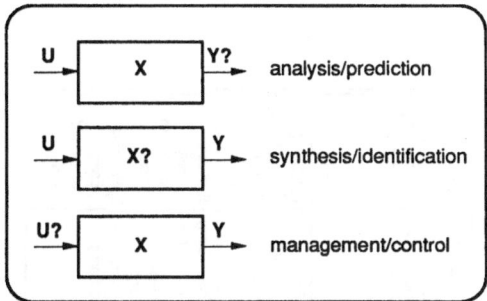

Figure 1.4. Systems concepts

When modeling physiological systems, the bioengineer mostly lacks the physicians profound knowledge of biology whereas the physician does not have the bioengineer's profound knowledge of systems theory. But using equivalent elements to describe a system is of great advantage both of them. These elements are, for instance, the replacement of the blood flow QB or the respiration flow QR by current i. Hence a table of correspondences is helpful in creating models of physiological systems, given as follows in Table 1.1.

With respect to this table it is easy to model a physiological process based on the physical laws that can be shown in detail by an example. We assume that the systemic pathway of the cardiovascular system can be represented by a compartment model of two elastic reservoirs connected with a column of blood in a vessel as shown in Fig. 1.5. This could be accepted as a rough description of the "Windkessel" effect of the aortic arch, the column of blood in the descending aorta and the abdominal aorta, and its branching into many parallel vessels of the systemic circulation [Ingram, 1985; Möller, 1981].

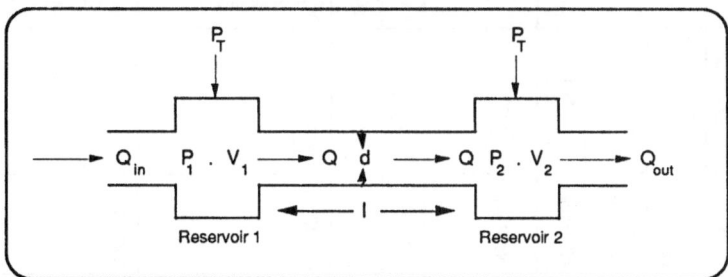

Figure 1.5. Simple mechanical model of the circulatory system

The description of the model is based on the elastic reservoir theory of the circulatory system. It represents the aorta and the main arteries as a compression chamber filled rhythmically, by the ejected blood-flow Q, in accordance with the heart contraction. Due to the variable distensibility of the walls of the blood vessels, the pulse-shaped blood flow caused by the rhythmical heart actions is converted

Table 1.1

Physically System	General Description	Electrical	Hydraulical	Pneumatical	Thermal	Translational	Rotational
Transversal Variable e(t)	Voltage, Pressure Velocity	$U(t)$; Voltage	$P(t)$; Pressure	$P(t)$; Pressure	$T(t)$; Temperature	$V(t)$; Velocity	$\omega(t)$; Angular Velocity
Transit Variable f(t)	Current, Flow Force, Momentum	$i(t)$; Current	$\dot{V}(t)$; Volume Flow	$\dot{m}(t)$; Mass Flow	$\dot{q}(t)$; Heat Flow	$f(t)$; Force	$M(t)$; Torque
e(t) Product	Power supplied to the element	$p(t)=u(t)\cdot i(t)$	$p(t)=P(t)\cdot \dot{V}(t)$	$p(t)=P(t)\cdot \dot{m}(t)$	$p(t)=\dot{q}(t)$	$p(t)=V(t)\cdot f(t)$	$p(t)=\omega(t)\cdot M(t)$
e(t) Relation	Power Consumption $e(t)=R\cdot f(t)$	R; Electrical Resistance	$R=\dfrac{8l\eta}{\pi r^4}$ Flow-resistance	identical to hydraulical	Thermal Resistance $R_F=\frac{1}{\lambda A}$ (Flow) $R_T=\frac{1}{\alpha A}$ (Transm.) $R_C=\frac{1}{\kappa c}$ (Convect.)	d^{-1}; Damping-factor	d_r^{-1}; Damping-factor
$\int e(t)\,dt$	$F(t)=1/L\cdot\int e(t)\,dt$	L; Inductor	$\dfrac{\rho l}{\pi r^2}$; Inertance	$\dfrac{\rho l}{\pi r^2}$; Inertance	———	c^{-1}; Spring-constant	c_r^{-1}; Spring-constant
$\int f(t)\,dt$	$e(t)=1/C\cdot\int f(t)\,dt$	C; Capacitor	$\dfrac{A}{\rho g}$; Hydraulic Capacity	$\dfrac{m_o}{\delta_o}=\dfrac{V}{R\cdot T}$; Pneumatic Capacity	$m\cdot c$; Thermal Capacity	M; Mass	θ; Moving Mass
$\int e(t)\,f(t)\,dt$	Energy done on system	E_M;Magnetic Energy of inductor E_E;Electric Energy of capacitor	E_K;Kinetic Energy of fluid flow E_P;Potential Energy of pressure head	E_K;Kinetic Energy of pneumatic flow E_P;Potential Energy of pressure	E_P;Thermal Potential Energy of stored heat	E_K;Kinetic Energy of moving mass E_P;Potential Energy of compressed Spring	E_K;Kinetic Energy of rotating mass E_P;Potential Energie of twisted spring
Symbols		R, L, C	R, L, C	C	T_1, $\frac{1}{g}R$, T_2 ; T_1, C, T_2	R, L, M, C	R, L, M, C

into a continuous one, i.e. the vascular volume flow Q and the peripheral flow rate Q_{out}. In Fig. 1.5 P_1, P_2 and V_1, V_2 represent the pressures and the volumes of the two reservoirs respectively. Supposing a linear relationship between the reservoir volume and the pressure difference to the surroundings, we come to the following conclusions for the two reservoirs:

$$V_1 = V_{10} + C_1 (P_1 - P_T) \tag{1.1}$$

$$V_2 = V_{20} + C_2 (P_2 - P_T); \quad P_T = \text{const.,} \tag{1.2}$$

where V_{10} and V_{20} are the unstretched reservoir volumes, C_1 and C_2 are the compartmental compliances, and P_1 and P_2 are the respective compartment pressures. P_T represent the transmural tissue pressure of the reservoirs.

The transmural tissue pressure, also called solid tissue pressure, is the pressure difference between the inner side and the outer side of the tissue wall ($P_T = P_1 - P_0$). Due to the fact that each vascular bed has its particular resistive properties, it is reasonable to introduce the peripheral resistance R_A, which connects the second reservoir segment with the systemic veins. Assuming for the venous pressure $P_3 = 0$, we come to the following conclusions:

$$R_A = \frac{P_2}{Q_{out}} \tag{1.3}$$

The differentiation of (1.1) and (1.2) results in

$$\dot{V}_1 = C_1 \dot{P}_1 - C_1 \dot{P}_T \tag{1.4}$$

and

$$\dot{V}_2 = C_2 \dot{P}_2 \tag{1.5}$$

Applying the principle of conservation of mass for the reservoirs, we get

$$Q_{in} - Q = C_1 \dot{P}_1 - Q_A , \tag{1.6}$$

and

$$Q_{in} - Q = C_1 \dot{P}_1 - Q_A , \tag{1.7}$$

whereby the definition

$$Q_A = C_1 \dot{P}_T \tag{1.8}$$

has been used.

Supposing that the blood flows in a rigid tube of length ℓ and of uniform cross-sectional area $A = 0{,}25\pi\, d_2$, the law of conservation of momentum:

$$m\dot{v} = F_1 - F_2 , \qquad (1.9)$$

with m, $\dot{v}$, F_1 and F_2 as mass, acceleration and forces acting on the system, can be written as

$$\rho \cdot A \cdot \ell \cdot \frac{d}{dt}\left(\frac{Q}{A}\right) = P_1 \cdot A - P_2 \cdot A$$

where ρ is the density of blood, and Q/A is the velocity of blood.

Introducing finally

$$M = \frac{\rho \cdot \ell}{A} \qquad (1.10)$$

the last equation can definitely be written as

$$M\dot{Q} = P_1 - P_2 . \qquad (1.11)$$

This equation, together with the equations (2.6) and (2.7), represents the mathematical model of the simplified circulatory system, as shown in Fig. 1.5. It is also possible to describe the physiological process shown in Fig. 1.5 on the basis of electrical equivalents. Hence we have to replace the flow rate Q by current i, the pressure P by voltage U, the mass M by inductance L, and the compliance C by capacitance C, to find the model equations of the circulatory system, described by (1.6), (1.7) and (1.11):

$$\dot{U}_{C_1} = \frac{1}{C_1}(i_{in} + i_A + i) \qquad (1.12)$$

$$\dot{U}_{C_2} = \frac{1}{C_2}(i - i_{out}) \qquad (1.13)$$

$$\dot{i} = \frac{1}{L}(U_{C_1} - U_{C_2}) \qquad (1.14)$$

the graphical representation of which is shown in Fig. 1.6. The equations (1.12), (1.13) and (1.14) can be further simplified by supposing that during the diastole $Q_{in}=Q_A=0$, which corresponds to $i_{in}=i_A=0$, and that the peripheral resistance of the system is R_A (and the system behaves linearly), so that the relation

$$Q_{out} = \frac{P_2}{R_A} , \qquad (1.15)$$

and correspondingly

$$i_{out} = \frac{U_{C_2}}{R_2} \ , \tag{1.16}$$

holds. By introducing the state vector

$$\underline{X} = \begin{bmatrix} i \\ U_{C_1} \\ U_{C_2} \end{bmatrix} \tag{1.17}$$

the analogous model equations (1.12), (1.13) and (1.14) can be written as

$$\dot{\underline{x}} = \underline{A} \cdot \underline{x} \ , \tag{1.18}$$

whereby the system matrix is given by

$$\underline{A} = \begin{bmatrix} 0 & \dfrac{1}{L} & -\dfrac{1}{L} \\[2mm] -\dfrac{1}{C_1} & 0 & 0 \\[2mm] \dfrac{1}{C_2} & 0 & -\dfrac{1}{R_A C_2} \end{bmatrix} \tag{1.19}$$

The electrical variables of the model i, U_{C_1}, U_{C_2} are called model state variables. They correspond to the system state variables Q, P_1, P_2.

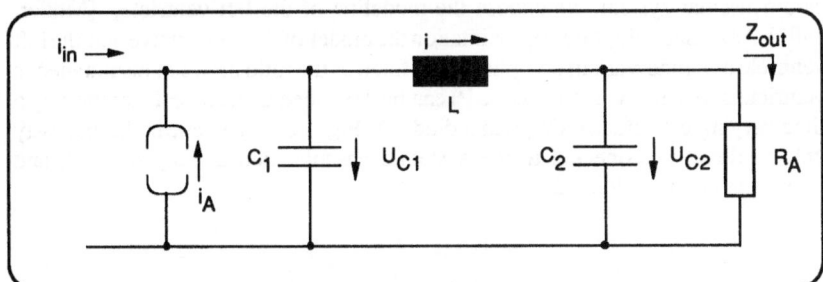

Figure 1.6. Electrical representation of a circulatory system

In a similar way, the values L, C_1, C_2 and R_A are known as model parameters. They correspond to the system parameters M, C_1, C_2, and R_A. Thus (1.18) is called the state-space model of the circulatory system, the system matrix $\underline{A}$ of which is given by (1.19), [Möller, 1981]. Simulating a physiological system as described in this section is generally no problem when using a powerful simulator. Using a block-oriented or an equation-oriented simulator is more a statement of the personal preference than a general one. The simulation of bio systems will be discussed in detail in the different case study examples.

1.2.3 Difficulties in Modeling Physiological Systems

There are some mathematical difficulties when deriving the model equations from a physiological system. One important point is the distributed nature of the system parameters. This is an essential property of some system parameters because many physiological phenomena are often describable by using, for example, the flow, which has to be locally described based on the principles of flow dynamics. This implies the use of partial differential equations and causes essential difficulties when simulating the model. For the purpose of simplification, dividing the system into small segments that can be treated as lumped-parameter systems is a useful tool.

A further difficulty in describing physiological systems mathematically is due to the well-known fact that the vast majority of physiological processes implies wide-range, intrinsic nonlinearities [Möller, 1981], that cannot be approximated by linearities in the full signal and parameter range of interest without loss of the essential process characteristics. Typical nonlinearities are: the O_2-saturation; the threshold of the nervous control; the rectification effect of the heart-valve mechanism; state-, signal- and frequency- dependency of some system parameters; nonlinear and adaptive feedback links; and state-dependent transport delays within the system.

As an example of the need to introduce nonlinearities into a simple model of the cardiovascular system, we choose the modeling of the left ventricle, [Möller, 1981]. As a relatively good approximation the model of the aortic valve and the left ventricle as a time-variant compliance (defined as the ratio between instantaneous ventricular volume V and pressure P) can be described in electrical notation by a time varying capacitance C(t) and a diode D (Fig. 1.7). The ventricular pathway will be directly connected to the systemic vascular bed already derived, and implemented in section 2.2.

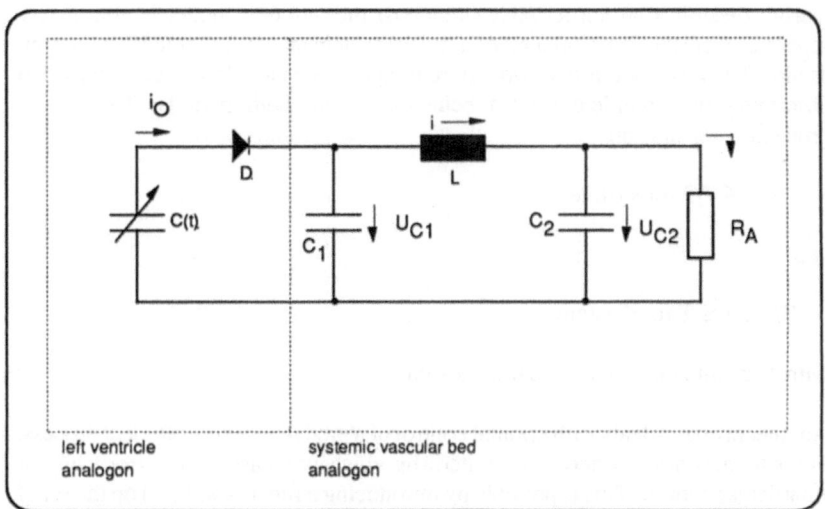

Figure 1.7.   Electrical model of the left ventricle and of the systemic vascular bed

During systole (the aortic valve is open) the model (left ventricle plus systemic arterial bed) is described by

$$\dot{i} = \frac{U_{C_1} - U_{C_2}}{L} \tag{1.20}$$

$$\dot{U}_{C1} = - \left(i + \dot{C}(t)\, \dot{U}_{C_1} \right) \cdot \frac{1}{C_1 + \dot{C}(t)} \tag{1.21}$$

$$\dot{U}_{C_2} = \left(i - \frac{U_{C_2}}{R_A} \right) \cdot \frac{1}{C_2} \tag{1.22}$$

so that the system matrix $\underline{A}$ is relative to the same state vector $\underline{X} = (i\ U_1\ U_2)^T$, as in section 2.2,

$$\underline{A} = \begin{bmatrix} 0 & \dfrac{1}{L} & -\dfrac{1}{L} \\[2ex] \dfrac{-1}{C_1 + C(t)} & \dfrac{\dot{C}(t)}{C_1 + C(t)} & 0 \\[2ex] \dfrac{1}{C_2} & 0 & -\dfrac{1}{C_2 R_A} \end{bmatrix} \tag{1.23}$$

During diastole (the aortic valve is closed) the vascular model is completely separated from the ventricular model, and the system matrix is defined by (1.19) in. section 2.2. It is obvious that two different state-space models are necessary when describing the complete system behavior during both periods. The timing procedure depends on

$$0 \leq t \leq T/2 \text{ for systole}$$

and

$$T/2 \leq t \leq T \text{ for diastole}$$

with T as time interval of one heart action.

Another problem deals with optimal control of a model of a physiological process. For optimal control we need some criteria by which we measure the performance of a particular control. This is possible by introducing a functional $J(.)$ on the set of admissable controls, v, whereby $J(U_1(.)) < J(U_2(.))$ means: $U_1(.)$ performs better than $U_2(.)$.
Finding a control $U^*(.) \in v$, with $J(U^*(.)) < J(U(.)) \forall U() \in v$, then, $U^*(.)$ brings the state from X_0 to S, and $U^*(.)$ is said to be an optimal control, which is difficult to solve with respect to the nonlinearities.

1.3   Models in Biomedical Engineering

1.3.1   General Aspects

As previously described, models are used to help elucidate biomedical mechanisms; they usually possess a simpler or more accessible structure or mechanism than the authentic biomedical system of research interest. Also they are needed because certain classes of experiments are not allowed to be carried out on human beings in order to avoid harm to them. Moreover, many scientists are feeling strong social pressure to use fewer animal experiments in biomedical research and, therefore, have become more interested in models. The models used today in biomedical engineering vary greatly in both size and complexity. Some of them are as simple as the analysis of a two or three compartment system, while others model complex biomedical systems of interrelated processes and hierarchies. Thus, the number of state variables in the biomedical stimulation models of today can range from one or two to several thousands or more. Due to this fact, stimulation as a third category apart from theory and experiment is of considerable importance in biomedical engineering. In the many cases where experiments cannot be realized, it is possible to obtain answers on extremely complex questions by means of simulation models which contribute to a better understanding of theory. Further, simulations provide information about the usability of hypotheses, e.g. temperature regulation or the growth of tumors and tumor therapy, cardiac assist devices, control of an artificial heart etc.

Within the framework of model building, the unstructured initial data of the research object are analyzed from certain points of view by means of functional decompensation and abstraction (in a certain sense, this could be called a filtering process) and are reproduced by clearly identified elements, including their attributes which are distinguished marks, characteristics of features and relations. As a result of this process, we get the "structural concept" of a model, which is a replacement for the real biomedical process, which we call an abstract model. From this, a generally valid limit of simulation immediately becomes evident: models in biomedical engineering only provide statements within the subset of elements of the whole, which have been important within the model building process. This is valid for the following model building principles:

-   Principle of physical similarity
-   Principle of physical isomorphism
-   Principle of mathematical reproduction.

If these explanations are abstracted by one more step, one can say that scientific research in modeling biomedical engineering problems takes place in subsequent semantic stages, which are shown in Table 1.2:

Semantic stage:

0:	*Internal model building*	=Model of the 1st semantic stage, based on: stimuli, impressions (material information), ideas
1:	*Perception model*	= f(stimuli, impressions)
	Cogitative model	= f(picture) this picture is based on the association of impressions with other impressions and ideas and on the combinations of the different pictures and ideas
2:	*Communication model*	Ideas are expressed in an intersubjectively understandable language
3:	*Graphic model*	Language form is described in graphic form
4:	*Fixing of the language*	
5:	*Formalized idea*	e.g. in computer language Generate reproduction of idea in the computer

---

Table 1.2. Scientific research in models and respective semantic stages

At semantic stage 0 there are the stimuli and impressions (material information), the objects and processes (also the information from other subjects) have on us.

These stimuli and influences create certain ideas inside us and convey a picture of the objects, etc. We will take this process as being internal model building and the picture as the model of semantic stage 1. This process can be differentiated in the actual picture producing the stimuli and impressions - this is called the perception model - and the extended picture resulting from an association of these impressions with other impressions and ideas by combining the different pictures and ideas. In this context we speak of the cogitative model.

These pictures are articulated at semantic stage 2, i.e. are expressed in an intersubjectively understandable language, this is the communication model. At higher semantic stages these language forms may be described in other symbolic forms. The next step will be to fix the language in writing.

In further steps one could think of expressing the ideas in formal languages, e.g. in a computer language and thus of generating a reproduction of the ideas in a computer. In this way the semantic stages of scientific research may be continued. Each stage, however, means a limitation of the information content, which at the same time is rendered more precisely: the language is more limiting but more precise than the association of ideas and feelings, the writing on the other hand is more limiting and more precise than the richness of linguistic possibilities but more precise than colloquial language, etc.

1.3.2   Modeling Biomedical Systems

An essential difficulty in modeling biomedical systems is that a relatively large number of system parameters has to be taken into account. Normally, not all essential system parameters are directly known or measurable and very frequently the values of some of them have to be guessed, and thus are introduced as constant values into the simulation model of the biomedical system. The last restriction is an artificial one, since in a real biomedical system there are seldom parameters of which the values are frozen within the whole wide range of operating the process. Often abstract models are used in order to simplify the complex biomedical systems, e.g. biomedical movement for studying human locomotion (walking, running). Based on anatomical reality, the degrees of freedom will be up to 250 for such a biomedical system analysis, which is really an unrealistic task to set. Therefore, one method will be to reduce the degrees of freedom, combining sub-elements, the "characteristic elements", for the reason that modeling is a compromise between model "goodness" and expenditure. Based on this analysis, a biomedical model with x degrees of freedom limbs can be derived and used for studying the transient behavior of limb prosthetic devices during walking and running. Based on such a study, attractors could be found out by system analysis, as well as stability margins of locomotion, which are important performance criteria for the optimal design of prosthetic devices as one biomedical task. Such differential equation systems can be solved properly with a software tool like Phaser [Kocak, 1986].

Thus, by using a mathematical model of the biomedical system and by applying parameter estimation techniques as biomedical engineering research tools, we can also "solve the unsolvable": the determination of some system parameters which cannot be measured directly. The main difficulty in parameter estimation of biomedical systems is the large number of relevant parameters and their inter-relation. This requires that some of the system parameters should a priori be known or directly be measured in vivo, a requirement that is a rather difficult task, as has been discussed above. Another essential problem in the parameter estimation of a biomedical system is the problem of identifiability, which means the possibility of system parameter estimation based on experiments. Here, the systems frequently show internal or structural non-identifiability, so that the identifiability of the given system should be proved. Usually this is done in the case of linear models by proving the system controllability and system observability. If algebraic methods for testing identifiability are not applicable, a graphical test is a proper proof. The philosophy behind this test is that it is important to check whether the parameter estimation based on a large number of initial values, taken from an expected neighborhood of the system parameter vector, will be an sufficiently accurate estimate of the minimized parameter vector. A further essential problem in the parameter estimation of biomedical systems is the problem of parameter estimation accuracy. Identifiability as discussed before does not imply good accuracy of the parameter estimates in the presence of measurement noise. If the variance of the parameter estimates is too big, conclusions from these estimates will be unreliable.
How does one get an idea of estimation accuracy?
As a measure of the parameter estimation error, the deviation of the minimum error functional for noise-free data was chosen, which for low-noise corrupted data gives:

$$COV\,(\Delta\underline{\theta}_{min}) = \left\{ [\,\underline{S}^{\underline{Y}}_{\underline{\theta}}\,(\underline{\theta}_s)]^T \,\underline{S}^{\underline{Y}}_{\underline{\theta}}\,(\underline{\theta}_s) \right\}^{-1} \cdot\, \sigma_v^2 \qquad (1.24)$$

where $\underline{S}$ is a sensitivity matrix of the model output $\underline{Y}$ with respect to $\underline{\theta}$, and σ_v^2 is the variance of the measurement noise. From this equation it follows that the variances of the estimation errors can become significant when $\underline{S}$ is ill-conditioned. For instance, for relatively small values of σ_v^2 the rough evaluation of the above equation by calculation of the diagonal elements gives the approximate values of the expected variances of estimation errors for the respective system parameters. Since the calculation results are valid only for a small neighbourhood of the nominal parameter values $\underline{\theta}_s$, they have to be tested over an area of expected true parameter values.

We will speak of the true parameter vector $\underline{\theta}_s$ and correspondingly of the 'true model' if the output sequence $\{Y_k(r)\}$ coincides with the output of the true model $\{Y_k\}$, and also the difference between the measured system output $\{Y_{meas,k}\}$ and the model output $\{V_k\}$ fits the output measurement error Y_k in such a way that:

$$\{ V_k(\underline{\theta}) \} = Y_{meas,k} - Y_k(\underline{\theta}) = \{V_k\} \tag{1.25}$$

as shown in Fig. 1.8:

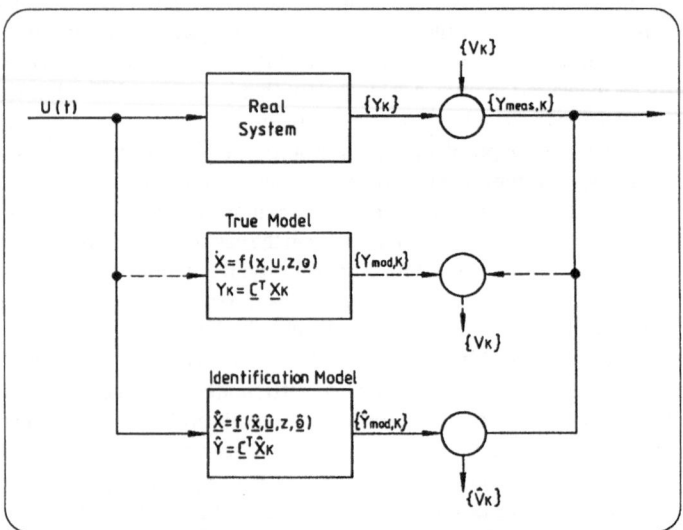

Figure 1.8.   Relationships between the real system, the true model, and the identification model [Möller, 1990]

The basic principle of parameter estimation is the adjustment of the parameter vector $\underline{\theta}$ of an identification model, with the same structure as the true model, in such a way that its output $\{Y_k(\underline{\theta})\}$ will coincide with the output $\{Y_k(\underline{\theta}_s)\}$ of the true model [Möller, 1983; Kaehr, 1989]. If $\{Y_k(\underline{\theta}_s)\}$ is not known it is, at most, possible to compare $\{Y_k(\underline{\theta})\}$. If the difference:

$$V_k(\underline{\theta}) = Y_{meas,k} - Y_k(\underline{\theta}) \tag{1.26}$$

is interpreted as the estimate of v_k, the task will be to adjust $\underline{\theta}$ in such a way as to impress on the sequence $\{v_k(\underline{\theta})^2\}$ some known statistical properties of $\{v_k\}$, e.g. its mean and its variance. Assuming $\{v_k\}$ would be white and stationary with mean zero and variance σ_v^2, this task can be done by minimizing the well-known output error least-squares functional:

$$J(\underline{\theta}) = \sum_{k=1}^{N} v^2(\underline{\theta}) = \sum_{k=1}^{N} (Y_{meas,k} - Y(\underline{\theta}))^2 \qquad (1.27)$$

the minimizing argument of which is a consistent estimate of the true parameter vector $\underline{\theta}_s$.

1.3.3 Block Diagram of Algebraic Representation of Biomedical Systems

In this section we will introduce the concept of block diagram algebraic representation, a powerful tool for simulation and parameter identification of biomedical systems. From biomedical engineering we know to consider a composite system as consisting of two or more subsystems. There are many forms of composite systems, however. Mostly they are built up on the following basic structures: parallel, sequential (or tandem), and feedback.

First we study in general the input-output description of composite systems. Consider the multivariable system S_i, which is described by

$$Y_i(t) = \int_{-\infty}^{\infty} G_i(t,\tau), U_i(\tau) d\tau \ , \ i = 1 \ , \ ... \ , \ n \qquad (1.28)$$

where U_i and Y_i are the input and the output, and G_i is the impulse response matrix of the system S_i (transfer function matrix). Hence we can rewrite with

$$G_i(s) = \frac{Y_i}{U_i}, \qquad\qquad i = 1 \ , \ ... \ , \ n \qquad (1.29)$$

the algebraic notation of the input-output description of blocks, representing the subsystems of a composite system. This general structure is shown in Fig. 1.9.

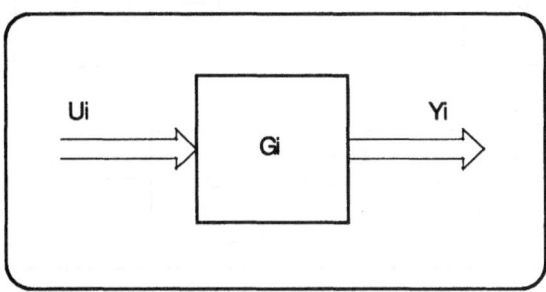

Figure 1.9. Block diagram structure of subsystems of a composite system

A classification of some elements, which are proper to use as blocks for subsystems, is shown in Table 1.3.

Block	Mathematical Description	Algebraic Operator
Proportional block	$Y(t) = K\,U(t)$	$G(s) = K$
Integral block	$Y(t) = K_{t_0}\,U(t)dt$ t_0: initial time t: time	$G(s) = \dfrac{K}{s}$
Differential block	$Y(t) = K\,U(t)$	$G(s) = Ks$
Exponential block	$Y(t) = e^x$	$G(s) = \exp(U)$

Table 1.3.   Classification scheme;   K = proportionality constant

The structure of general composite systems is now shown in Fig. 1.10.

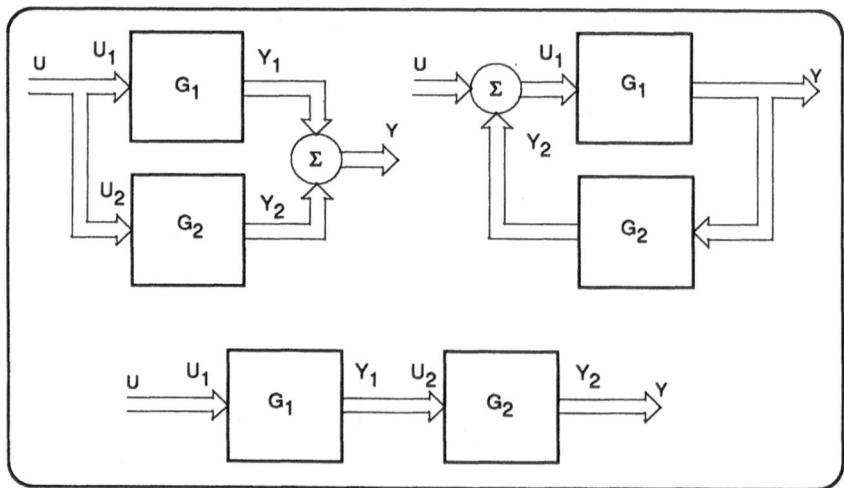

Figure 1.10.   Composite connections of two systems
        a) parallel, b) feedback, c) sequential (tandem)

From Fig. 1.10a we see that in the parallel connection we have $U=U_1=U_2$, $Y=Y_1+Y_2$, in the feedback system structure (Fig. 1.10b), we have $U_1=U-Y_2$, $Y=Y_1$, and the sequential system structure (Fig. 1.10c) shows that we have $U=U_1$, $Y_1=U_2$, $Y=Y_2$. Note that it was assumed that the systems G_1 and G_2 have compatible numbers of inputs and outputs.

It is easy to show that the impulse response matrix of the parallel connection (Fig. 1.10a) is:

$$G(t,\tau) = G_1(t,\tau) + G_2(t,\tau) \tag{1.30}$$

For the feedback connection, shown in Fig. 1.10b, the impulse response matrix is the solution of the integral

$$G(t,\tau) = G_1(t,\tau) - \int_\tau^t G_1(t_1,U)\int_t^U G_2(U,V)G(V,\tau)du\,dv \tag{1.31}$$

For the sequential solution, Fig. 1.10c, we have

$$G(t,\tau) = \int_\tau^t G_1(t,U)\, G_2(U,\tau)dU \tag{1.32}$$

Based on these general remarks, now we can introduce the relevant laws for rearranging blocks of linear and nonlinear composite systems, which are based on block diagram algebra. The laws are described with the respective algebraic equations, showing the original block diagram and the equivalent block diagram.

Combining parallel blocks:

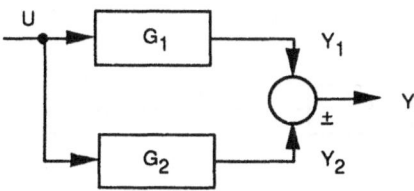

The output variables Y_1 and Y_2 are

$$Y_1 = U\ G_1$$
$$Y_2 = U\ G_2$$

Adding Y_1 and Y_2 gives

$$Y = Y_1 \mp Y_2 \; U(G_1 \pm G_2)$$

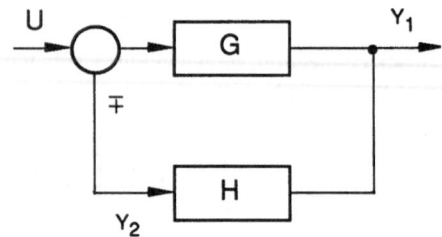

Combining feedback loops:

The output variable is

$$\begin{aligned} Y &= U + Y_2 G \\ &= U + Y_1 HG \end{aligned}$$

The algebraic description is given by

$$Y_1 = \frac{G}{1 \mp GH} U$$

Combining sequential blocks:

The output variables are

$$\begin{aligned} Y_1 &= G_1 U \\ Y_2 &= G_2 Y_1 \\ &= G_1 G_2 U \end{aligned}$$

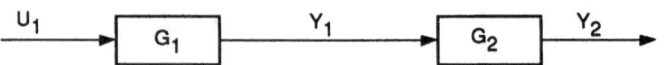

Permutation of blocks:

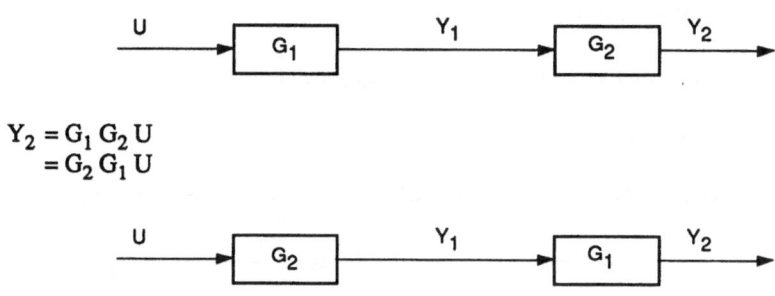

$$Y_2 = G_1 G_2 U$$
$$\quad = G_2 G_1 U$$

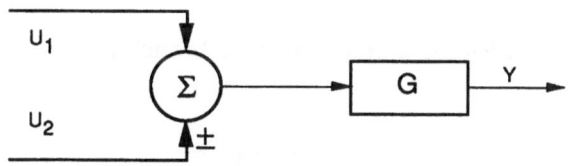

Moving a block before a summing point:
The original situation is shown as follows

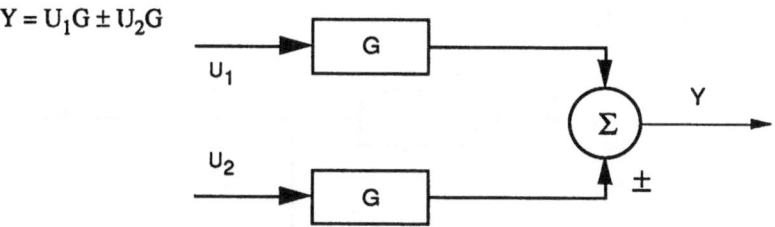

The output variable is

$$Y = (U_1 \pm U_2)\, G$$

and for the rearranged case

$$Y = U_1 G \pm U_2 G$$

Moving a block after a summing point:
The starting structure is as follows

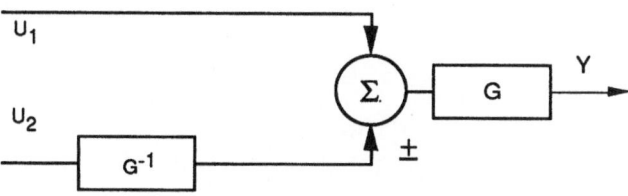

with

$$Y = U_1\, G \pm U_2$$

The output variable after the moving procedure is

$$Y = G(U_1 \pm G^{-1}\, U_2)$$

which yields a multiplication with the inverse transfer function.

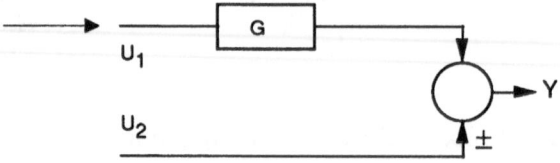

Moving a block before a branch point:

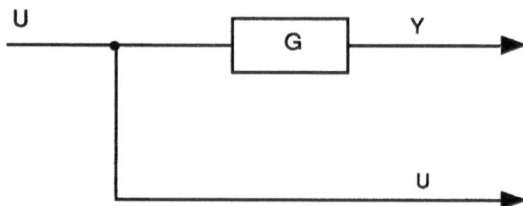

$$Y = G \cdot U$$

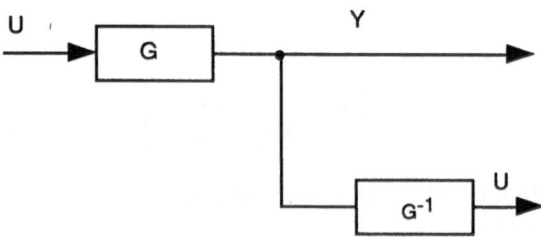

Rearranging summing points:

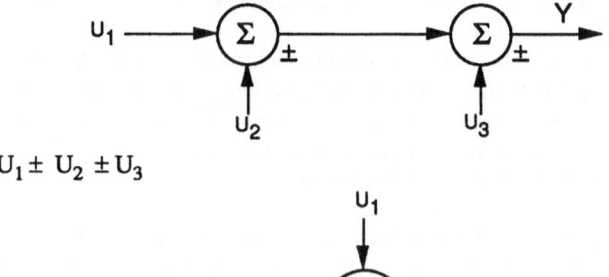

$$Y = U_1 \pm U_2 \pm U_3$$

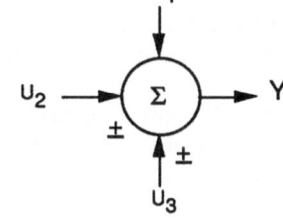

Inversion:

$$Y = G\,U \qquad U \longrightarrow \boxed{G} \longrightarrow Y$$

and the inverse function

$$U = G^{-1}\,Y \qquad Y \longrightarrow \boxed{G^{-1}} \longrightarrow U$$

The block diagram algebraic representation rules give a brief overview of how to model and simplify biomedical systems in order to be used for model-based identification of unknown parameters, which has direct clinical importance.

1.4 Sensitivity Analysis

1.4.1 Introduction

Investigation of models by simulation shows that the state variables that describe the physiological system are sensitive to model parameter deviations. This is because the state vector $\underline{X}(.)$, as the solution of the vector differential equation by which the physiological system is mathematically described, generally depends on the parameter vector $\underline{\theta}$ and initial value vector $\underline{X}_0$, i.e. that

$$\dot{\underline{X}} = \underline{X}\,(t\,,\underline{\theta}\,,\underline{X}_0) \tag{1.33}$$

Equation (1.33) is of great importance when estimating the system parameters, especially when using an adjustable system model. With significant errors in the parameter estimates, the model of the physiological system to be identified can be physiologically inadequate.

On the other hand, a mathematical model of a physiological system that is sensitive to relatively small deviations of its parameters is inappropriate for system analysis [Möller, 1981]. The sensitivity analysis of the model has to be carried out before accepting it as an adequate description for the real system, which will be a rather difficult task in the case of complex models.

The sensitivity analysis of a model consists in the comparative study of its time behaviour for different sets of parameter values with the aid of a computer, on which the model is implemented.

As an example of how to directly apply the methods of sensitivity analysis to the study of parameter influence on the model behaviour, we will give a brief introduction into the method.

1.4.2   Methods of Sensitivity Analysis

Modern methods of sensitivity analysis can successfully be applied to study the parameter sensitivity of physiological or biomedical systems, which, in general, can be described by a nonlinear state-space differential equation [Möller, 1981]

$$\dot{\underline{X}}(t) = \underline{f}(\underline{X}(t), \underline{U}(t), \underline{\theta}(t), t) \; ; \; \underline{X}(t_0) = \underline{X}_0 \tag{1.34}$$

and an output-equation

$$\underline{Y}(t) = \underline{g}(\underline{X}(t), \underline{U}(t), \underline{\theta}(t), t), \tag{1.35}$$

where $\underline{X}(t)$, $\underline{Y}(t)$ and $\underline{U}(t)$ are the state, output and input vector (of order k, n and m), $\underline{\theta}(t)$ the parameter vector (of order p), and $\underline{f}(.)$ and $g(.)$ are vector-valued functions. The solution of the first equation, neglecting the dependence on x_0 in (1.23), is time and parameter dependent, i.e.

$$\dot{\underline{X}} = \underline{X}(\underline{\theta}(t), t),$$

where in general $\underline{\theta}$ could be a time-variant parameter vector. Thus, the time-derivative of $\underline{X}$ will be given by

$$\frac{d\underline{X}(\theta, t)}{dt} = \frac{\partial\underline{X}(\theta, t)}{\partial\underline{\theta}} \cdot \frac{\partial\underline{\theta}}{\partial t} + \frac{\partial\underline{X}(\theta, t)}{\partial t} \tag{1.36}$$

where

$$\frac{\partial \underline{X}(\underline{\theta}, t)}{\partial \underline{\theta}} = \begin{bmatrix} \dfrac{\partial x_1}{\partial \theta_1} & \cdots & \dfrac{\partial x_1}{\partial \theta_p} \\ \cdot & & \cdot \\ \cdot & & \cdot \\ \cdot & & \cdot \\ \dfrac{\partial x_n}{\partial \theta_1} & \cdots & \dfrac{\partial x_n}{\partial \theta_p} \end{bmatrix} \qquad (1.37)$$

is the corresponding Jacobian matrix of partial derivatives, which is assumed to exist and to be unique. The sensitivity of the equations (1.24) and (1.25) to any specific component θj of the parameter vector $\underline{\theta}$ can be evaluated by the use of the partial derivatives

$$\frac{\partial \underline{X}(\underline{\theta}, t)}{\partial \theta_j} = \frac{\partial \underline{f}(\underline{X}, \underline{\theta}, \underline{u}, t)}{\partial \theta_j} \qquad (1.38)$$

and

$$\frac{\partial \underline{Y}(\underline{\theta}, t)}{\partial \theta_j} = \frac{\partial \underline{g}(\underline{X}, \underline{\theta}, \underline{u}, t)}{\partial \theta_j} \qquad (1.39)$$

which, in the case of a constant parameter vector, are equivalent to

$$\frac{\partial \underline{X}}{\partial \theta_j} = \frac{\partial \underline{f}}{\partial \underline{X}} \cdot \frac{\partial \underline{X}}{\partial \theta_j} + \frac{\partial \underline{f}}{\partial \theta_j} \qquad (1.40)$$

and

$$\frac{\partial \underline{Y}}{\partial \theta_j} = \frac{\partial \underline{g}}{\partial \underline{X}} \cdot \frac{\partial \underline{X}}{\partial \theta_j} + \frac{\partial \underline{g}}{\partial \theta_j} \qquad (1.41)$$

with partial derivatives $\partial \underline{f}/\partial \underline{X}$ and $\partial \underline{g}/\partial \underline{X}$ as Jacobian matrices defined in a similar way as $\partial \underline{X}/\partial \underline{\theta}$ in (1.27). Defining the vector sensitivity functions $\underline{s}_{\underline{X}, j}$ and $\underline{s}_{\underline{Y}, j}$ for θ_j as

$$\underline{s}_{\underline{X}, j}^{+} = \frac{\partial \underline{X}}{\partial \theta_j} \qquad (1.42)$$

and

$$\underline{s}_{\underline{Y}, j}^{+} = \frac{\partial \underline{Y}}{\partial \theta_j} \qquad (1.43)$$

finally we can write the general sensitivity equations, corresponding to (1.40) and (1.41) as

$$\dot{\underline{s}}_{X,j} = \frac{\partial \underline{f}}{\partial \underline{X}} \underline{s}_{X,j} + \frac{\partial \underline{f}}{\partial \theta_j} \qquad (1.44)$$

$$\dot{\underline{s}}_{Y,j} = \frac{\partial \underline{g}}{\partial \underline{Y}} \underline{s}_{Y,j} + \frac{\partial \underline{g}}{\partial \theta_j} \qquad (1.45)$$

In order to express equations (1.28) and (1.29) for all components of the parameter vector $\underline{\theta}$ we need the definition of the parameter sensitivity matrices

$$\underline{S}_X = [\,\underline{s}_{X,1}\ \ \underline{s}_{X,2}\ \cdots\ \underline{s}_{X,p}\,] \qquad (1.46)$$

and

$$\underline{S}_Y = [\,\underline{s}_{Y,1}\ \ \underline{s}_{Y,2}\ \cdots\ \underline{s}_{Y,p}\,] \qquad (1.47)$$

By the use of the sensitivity vector one can now approximate the deviation $\Delta \underline{Y}$ caused by a small deviation $\Delta \underline{\theta}$ of the parameter vector $\underline{\theta}$ as

$$\Delta \underline{Y} = \underline{Y}(\underline{\theta} + \Delta \underline{\theta} t) - \underline{Y}(\underline{\theta},t) \approx \underline{S}_Y^T \cdot \Delta \underline{\theta} \quad . \qquad (1.48)$$

For the linear system defined by

$$\dot{\underline{X}}(\underline{\theta}, t) = \underline{A}(\theta)\underline{X}(\underline{\theta}, t) + \underline{B}(\underline{\theta}) \cdot \underline{u}(t) \qquad (1.49)$$

and

$$\underline{Y}(\underline{\theta}, t) = \underline{C}(\theta)\underline{X}(\underline{\theta}, t) \qquad (1.50)$$

it follows from (1.34) and (1.35) that

$$\frac{\partial \underline{f}}{\partial \theta_j} = \frac{\partial \underline{A}(\theta)}{\partial \theta_j}\, \underline{X}(\underline{\theta}, t) + \frac{\partial \underline{B}(\theta)}{\partial \theta_j}\, \underline{u}(t) \quad (1.51)$$

and

$$\frac{\partial \underline{g}}{\partial \theta_j} = \frac{\partial \underline{C}(\theta)}{\partial \theta_j}\, \underline{X}(\underline{\theta}, t) \qquad (1.52)$$

so that the j-th sensitivity equation can be written as

$$\dot{\underline{s}}_{X,j}(\underline{\theta},t) = \underline{A}(\underline{\theta})\ \underline{s}_{X,j}\ (\underline{\theta},t) + \frac{\partial \underline{A}(\theta)}{\partial \theta_j} \cdot \underline{X}(\underline{\theta},t) + \frac{\partial \underline{B}(\theta)}{\partial \theta_j}\, \underline{u}(t) \quad (1.53)$$

and

$$\dot{\underline{s}}_{Y,j}(\underline{\theta},t) = \underline{C}(\underline{\theta}) \cdot \underline{s}_{Y,j}\ (\underline{\theta},t) + \frac{\partial \underline{C}(\theta)}{\partial \theta_j}\, \underline{X}(t) \qquad (1.54)$$

It should be noticed that the sensitivity equations (1.53) and (1.54) have the same system matrix $\underline{A}$ and the same measurement matrix $\underline{C}$ as the original system represented by (1.49) and (1.50) [Möller, 1981].

1.5 Outlook

Modeling and simulation of physiological and biomedical systems has been the subject of interdisciplinary research for many years. Accurate clinical measurements and the availability of electronic computers or powerful personal computers has allowed more quantitative hypotheses concerning physiological and/or pathological mechanisms to achieve considerable insight into the mechanisms of the system itself.

The literature of application of modeling and simulation in physiological research spans more than two decades, and shows there is a gap, for now, between desire and promise and the actual realizable achievement.

The use of physiological and biomedical systems models as an educational tool has been the subject of much experimentation over recent years. One of the most powerful physiological system simulator is HUMAN, developed by T.G. Coleman and J.E. Randall.

These educational systems models may be divided into four categories:

- statistical models for clinical decision making
- case study simulations for teaching clinically relevant processes
- dynamic (parametrized) models of physiological and/or pathophysiological systems
- computer-based medical consultations (expert systems)

One medical expert system known worldwide is MYCIN, developed by E.H. Shortliffe.
The introduction of microcomputers influenced computer aided education on many aspects and at many levels. An excellent review of this topic is given in [Ingram, 1985]. Based on the guide to the level of computer resource required to implement the teaching aid described in that work, table 1.4 gives an impression of the respective computer resources to run educational system models described in four categories.

	Hardware Configuration		
Teaching Application	8 bit CPU 64 kb RAM Floppy Disk	16 bit CPU 256 kb RAM Floppy and Hard Disk	32 bit CPU 512 kb RAM Floppy and Compact Disk
Statistical models for clinical decision aids	X		
Case study simulators	X	X	
Dynamic models of physiological and patho-physiological systems	X	X	X
Computer based medical consultation		X	X

Table 1.4. Guide to level of computer resource required to implement educational system models, modified after [Ingram, 1985]

The complexity of real physiological systems requires the computer aided analysis via simulation such as experimental design of clinical experiments with animals, to allow the study of individual system variables with respect to the normally varying conditions of a living organism. An essential difficulty in modeling physiological systems is a relatively large number of system parameters to be taken into account. This is the main obstacle when identifying the system experimentally. Here, the values of some system parameters should be known, roughly estimated or measured in vivo. However, this is the weakest point of a system identification procedure, since the errors contained in the values of such parameters influence the accuracy with which the values of other parameters can be experimentally estimated. Furthermore, as a rule, this estimation has to be carried out based on a relatively small set of noise corrupted experimental data. Thus, some considerations should be given to getting good estimated parameter values from a poor data base. An important aspect to be considered in connection with the accuracy of the system parameters to be estimated experimentally is the parameter sensitivity of the system model. The ideal case would be where sensitivities were very high and linearly independent for the parameters to be estimated, and very low to be taken as known. In this case the errors due to the incorrect values of parameters which are assumed to be known will have a very low influence on the estimated values of other parameters, i.e. the estimation will be relatively robust in this sense. However, in the opposite case the initial errors will amplify the estimation errors of other parameters, the estimated values of which will not be acceptable from the application point of view [Möller, 1985].

From these brief remarks we find that the better the computer models fit the realistic scenario, the better the educational tool. It seems inevitable that computer-based educational material of physiological systems become more and more important, just as textbooks did decades ago. But it must be taken into account that computer models of physiological processes are never a substitute for experience with the real human processes, and computer simulation is never an equivalent for animal or human based experiments. One of the powers of computer models and computer simulation is the possibility of experimental design.

1.6   Learnability and Machine Learning

In order to extend the possibilities of applying the tools of modeling and simulation, presently people worldwide try to extend the classical description levels such as the information processing method by the metaphrasis of the knowledge- rule-processing method, e.g. by inclusion into or combination with an expert system, as shown in Fig. 1.11:

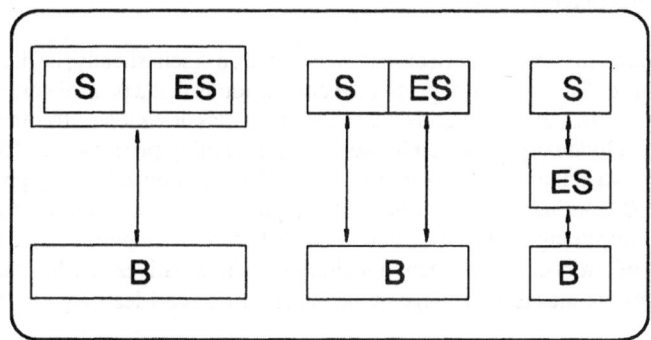

Figure 1.11.   Possibilities to combine simulation and expert system
               (For details see text).

The knowledge-based system (expert system) is included in the simulation system, in the case of combination, the simultaneous use of the knowledge-based system and the simulation system is possible, whereas in the case of the intelligent man-machine interface communication with the user is carried out via the knowledge-based system.

A knowledge-based system is a level of description for making diagnoses, generating plans, giving information on the basis of data and explicitly represented knowledge according to given algorithms and heuristics. These are currently not the limits of simulation, although this field of research is relatively new, because knowledge-based systems only produce a purely syntactical correlation of entered strings, with certain other strings which are given by the knowledge-based system without understanding the meaning of the characters.

Therefore, knowledge-based programs of today do not know anything; the term "knowledge" is not applicable to them, not only temporarily but in principle. Due to this fact, an exploration of the principles according to which man "thinks" and interacts is necessary.

Recently, old ideas in which knowledge is described as being distributed in patterns similar to a neuronal network have been taken up again. Thus the described knowledge is not allocated to specific symbols; it is, rather, distributed in the form of activity patterns via knots and linkages of the complete network connecting these knots. These are also called sub-symbolic representations. The connectional models derived from these representations therefore receive special scientific attention.

The ability to learn is one of the fundamental characteristics of human intelligence, and thus the complete field of machine learning is an integral part of the research of "artificial intelligence" (AI).

If one assumes that learning in a certain sense goes on stochastically, i.e. it contains components of "attempt" and "error", the conclusion is that a possible order of the learning processes may be based on a hierarchical classification of the error types which are individually corrected in the multiple learning processes. In [Bateson, 1972] the descriptive type theory is used for the classification of learning processes. The stage 0 learning characterizes all those processes which are not subject to correction by attempt and error, whereas stage 1 learning implies change through correction of errors through selection within a set of alternatives. On the other hand, stage 2 learning means the change in the process of stage 1 learning.

Therefore, a correcting change in the set of alternatives from which a choice is made takes place, which corresponds with and is related to a diagnostic expert system which is able to learn, and the system itself to change the rules/algorithms.

This means that the change does not concern the data base (data of the algorithm), or the operands (that would be first stage learning), but that it concerns second stage learning. In all higher learning types, the system itself changes the operands, the learning rules. This is the same as a context change [Bateson, 1972].

It should be mentioned here that a technical system for second stage learning does not yet exist, since a machine which draws up its own algorithms (without a "teacher/constructing engineer") has not been developed yet [Kaehr, 1989].

All machine learning which has been described in AI-research up to now is either stage 0 or 1. However, especially in the case of machine learning simulators, a stage 2 learning would be desirable, because the simulator itself could manage a change of its context from its "experience", a proceeding carried out by AI-tools which is similar to human intelligence, corresponding to an autoreferentiality. Dissipative structures, including among others vital processes, and thus mental processes in human beings, only arise in open systems, far from the thermodynamic equilibrium, through autoreferentiality.

This means, on the other hand, that the contexture of the bivalent classical logic has to be extended by the correlation of a polyvalent logic, the "polycontexturality logic" [Kaehr, 1989]. This originates from the fact that in the case of three values - the third value not being between zero and one, or true and false as in the bivalent logic - three bivalent logics result to which three contextures are allocated. Contexture in this case means a logic domain, where all logic rules are fully valid.

With respect to the above, the realization principles of simulators should be considered again, and we should ask how a simulator which can reproduce a process like stage 2 learning should be designed. It has to be able to carry out logic processes (logic operations) and in parallel to analize every individual step of such a process and to set the results of the analysis in correlation to the steps of the processes in order to correct them by control. It has to change it, i.e. it has to possess autoreferentiality. Thus it comes down to the conception of a machine (simulator) for which the operators of one process simultaneously appear as operands of operands of the other process.
The problem arising here is to formulate the problem itself in a logically adequate manner, i.e. if learning is not part of the machine - that is to say it is not to be pro-grammed - the machine must be able to differentiate between itself and the simulation which, however, is an autoreferential process. This has already been pointed out by philosophy since Immanuel Kant with respect to the existing limitation of logic descriptive modes.

References

Ingram, D. and C.J. Dickinson: "A Review of Modelling and Simulation Techniques in Medical Education", In: *Computers and Control in Clinical Medicine*, Eds.: E.R. Carson and D.G. Cramp,
Plenum Press, New York, 1985.

Möller D., D. Popovic and G. Thiele: *Modelling, Simulation and Parameter-Estimation of the Human Cardiovascular System*,
Vieweg Verlag, Braunschweig-Wiesbaden, 1981.

Möller D., D. Popovic, G. Thiele and W. Barnikol: "Parameter-Estimation of the Shortterm-Behaviour of the
Arterial Pressure Regulation System by Modelling and
Simulation", In : *Applied Modelling and Simulation*, pp. 1-4,
Ed.: M. Hamza, Acta Press, Anaheim, 1983.

Möller D., D. Popovic and G. Thiele: "Reliability of Parameter Estimation Methods applied to the Identification of Biomedical Multicompartment Systems", In: *Identification and System Parameter Estimation*, pp. 1385-1390, Eds.: H.A. Barker, P.C. Young,
Pergamon Press, Oxford, 1985.

Möller D.: "Parameter Estimation: An Advanced Simulation Tool in Biomedicine", In: *Advanced Simulation in Biomedicine*, pp. 71-82.
Ed.: D.P.F. Möller, Springer Verlag, New York, 1990

Kocak H.: *Differential and Difference Equations through Computer Experiments*.
Springer Verlag, New York, 1986

Bateson G.: *Steps to an Ecology of Mind*.
Chandler Publishing Comp., 1972

Kaehr R. and E. v. Goldammer: "Poly-contextural modelling of heterarchies in brain functions". In: *Models of Brain Function*,
Ed.: R.M.J. Cottervill, Cambridge Univ.Press, Cambridge, 1989

2
Computer Simulation as an Educational Tool

Fredericus B.M. Min

Abstract

- Why use computer simulation as an educational tool?
- Why is computer simulation based on mathematical models?
- What kinds of computer simulations are there?
- What is the difference between computer simulation and modeling?
- What is the difference between a model and a program?
- What are some advantages and disadvantages of computer simulation programs?
- What is the difference between learning through discovery and various kinds of coached learning?
- In which five ways can computer simulation programs be used?

2.1 Computer Simulation and Modeling

2.1.1 Introduction

Computer simulation is a form of computer-assisted learning (CAL) in which the user may experiment with a simulated situation. This situation strongly resembles reality or is a deliberate simplification of it. Computer simulation enables students to make decisions that entail few risks. As a result of the decisions made by the user, the computer reacts with informational feedback. This feedback is almost always visual. Visual feedback is an important characteristic of computer simulation. Therefore a computer simulation program often also has the characteristics of an animation program. Computer simulation in the instructional situation offers the teacher possibilities for providing experimentation with the subject matter in a well ordered way. Moreover it can make it easier to fully realize preset goals in certain situations. Many definitions of computer simulation exist. Unfortunately many authors do not make a distinction between simulation and modeling. A number of definitions collected from the literature can illustrate this:

Hinton, 1978: "A simulation package is based on a known model of physical phenomena. The model, usually in the form of a mathematical relationship, can be set up within a computer program and the student can simulate the phenomena or process by controlling and observing the output."

MacArthur, 1984: "A simulation is simply a model of some aspects of reality to focus on points of interest."

Manning and Potter, 1984: "The objective of simulation models is to present a simplified version of reality, whereby a complex system is distilled to only its most important elements or variables."

Shaw, 1984: "A computer simulation is a simplified representation or a real event or thing that recreates pertinent characteristics."

In view of the perspective of this chapter and the classifications used in it, all these definitions are incomplete. Hinton is the most complete, since MacArthur, for example, does not make a clear distinction between a simulation and a model. The latter definition would have been better if it had been reworded to say: "A simulation is simply based on a model ...". Shaw is also not precise in the distinction of the model as opposed to simulation, for a simulation is not a representation: a model is a representation and with a model simulation becomes possible.

It must be stressed, as many other authors have noted, that in the field of computer-assisted learning a distinction has to be made between simulation and modeling. For example, Hinton makes the following remarks:

"In the area of computer-assisted learning it is useful to make an operational distinction between simulation and modeling. It will be assumed that a simulation allows the student to control input parameters and observe the output. The model, usually in the form of a mathematical relationship, is set up within the computer program and the student can simulate the phenomena or process by controlling the input and observing the output. In modeling the student is able to control or change the model upon which the simulation is based." The student learns to change the model and the model relations until the model concurs with the mental model of what the student has envisioned.

In teaching there are a number of definitions of computer simulation, but we seldom find a universally accepted definition in the literature. Frequently, as has been demonstrated here, this is because a distinction between simulation and modeling as separate forms is often not made. In addition, apart from the difference between simulation and modeling as CAL forms, gaming also has to be distinguished.

2.1.2 Simulation

A definition of computer simulation that is too broad or inexact can lead to misunderstanding when, for example, a discussion of specific advantages of computer simulation occurs. As an example, the following question requires a common understanding of what is meant by a simulation: Can a simulation in the form of an inquiry CAL-program, in which the student has to think up and pose questions himself, also be used in the same way as a simulation of a certain phenomenon?

No, it is necessary to distinguish at least four different kinds of computer simulation, each with its specific, internal characteristics and without typical graphic feedback with respect to use as an educational tool, namely:

- *Simulation of conversations,* for example the physician-patient conversation involving history taking. This simulation is classified in this chapter as "inquiry CAL". Examples of these are programs by Verbeek [1986] and the training program by Min and Ephraim [1978] ANAMNESE. The student has to be able to ask a series of questions by himself and is minimally coached. The simulated "patient" answers the questions of the (medical) student, the "prospective physician". Below is an example of such a simulated conversation from the program ANAMNESE.
On the screen a list of questions appears:
  1. Do you suffer from listlessness?
  2. Do you suffer from fatigue?
  3. Do you suffer from pain in the chest?
  4. Does the pain radiate to your arms?

The student chooses one or more questions and receives, after typing in the numbers of the questions, an answer from the simulated patient.
  3. No, I do not suffer from pain in the chest.
  1. Yes, I have been feeling listless for the past few weeks.

- *Simulation of phenomena.* This kind of simulation almost always involves phenomena that can be described with a mathematical model. In this chapter we call this form "real" simulation in the sense of the classical classification. Here it is defined as: "simulation CAL". Examples of this are the computer simulation programs (CS-programs) CARDIO and FLUIDS, which are among those described later in this book. An example of a simulated physiological process can be seen in the following fragment of the CS-program CARDIO using text instead of the graphic display [Min, 1982]:
The CS-program shows on the screen:
  T=0 hour; mean arterial pressure: 100 mmHg.
  etc.

To enable the student to intervene in the process, the following questions appear on the screen:

Which intervention do you want to do?
1. Contraction strength of heart muscle
2. Renal arterial constriction
3. Vasoconstriction

The student chooses 1. On the screen appears:

This parameter is now fixed at 1.0 units.
How many units do you want?

The students types: 0.4. The program proceeds with:

T=1 hour; mean arterial pressure: 75 mmHg.
Doctor, I feel so dizzy.
T=6 hours; mean arterial pressure 80 mmHg.
Doctor, my shoes pinch, my feet are swollen.

- *Simulation of the 'conduct' of an expert.* In this form of computer simulation a (mathematically described) model is not the 'nucleus' of the CS-program; instead an expert system is used. This form is usually called "intelligent CAL" or CAI and is usually abbreviated to ICAI.

Increasingly, among the social sciences and particularly the medical sciences, expert systems are assuming the role that mathematical models have already played for a long time in the natural sciences. Both kinds of systems (models and expert systems) are based on rules and knowledge of experts that, with the help of a computer, can be used by a third party, such as the student.

The following is an example of an expert's computer simulation: The simulated "expert" answers after the student has introduced a series of data. The example fragment is from the program ARTS, and it works with conditional chances [Min, 1987].

On the screen appears:

Do you feel pain yourself while straining?

The student answers: YES. On the screen appears:

Do you have pain on the middle of your breastbone?

The student answers: YES. On the screen appears:

Do your legs hurt?

The student answers: NO.

etc.

Finally the simulated "expert" answers:

The most probable diagnosis, with a chance of 89%, is in 'my' opinion:
Angina Pectoris.

The argument can also be made that there exists a fourth form of computer simulation:

- *Simulation with moving pictures.* This kind of simulation, with many fragments of moving pictures giving the impression that each picture is unique, almost always occurs within the framework of ordinary "tutorial CAL". In this chapter it is called: "quasi simulation". Some examples of this are simulation programs on the DIDACDISC of the company QBit and Verhagen and Pals [1986]. These programs are "tutorial CAL with moving pictures" with "visible multiple choice moments". In the literature many programs labelled as simulations appear to be, in fact, only tutorial CAL and are, therefore subject to all the disadvantages of tutorial CAL rather than the advantages of "real simulation CAL".

For educational use, the University of Twente works mainly on computer simulations in which:

- a phenomenon or process from reality is simulated,
- a previously developed mathematical model is implemented in a computer simulation program, and
- vividly visualized forms of representation (animation) are used.

However, the model underlying the simulations is always regarded as invariable. In this way the model represents the reality, allowing it to be studied by experimenting with the computer simulation program.

According to Hinton, CS-programs have a special value in teaching, particularly when used for subjects that are mathematically complicated and make it very difficult for the student to see the complete pattern of relationships because of mathematical formalization. Situations such as these appear frequently in physics, chemistry, biology and economics.

CS-programs based on mathematical models should be designed in such a way that the student can easily use and easily intervene in the model. The model and the location of the interventions can, for example, be represented visually as shown in Figure 2.1. A parameter can be changed by using a mouse and clicking in an "inclick region". This form of intervention in a CS-program is called an input-animation technique here. Programs utilizing this technique will be discussed in later chapters. In particular, CS-programs made with the design system MacTHESIS use this technique, enabling the simulation programs to have the appearance of a real educational appliance instead of simply a computer with a keyboard.

2.1.3 Modeling

Modeling is one of the eight forms of CAL, at least according to the "classical classi-fication" used in this book. As already noted, modeling is distinguished from simu-lation in that the student develops a model on his own that represents a situation in re-ality. The model itself in this form of CAL is continually subject to discussion and is not regarded as a given, as occurs in simulation. Modeling has as its educational aim the development of the systematic ability to represent complex relations within a system. Modeling involves executing certain purposeful tasks in order to record the interrelationships within a certain system. This happens in two steps:

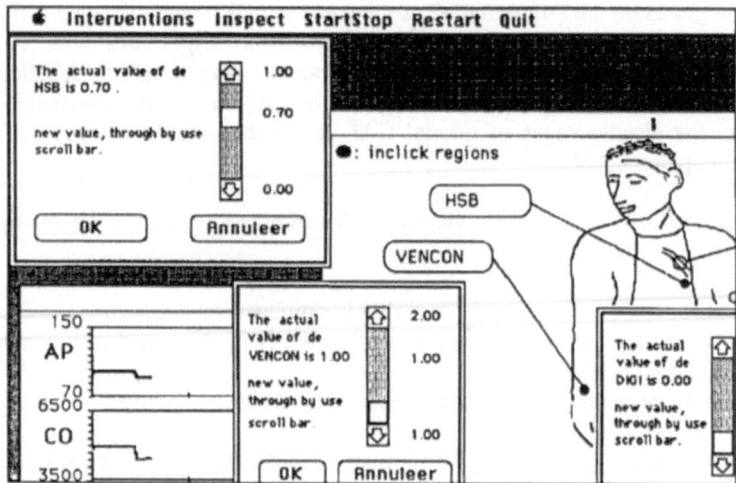

Figure 2.1. An example of what is called 'input-animation-technique'. This program consists of two windows (the animation window and the representation or output
window). These windows can be placed one over the other, or next to each other,
for examination in relation to each other. (In Macintosh terms this is called 'desk
top'.) In the animation window are 'click-in' regions by which an auxiliary window with a scrollbar can be popped up. A parameter of an underlying model can
be adjusted with this scrollbar. (CS-program CARDIO, MacTHESIS system,
Min 1982)

- defining a system from reality,
- representing this system in a model.

Modeling involves learning to make a model of something, such as making a
mathematical model, making something (drawing a flower) in LOGO, or making a
program for a "programmable logical controller" (PLC) that resembles programming a robot. Programming in its widest sense is also included as modeling in the
structure of this chapter.

Simulation involves learning to experiment with a simulated reality. Although
models are used while doing this, simulation programs are "discovery
environments" for bringing reality into the classroom. The student trains or does
exercises while learning to see connections between quantities and components of a
system.

Paradoxically, CS-programs appear to be very useful in learning how to make a
model. As an example, a CS-program of the blood pressure regulation mechanism of
the human body (CARDIO or FLUIDS) allows a medical student to construct for
himself a (conceptual) model of the relation between blood pressure, cardiac output,
and peripheral resistance and find that there is a linearity among them.

2.1.4 Gaming

Gaming is different from simulation. In this form of CAL there is usually competition among various participants or between the participant and the computer. Moreover there are preset rules to which participants are bound. Each simulation or modeling program has some of the characteristics of gaming that the student himself is experiencing while using the program. Competition is not necessary in gaming; to have a goal is enough in itself, as for example, in games of skill development.

2.1.5 Reality, Model and Program

The educational value of simulations is the development in the user of a feeling for the relationships between different variables within a model. This feeling, however, cannot be verified because it cannot be measured. It is questionable whether every user develops this feeling. Moreover not all relationships in the real system can be distinguished by the user, because the designer will necessarily have decided to make certain simplifications out of necessity. Also, certain important elements, from an educational point of view, are difficult to represent in a mathematical model, so these elements cannot be transposed into a CS-program. The danger exists, then, that too much attention will be given to things that can easily be condensed into a model and which, from an educational point of view, may be of less or even of no importance.

In computer simulation, purposeful actions are executed with a model fed into the computer. With gaming, how well the preset goal is reached is also calculated and displayed. The model is a substitute of a phenomenon or happening from reality. Instead of the terms "happening", "phenomenon", or "reality", the general term "system" can be used. This requires an examination of the relationships and the differences among reality, a system, a model, and a CS-program.

The "Really Existing" System

A system is defined as the connection of parts in an existing reality that are related in a certain way to each other, to the whole, and to their surroundings, and that interact and influence each other. Kronbluht and Little [1976] remark on this:

"A system means a grouping of parts that operate together for a common purpose. An automobile is a system of components that work together to provide transportation."

Another example of a complex system is a national economy. The separate components of the economy influence each other in the production, distribution and consumption of goods and services.

A number of specific characteristics are, according to Kornbluht and Little, connected with the term system. The most important of these is the state of the system. This state is related to the value of the state of each of the system's parts at any given moment. In some cases this state can vary from time to time, in other cases it remains stable or practically stable.

A system is described as being stable when the overall measured state of the system varies only within an accepted margin, in spite of the occurrence of changes among the separate parts. If the measured state alters beyond the accepted margin then the system is described as unstable. When we want to examine a problem that is specifically related to a certain system, we first have to determine where the margins of the system are: what we do and what we do not want to represent in the model. Next we try to make a representation of this system. This happens in the form of a model. The model can then be regarded as a substitute of the system. With the help of such a model it is possible to indicate the important, often complex relations within the system.

The Model

The construction of models representing a system is a complicated, specialized, and labor intensive activity. That is why the production of CS-programs in education generally relies on models already developed from research in other discipline. Modeling is most advanced in physics, chemistry, biology, and economics. It is therefore not surprising that we find the most educational CS-programs within these disciplines.

All types of mathematical models can be translated into a computer language without particular difficulty by individuals experienced with designing CS-programs. For example, students of the University of Twente, in the context of their training with the MacTHESIS system, learn to translate each kind of well-defined mathematical model into a Pascal source for implementation in the MacTHESIS simulation system [Min, 1987].

The Program

The ultimate goal of working with a CS-program is to get insight into a process, phenomenon or system that is represented through a model in a computer simulation. One aspect of this goal is that the student tries to detect the structure, interconnectivity, or the conduct of the entity being simulated. This kind of discovery is made possible because the program enables students to vary parameter values in the model and to analyze the outcomes of these variations. The benefits of these discovery experiences increase as the educational and technical aspects of CS-programs are enhanced.

With such CS-programs the student gets efficient and direct feedback when he changes the model parameters. This feedback is related to the consequences of the input relative to the underlying model. It may appear graphically, in table form, in animation form, or in another form of representation, often also furnished with some sort of comment. The student formulates for himself a hypothesis that he tests and to which he gets an answer or material from which he may draw a conclusion. The correctness of the hypothesis is examined by intervening in the model and noting the results of the intervention. On the basis of these results or other feedback, the student can adjust his hypothesis. This process of intervening and adjusting continues until a certain kind and a certain level of insight, relative to the simulated system, has come about.

2.2 Computer Simulation and Learning

2.2.1 Introduction

Many advantages can be cited with respect to the use of CS-programs in education. A number of these do not specifically relate to simulation, but are valid for practical laboratories or the general use of computers in education. There are also some specific advantages connected with computer simulation. First, some general advantageous aspects of simulation as a form and method of learning can be indicated.

- Computer simulation offers the opportunity to experiment with phenomena or events which, for a number of reasons, cannot be experimented with in the traditional way. Bork [1981] remarks: "Simulations provide students with experience that may be difficult or impossible to obtain in every day life."
- CS-programs can be used in education to give the student more feeling for reality in some abstract fields of learning. Foster [1984] remarks on this: "Simulations can be entertaining because of dramatic and game-like components."
- When a teacher tries to explain a difficult interrelationship, such as a hybridization experiment with fruit flies, there is a likelihood that part of the class will fail to fully understand. Executing the real experiment is impossible because this will take a number of weeks and cannot be integrated as such within a lesson. When a simulation experiment follows the necessary theoretical discussion of the material, there will be a greater chance that more students will understand a complete relationship.

According to Elron [1983] the best simulation does not have to resemble reality in the most accurate way. The power of simulation often lies, according to him, in the simplification of the reality. Good simplifications can give students a better insight into reality than could be obtained by examining all components of a complex situation.

While working with a CS-program the student is experimenting, so he is playing an active rather than passive role. This active engagement contrasts with the situation students often have during "frontal" teaching as passive listeners. Simulation creates, according to Foster, an interactive educational setting that offers the opportunity to effect changes in the learning experience in a more efficient way than is ordinarily possible through the use of other didactic methods.

It generally follows that computer simulation does not work to its intended advantage with frontal teaching where students can only listen as a group rather than interact as individuals. Nor does it stand alone. Only when computer simulation is appropriately alternated with other didactic forms will its use render a positive result. Working with a CS-program is often coupled with student enthusiasm and as such has a positive influence on his motivation. Spitzer remarks on this: "Simulations are highly motivating, both intrinsically and extrinsically." However, no educational tool is effective for everyone and a differentiated supply of educational support tools is therefore important and a CS-program is one of them. [Min, 1987].

Working with a CS-program can increase a student's interest in a subject. [Min, 1982]. This shows, for example, in the fact that students will often look more into relevant literature concerning the subject after using a simulation than they would have done using traditional approaches to learning. The subject may also be discussed more among students and special experiences may be shared.

The computer can be used as a didactic medium and it can serve as a tool to realize a chosen educational strategy and to reach set goals in a better and fuller way than would have been expected otherwise. Yet, what justifies the use of computer simulation in the classroom? Which specific didactic functions can computer simulation fulfill in education? Often the technical possibilities and the particularly effective calculating capacity of the computer are advanced in order to justify a switch to the use of computer simulations in the classroom. But these types of reasons are not enough, according to Wedekind [1979], especially from the educational point of view. The didactic functions that can be fulfilled with computer simulation are more important to cite than the technical ones.

The use of computer simulation should not, however, lead to the displacement of the practical laboratory in the range of student experiences. When experience with aspects of a real experiment is important, but the practical laboratory has a limited capacity, then working with a CS-program can increase the impact of practical work.

The value of experimenting and training with a model of reality is frequently recognized in the literature, as the following citations indicate:

- *Martens* [1979]: "Computersimulation als Forschungsstrategie ist seit Jahren in der Anwendung und hat sich aus der natur-wissenschaftlichen Forschung und Problemstellung heraus entwickelt."
- *Manning & Porter*[1980]: "Simulation can present a simplified version of reality, whereby a complex system is distilled to only its most important elements or variables."

- *MacArthur*[1984]: "This is opposed to a laboratory where many superfluous 'circumstances' exist." MacArthur describes the experimental function which computer simulation can fulfill as follows: "... using simulations to understand systems." He proceeds by suggesting that "this form has as its aim that the student learns to see the existing relations between the various factors within a system through experimenting."

As already noted, computer simulation can be a good educational alternative to elaborate experiments. In the following sections we consider some of the advantages and disadvantages of using computer simulation as an educational tool.

2.2.2 Advantages

Some advantages of computer simulation as an educational tool are:

- The apparatus necessary to do an experiment in reality is too expensive, and it is often the case that such apparatus could only be operated by specialists, even if it could be obtained.
- The process to be examined occurs too quickly in reality to be examined through the traditional experiment. Here we can think of certain chemical processes. Changes in a chemical reaction have to be presented at an observable pace in educational situations. In reality these changes are sometimes hard to discern and involve tedious calculations not interesting in themselves, but necessary for the subsequent acquisition of insight.
- The process to be examined can proceed too slowly in reality, such as biological growing processes.
- The system to be examined can be too complex for traditional research, such as economical systems.
- The system to be examined can be too large a scale, for example, planetary movements in space.
- The system to be examined can be too small a scale, such as molecular movements.
- The system to be examined can be dangerous to manipulate, as, for example, a nuclear reactor or the human body.
- It can be irresponsible from an ethical point of view to do research through traditional experiments as with certain diseases.
- Simulation experiments can be used to good advantage prior to a course as an introduction to a new subject or certain parts of it. After a lesson they can be used again for reinforcement of concepts.
- Simulation often goes together with visualization. The results of changes that a student puts into a model are directly shown on the screen. This often appeals highly to the student.
- Simulation can be very purposeful and for certain students very useful, such as those who need insight before they are able to learn and understand a new concept.

- The student can insert those parameter values that may produce a result that is interesting for him. He can then choose to focus his attention on parts that interest him and skip other parts or aspects. In this way he learns how to experiment systematically.
- A student can choose in which way he wants to approach a simulation experiment, how many times he wants to repeat the experiment and to what degree he wants to intervene. In computer simulation there are usually many ways that can lead to the student's goals.
- If the simulation is well designed, learning how to operate a CS-program generally takes little effort. A short introduction by the teacher is often enough to enable the student to work with the program.
- It can be an advantage for the student to see that not everything can be used as input. He experiences that variables and parameters have their limits, what is a reasonable input for a particular variable and what inputs produce relevant information.

Figure 2.2. An example of a complex computer simulation system for medical education (the RLCS-system, Min, 1982; VAX version 1984)

2.2.3 Disadvantages

There are also disadvantages or limitations connected with the use of CS-programs in education. These limitations are in some cases the result of misuse or inappropriate input into such programs. Possible limitations of computer simulations for general educational use are:

- Simulation concerns the manipulation of a number of variables of a model representing a real system. However, manipulation of a single variable often means that the reality of the system as a whole can be lost. Also, certain systems or components of a realistic situation are not transparent. Some factors have a great influence on the whole, but have indistinct influences and are therefore not representable in the model of the system. These factors, however, cannot be forgotten in the learning process.

- A CS-program cannot develop the students' emotional and intuitive awareness; the goal of a simulation is to better understand the relationship among variables in a model. Therefore, student intuition has to be developed in another way.
- Computer simulation cannot leap into unexpected "sub-goals" that arise (for the student) during a learning process. These sub-goals can be discussed, during a teacher-student interaction, but they remain unnoted during the student's use of a simulation.
- CS-programs may function well from a technical point of view, but they are not easy to fit into a curriculum.
- A CS-program often cannot be adapted to take into account differences in student level within a group or class. A CS-program can certainly be made to adapt to different circumstances if the designer has taken this into account; however, for many CS-programs this has not happened.
- During the experience of interaction with a CS-program, problem-solving learning is frequently demanded of the student and creativity is often the decisive factor in the success of this problem-solving. The fact that this creativity is less present in some pupils than in others is not taken into account by the simulation. Mutual collaboration and discussion among students while using the software could be a solution to this.

2.2.4 Working Inductively

There can be an inductive element associated with working with a CS-program, but requires effort from the student. While going through the program the student has to work according to a certain strategy in order to solve the set problems. This problem is first described and presented as a "case". A case can be a direct part of the program but may also appear as a text on the screen. Furthermore the case can also be included in the course workbook.

After the case is presented the student has to interrupt the computer simulation in order to change the values of the parameters he thinks are relevant. The results of these changes are then presented on the screen in graphic or table form, in computer voice or through animation. On the basis of the feedback, the student can again take action if necessary. This course of action is repeated until the problem is solved. After the session the teacher can check the student worksheets.

Because of this inductive strategy, computer simulation shows a strong resemblance to all other inductive forms of learning. We find induction used as a strategy by others involved in research about discovery learning. Foster [1984] writes about the relation between simulation and discovery learning: "Simulations are excellent discovery learning techniques that often offer insight or gestalts not gained through more traditional didactic methods."

2.2.5 Discovery Learning and Problem Solving

In discovery learning, an unsolved problem is presented or the student is only exposed to an "environment" in which he can pose a problem himself. For the formulation of the problem and for its solution, a large amount of free activity by the student is required. The goal is for the student to discover the characteristics of a concept, object, or system. Bruner [1976] links to this discovery method the following advantages [Min, 1982]:

- By applying the discovery method, the intellectual possibilities of the students increase. They learn to see that there are many possibilities for solving problems and that information has to be translated in a certain way.
- The discovery method heightens the student's intrinsic motivation. The student can take initiative himself.
- The student learns to handle approaches to problem-solving which can be of use in solving new problems.
- Knowledge acquired by what you discover (or make) yourself stays in your mind better.

These advantages are also relevant when working with a computer simulation. Education by means of computer-augmented discovery learning could even be increased, but time and materials often play a decisive role in limiting the opportunity for discovery learning to occur.

Kolb [1973] has developed a "learning model" with regard to the discovery method. He says:

> "Learning should be regarded as a cycle which consists of four phases. The concrete immediate experiences make up the basis of the perception and the following reflection. The perception is embodied into theory from where new assumptions can be deducted. These implications serve as an induction to action, to collect new experiences."

For learning there are therefore at least three different skills required: the skill to learn from concrete experiences, the perceptual skills needed to form abstract ideas, and the skill to implement experimental activity. Perception and experimenting are made easy by simulation. Freidbichter [1981] suggests with reference to Kolb's model:

> "Wenn man das Kolbmodel mit Beschreibungen von simulative Verfahren aus didactischen Sicht vergleicht, wird man aus Anhieb überraschende Ahnlichkeiten entdecken. Eben die Fahigkeiten sollen mit simulative Verfahren gefordert werden können ... besser als mit anderen Methoden und Medien. Das Model des Er-

fahrungslernens nach Kolb scheint dagegen geeignet zu sein, zentrale Aspekte von simulative Verfahren herauszustellen: Was Simulation auszeichnet, ist u.a. die Ermöglichung konkrete Erfahrung (und experimentieren)."

The assertion that computer simulation has an educational advantage by stimulating discovery learning is often advanced. But it should be remembered that this happens only when the necessary prerequisite learning for the subsequent concepts has taken place. This kind of learning leads to the student's discovery of certain characteristics of a process or phenomenon. In contrast, what is learned in frontal teaching is often provided by the teacher.

Much of the information that is transmitted in a frontal situation will not reach the student. Telling, without confrontation or experiments, can lead to a failing motivation and less enthusiasm in the student. So often our education is rich with stimuli and deficient in student response. Too often we encounter students who have the attitude of a passive listener. With discovery learning, when conducted in the right way, the student will acquire knowledge that he is more likely to retain. This knowledge stays, we assume, better available for reproduction.

The process of learning with a computer simulation can be transformed into a problem-solving process. A problem solving process requires student involvement. During the problem solving process adequate insight in the field of the problem will be available. During the process of learning there arises in the student an insight into means to reach the goals of the problem. This is usually supported by the subsequent discovery of a certain way of structuring, and of a certain type of reorganization of the problem situation, which leads to the solution. [Min, 1987].

The total problem has to be split up by the student into a number of small problems. During the learning activity the student is motivated to find the right solutions for these small problems. By this strategy the student often needs less urging. It may be that a valuable effect of working with a CS-program is that the student's mental flexibility is stimulated and thus analytical solution performance is increased.

The problems that we want to solve with the help of a CS-program must be well structured. A well structured problem has to meet the following conditions:

- The set of alternative solutions (hypothesis) is limited and finite.
- The solutions are consistently deducted from the model, which in turn corresponds correctly with reality.
- The effectiveness and efficiency of the solutions must be evident.
- The solutions (hypotheses) have to be controllable by the program.

Problem solving happens because the student becomes able to make a representation of the real world in which the problem manifests itself, in the form of testable hypothesis.

2.2.6 Five Different Ways to Use Computer Simulation

A CS-program can be used in education for instruction, training, or learning in at least five different ways [Min, 1987]:

- *(Free) discovery learning.* The student can do experiments which he himself finds useful, think up a problem himself, and try to solve it.
- *Learning through making an assignment.* A problem of free discovery learning is that the student will soon stop the simulation program when help stimulating him to go on is not offered. CS-programs do not include lesson material but instead are discovery environments. An important adjunct to simulation use is to furnish assignments (exercises) on paper. It is necessary that before the student uses the simulation he writes down his hypothesis about what is going to happen with the underlying model if he, for example, changes a parameter. The experiment then determines if his hypothesis will be supported or rejected.
- *Coached (discovery) learning.* The method of learning through the giving of assignments can also be applied to an assignment generated by the CS-program itself the moment that the CS-program 'thinks' that this, in view of the previous history, is possible or necessary. This method is complicated. Experiments have been done with the help of an expert system, the so-called computer coaches. For example, the training program ANAMNESE generates multiple-choice questions at unexpected moments, depending on how the student goes through the program.
- *Problem guided (discovery) learning.* This manner of using a CS-program shows a certain way of conducting the model on the screen. Certain variables do something "abnormal" to the system and the student has to make an analysis of the model's conduct in order to come to a diagnosis of the "problem". Figure 2.3 schematically shows how the educational learning process flows while using a CS-program this way.
- *Learning by doing real (scientific) experiments.* The last application of a CS-program is almost a replication of real practical laboratory work in which something must be measured. Measuring is typically done at the values 0.1, 0.2, 0.5 and 1.0 units of a parameter, and the pertinent values of a variable are read and written down. These data then have to be put into a graph either on screen or on paper during such practical work. The student must then come to know something about the relation between two quantities.

2.3 Outlook

Simulation can be used to considerable advantage for training (vocational) skills, for skills in solving intellectual problems in a great number of fields, and for giving insight into the complexity of reality. And, from an educational point of view, computer simulation of some human body systems is better than using a rat as a model of the human body!

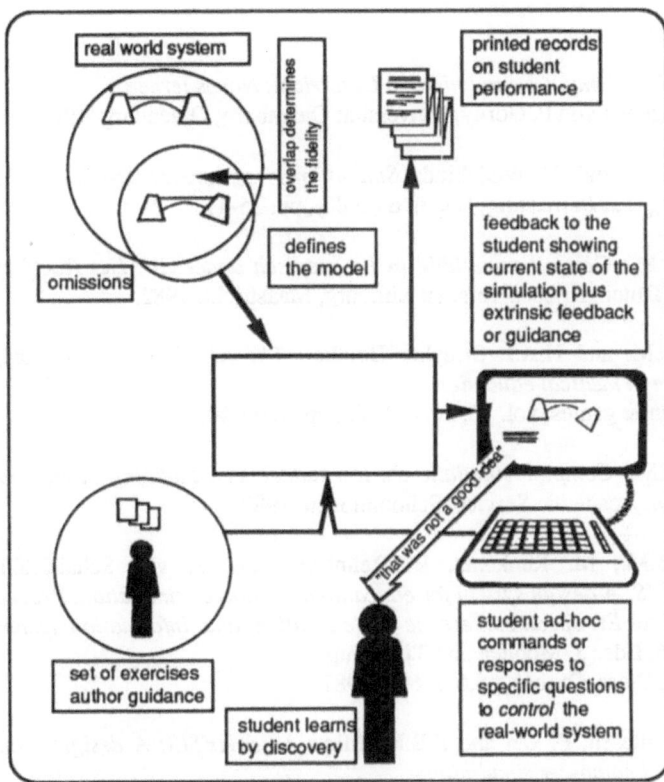

Figure 2.3. Schematic representation of aspects in a computer simulation experiment. Observable are the reality, a model of reality, the casual relationships, and feedback referring to an experiment (source unknown)

References

Daldrup, U.: *Computersimulation im Unterricht: Neues lernen,*
Herausgegeben von P. Gorny, Universität Oldenburg, Oldenburg, 1987.

Latzina, M. and J. Wedekind: *Simulationsprogramma: Systematische Be-
schreibung und Bewertung,* Log in 6 (1986), pp. 35-41.

Min, F.B.M.: *Computersimulatie in het medisch onderwijs, Het RLCS-system,*
Thesis in Dutch, Rijksuniversiteit Limburg, Maastricht, 1982.

Min, F.B.M. and H.A.J. Struyker Boudier: *The RLCS-system for computersimulation in medical education,*
Simulation & games, vol. 16, no. 4, 1982, pp. 429-440.

Min, F.B.M.: *Computersimulatie als leermiddel: een inleiding in methoden en
technieken,* Academic Service, Schoonhoven, 1987.

Min, F.B.M., M. Renkema, B. Reimerink and P. van Schaick Zillesen:
*MacTHESIS: A design system for educational computer simulation, Proceedings
of the First European conference on education and information technology,*
EURIT 86, Eds.: J. Moonen and Tj. Plomp,
Pergamon Press, Oxford, pp.689-691, 1987.

Schaick Zillesen, P. van and F.B.M. Min : *MacTHESIS: A design system for
educational computer simulation.*
Wheels for Europe, no. 2, 1987, pp. 23-33.

Schaick Zillesen, P. van and F.B.M. Min: *A design system for   educational
computer simulation programs used in secondary education in The Netherlands.*
Computers in Education, Elsevier Science Publ, North Holland, IFIP, Eds.: F. Lovis
and E.D. Tagg, 1988.

Wedekind, J.P.E.: "The Instructional use of Computer Simulation in the Teaching
of Biology: Three Examples", In: *Computer simulation in university teaching,* Ed.:
D. Wildenberg, North-Holland Publishing Company, 1981.

3
Artificial Intelligence and Simulation: An Introductory Review

Eugene J.H. Kerckhoffs and Henk Koppelaar

Abstract

Historically, two distinct approaches to modeling reality have been developed: numerical (quantitative) and symbolic (qualitative) modeling. Recently, the gap separating these two schools has narrowed. Both the numerical modeling community and artificial intelligence (AI) community have found that AI can contribute to simulation and vice versa. In this contribution an introductory review is presented on AI as a tool in simulation and, another addresses simulation as a tool in AI. As an example of a combined AI/simulation approach, a biomental application is briefly discussed.

3.1 Introduction and Overview

Computer simulation of systems comprises modeling real world systems, implementing the resulting models on computers and experimenting on these computerized models. The basis of the modeling problem is whether or not our knowledge about reality is expressible in axiomatic terms. If we succeed in transforming our knowledge into a computer program, then we can subject this knowledge to execution. The latter step implicitly introduces a formal way of testing the knowledge and finding the utmost consequences of the implied knowledge.

Depending on the kind of application, the aims of a simulation study can be:

- analysis and prediction of the dynamic system's behavior (for instance, for decision making purposes)
- accessibility and documentation of knowledge about specific systems
- education and training

Simulation studies normally follow some well-defined steps (see figure 3.1):

1. problem specification (resulting in a detailed abstract problem description)
2. selection of modeling method, conceptual model construction (resulting in a tool independent model description)
3. selection of solution techniques and tools, realization of an executable model (resulting in a tool/computer dependent model)
4. model validation (resulting in a validated model)
5. experiment planning, performing model experiments (resulting in simulation results)
6. analysis and interpretation of results.

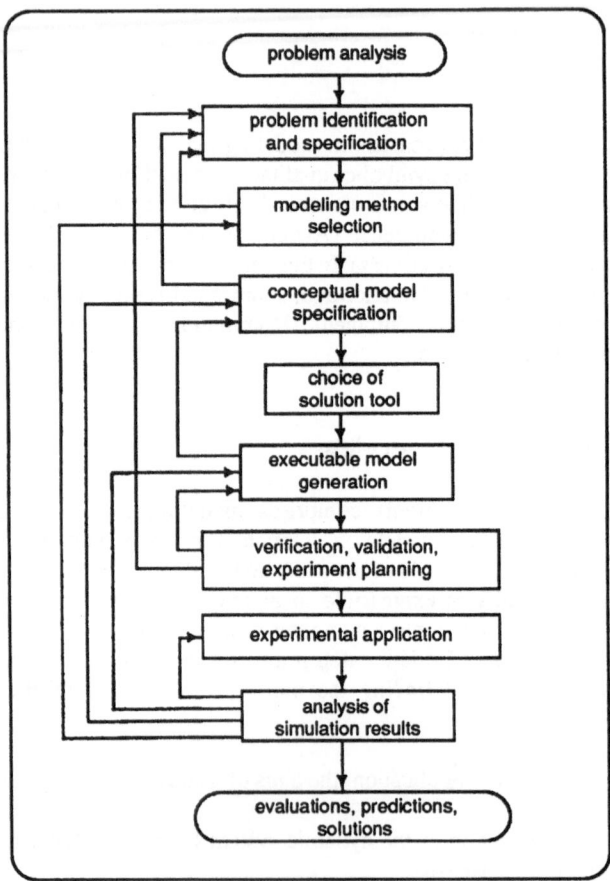

Figure 3.1. Stages in a typical modeling and simulation process

In this contribution, first we globally analyze some needs in complex simulation (section 3.2). It is concluded that, in addition to sophisticated methodologies and methods, advanced information processing techniques (such as parallel processing, AI, data base management) are needed and should be combined to create advanced simulation environments. Then, in the subsequent part of this paper we focus on AI (and especially its applied branch, "knowledge engineering"). After a short introduction to this domain (section 3.3) the application of AI concepts to enrich simulation environments (section 3.4) and the impact of simulation on AI (section 3.5) are reviewed. Finally, in section 3.6 we briefly discuss a biomental application as an example of a mixed AI/simulation approach.

3.2   Some Needs in Complex Simulation

3.2.1 Methodologies and Techniques

More advanced methodologies and new techniques in modeling and simulation are certainly in need of development. More effort is required in integrating various separate techniques, such as statistical aids in model construction and the extension of deterministic models through the use of stochastic models. It should be noted that pattern recognition appears in one form or another in many sophisticated software projects, as well as in mathematical modeling. Certainly, conceptual models involve at their very essence the problem of pattern recognition. Hence, a breakthrough from human operator pattern interpretation to the automated version based on powerful statistical packages would have a high influence on the construction of abstract conceptual models.

Improvements have to be made on mathematics as a tool for modeling of complex, especially non-engineering, systems. A mathematical basis for the concept of complexity is needed [Deland, 1983]. Frequently, there is a gap in understanding how parts of systems interact to produce the derived characteristics. New techniques could deal with the fundamental primitives of complex non-engineering systems, just as the equations relating to mass, force and velocity deal with the primitives in engineering. This would allow the user to interact with the system in the jargon of his discipline, so that his intuition and training could be directly applied to the problem. In biological systems, for instance, natural primitives could be diffusion, mass balance, dilution, delay, etc.; they could comprise a model hidden from the user except for its input, output and calibration parameters [Deland, 1983].

3.2.2 Simulation Tools

An increase of computing power and advanced software tools should allow research workers in scientific and engineering fields to concentrate more and more on the disciplinary aspects of the implemented problems, rather than on implementation, programming and numerical aspects. High speed computers, such as arrays of (identical) processing elements and pipeline-oriented processors, permit the (real-time) simulation of more detailed and large-scale models [Hockney, 1981], [Karplus, 1984, Karplus, 1987], [Kerckhoffs, 1985], [Spriet, 1982]. Advanced interactive environments should change fundamentally the way data are perceived, and the way the user behaves during problem analysis and model design; the machine should behave at the user's speed through the process of trial and invention, restructuring the model several times while simultaneously testing against real data.

Sophisticated methodology-oriented general simulation languages and packages are certainly in need (including versions to run on multicomputers); on the other hand, highly specialized macro-languages for systems analysis could allow the construction of complicated models written entirely in the language of, for instance, physiology and chemistry [Deland, 1983]. Simulation-oriented data base management systems should facilitate the interactive managing of both descriptive and quantitative information about real world systems and their simulation models as well as the experiments on these entities [Standridge, 1981], [Standridge, 1982], [Vansteenkiste, 1984]. Model bases may allow users to either search for a model with prescribed features that optimally fit their wishes, or to create new models from models previously stored (the latter on the basis of consulting modeling expert systems accessing these model bases; see section 3.4.1).

Inter-computer and inter-network communications can considerably affect the way simulation is performed. Files from data base management systems can be queried by other remote data base systems. (It should be noted in this respect, that a validated model can be used as a data base instead of the source research documents; the concept of a model as a data base has not yet been exploited to its full potential.) Another investigator's model on a remote machine can be accessed to test research questions or to be linked to a locally developed model to create a network for testing a more complicated model.

3.3 Knowledge-based (Expert) Systems (KBS)

A knowledge-based system is an intelligent computer program that uses knowledge and inference procedures to solve problems that are difficult enough to require significant human expertise for their solution; if the human expertise in a specific narrow domain is emulated, we speak of a (knowledge-based) expert system [Harmon, 1985], [Johnson, 1984].

Knowledge can be represented at different conceptual levels: consultation (inferred facts), procedural (goals and sets of rules), descriptive (context, objects and relationships) and meta level (rules that examine the other levels). Depending on the knowledge representation technique used, knowledge-based systems can be distinguished between:

- rule-based systems (knowledge at procedural and – if the system is being used – at consultation level), embodying independent chunks of knowledge (rules), and
- object-oriented or frame-based systems (knowledge at descriptive, procedural and – if being used – consultation level).

In the latter case, the developer begins by describing objects and their relationships; some of the objects might have rules associated with them, and these rules would create a procedural level. Structured rule systems and context trees fall within the above-mentioned extrema.

A "second generation" AI approach to knowledge representation is oriented around so-called "deep models", which capture explicitly causal relationships in the system being modelled [Chandrasekaran, 1983]. In contrast, production rules capture the empiric mapping from causes to effects, without asserting causality. As these deep models have become more complex, their application has become progressively more difficult.

From the methodological point of view there are extensive similarities between simulation systems and knowledge-based systems:

- Both are meant to represent knowledge and expertise about a system and its behavior in a specific domain. In simulation the knowledge is represented by events, processes, activities, frames and state transition descriptions, and in knowledge-based systems by frames, rules, semantic nets, etc.
- Both know the demand to describe and process uncertainty. Uncertainty in simulation systems is normally expressed by probabilities, and in expert systems by fuzzy sets, certainty factors, degrees of belief, etc.
- Both in simulation programs and expert system programs the flow control depends on logical decisions. Forward and backward reasoning in knowledge-based systems are methodologically comparable with, for example, next event/ previous event scheduling in discrete event simulation.

The methodological similarities offer good possibilities to integrate simulation systems and knowledge-based systems (see section 3.4 and 3.5). For example, one could think about (real-world system and simulation model) experimentation in terms of series of frames. It may be possible to exploit the similarity between knowledge representation frames and experimental frames, the latter having been defined to characterize the constraints in experimentation [Ören, 1979], [Zeigler, 1976], [Zeigler, 1984].

The current trends in AI are integration and enhancing user interfaces. Both reflect the strategic business value of expert systems. The many expert system shells available on the PC prove this. Most of the available expert system shells are applicable for shallow reasoning. This type of reasoning occurs in knowledge domains sustained by verbal information. For biological knowledge a deep reasoning tool is better. For biomedical and biomental modeling we need hybrid reasoning. Suitable expert system shells on the PC for this are GoldWorks, Guru and Nexpert.

3.4   Applying AI to Enrich Simulation Environments

3.4.1 How Could KBS Help in Continuous Systems Simulation?

Knowledge-based (expert) systems can be used in modeling and simulation both as advisory systems (see section 3.4.2) and to be linked to conventional simulation models for (on the fly) decision making and for intelligent control purposes [Birtwistle, 1985], [Holmes, 1985], [Kerckhoffs, 1986a], [Kerckhoffs, 1986b], [Luker, 1986], [Luker, 1987]. The task of knowledge-based advisory systems in, for instance, continuous systems simulation could be manifold [Kerckhoffs, 1986c]:

- user's support in the synthesis of differential models (general knowledge on the modeling process; candidate model choice and elaboration; selecting and lumping domain   primitives; defining differential models from schematic representations such as flow diagrams used in compartmental analysis, symbolic notations of chemical reactions, block diagrams used in engineering, Bond graphs, etc.)
- providing knowledge of the known mathematical properties of a model
- formal manipulations, such as symbolic differentiation or symbolic solving of equations (to calculate stationary solutions, sensitivity functions, etc.)
- user's support in the choice of algorithms, such as numerical integration algorithms, parameter estimation algorithms, validation procedures, etc.
- user's assistance in the interpretation of mathematical expressions. (To overcome the drawback that equations are often semantically disconnected from the application domain and are much less suggestive than, for instance, schematic representations, mathematical equations could be translated in such schematic representations.)

The advisory system should be able to simultaneously manage three kinds of bases, and multiple inference processes should operate on these bases:

- a data and objects base (consisting of a system base, model base, experimental frame base, real world experiment base and simulation experiment base [Kerckhoffs, 1986c], [Vansteenkiste, 1984])
- an algorithms base (containing the numerical algorithms, parameter identification algorithms, formal manipulations, etc.)
- knowledge base (including general modeling knowledge, knowledge about the conditions of applying methods and algorithms, mathematical properties

knowledge, model interpretation knowledge and experiment analysis knowledge).

In the subsequent section 3.4.2 we give general consideration on KBS as advisory systems in simulation, in which we do not purely focus on continuous systems simulation.

3.4.2 KBS as Advisory Systems

An advisory system or advice giving system is an expert system that gives coherent useful advice on a particular topic following a short consultation. Advisory systems can be viewed as alternative methods of maintaining and transferring knowledge; more traditional methods of doing this include documentation and teaching. Advisory systems for simulation (see figure 3.2) can contribute to improved performance with cost savings, important considerations as the complexity of simulation models grows!

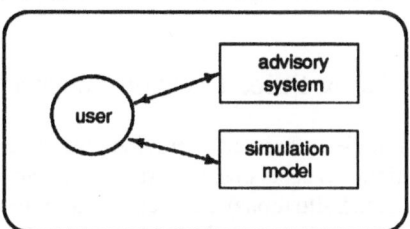

Figure 3.2. Advisory KBS in simulation

Since simulation itself is domain independent, expert systems in these areas may have to address an application domain as well as the "domain" of simulation. This generates a new set of problems not encountered by those expert systems that address a particular problem in a limited domain (e.g., medical diagnostic systems). According to [O'Keefe, 1986] the above suggests possibilities for:

- simulation advisory systems for domain experts
- domain advisory systems for simulation experts
- specialist simulation advisory systems for simulation experts (without the required specialism)
- simulation and domain advisory systems

Advisory systems for simulation are developed for application in various stages of the modeling and simulation process. Currently, available expert systems are only applicable for isolated problem areas, such as the selection of a model-adapted and computer-adapted simulation language [Elmaghraby, 1985], model validation

[Levine, 1984], or analysis of results for presentation to the user [Lehnert, 1983], [Swartout, 1983]. Also conceptual modeling is a particular area of the simulation process, where users may benefit from advice on how to structure their thoughts. An example of an expert system in this vein is given in [Doukidis, 1985], which allows natural language input of model definitions (based on activity-cycle diagrams) and prompts the user to extend these definitions, pointing out inaccuracies and inconsistencies.

There are only a few products supporting a user in several phases (hence more than only one) of the modeling and simulation process [Fox, 1986], although current research projects in this area typically address more than one of these phases [Ford, 1987] , [Haddock, 1987], [Hill, 1987], [Lehmann, 1986]. Since simulation suffers from a diversity of terminology and methodology, the production of general purpose simulation advisory systems will certainly be dependent upon the integration of methodologies.

Commercially available expert system shells are powerful enough for the production of simple advisory systems and are comparatively easy to use. Although rule based systems are adequate for simple advisory systems, more sophisticated simulation advisory systems may require deeper knowledge representation.

3.4.3 KBS as Simulation-Based Decision-Support Systems

Frequently, simulation studies are aimed to provide predictive information, which ultimately should contribute to decisions. Expert systems can select pertinent data (especially in data overload situations) and help in the interpretation of trends, provide perspectives from similar cases in its data base, and help define further questions and new simulation experiments addressing unresolved issues [Reddy, 1987], [Stefik, 1981]. Similar to the advisory systems mentioned in section 3.4.2, decision support systems and their related simulation systems are separately accessible to the analyst (see figure 3.3). Applications have been reported with regard to military systems [Stewart, 1986], complex manufacturing systems [Mellichamp, 1987], management information systems [Moser, 1986] and medical systems [Dassen, 1986].

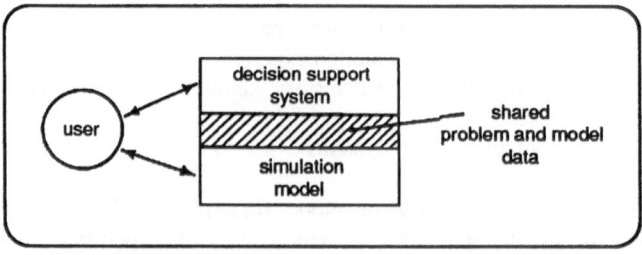

Figure 3.3. KBS in simulation-based decision support

3.4.4 KBS as Intelligent Front-End Systems

An intelligent front-end (IFE) system is a user-friendly interface to a (in this case: simulation) software package that would otherwise be technically incomprehensible and/or too complex to be accessible to many potential users [Bundy, 1984]. Front-ends are directly accessible for the analyst, while access to the simulation system is via these front-ends (see figure 3.4). This class of expert systems is used to bridge the gap between problem domains and a certain modeling tool. The meaning and description of items of a specific application domain are internally mapped into terms used and needed by the modeling tool.

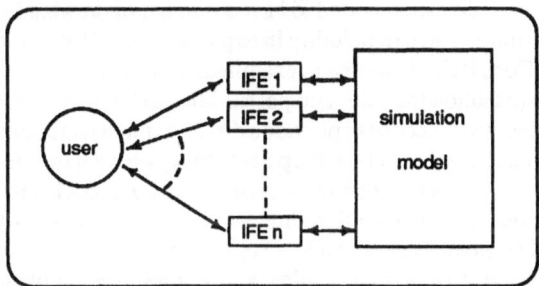

Figure 3.4. KBS as intelligent front-ends (IFEs) to a simulation software package

Motivations for using IFEs are:

- many analysts do not possess the mathematical or programming expertise to construct reasonably complex and realistic simulation models
- programs in high level programming languages or simulation languages are difficult to understand, modify or share with others.

In general, the main modules of an IFE are:

- "user interface" (or "dialogue handler"), capable of accepting unprompted input from the user, but also of guiding the user through the model building process
- "input file generator" or "task specification module", creating a set of instructions for the particular simulation package [Fjellheim, 1986] or supplying a formal representation of the model's mathematical structure that can be used to control a master simulation program [Muetzelfeldt, 1986]
- "knowledge base", storing domain specific knowledge dealing with, among others, the basic relations and parameter values that relate to the domain objects.

3.5 The Impact of Simulation on AI

3.5.1 Object-Oriented Techniques

The current so-called object-oriented modeling techniques originated from simulation, but have been very influential to modern knowledge representation developments in AI. These model-based knowledge representation techniques are indeed an example how simulation has contributed to the AI field. In the following, we briefly summarize object-oriented modeling and simulation.

Due to the used representations of knowledge in simulation models, traditional modeling and simulation are reported to have a number of drawbacks, such as lack of explicity, flexibility and extendibility in expressing model structures, extensive programming effort, lack of automatic examination of consistency and completeness, missing explanation facilities and possibilities to heuristic solutions. Consequently, techniques were needed to provide a direct, explicit and natural representation of the knowledge, we have about the part of reality we want to model. Compared to the conventional simulation models, we wish to make models more transparent with regard to our own knowledge, but we would also like to test them on consistency and completeness, and allow solutions on a heuristic level.

A sophisticated way of representing knowledge in a simulation model is offered by the object-oriented modeling tools [Middleton, 1986], [Klahr, 1980], [Klahr, 1986]. An object (or frame) is like a record: it has a name and contains information about the entity that is named. In fact, object-oriented languages (such as Simula and Smalltalk) assume that there are root or class objects that underlie more specific objects. A specific object may inherit information from other more generic objects in the domain knowledge base. An object has a number of slots (class properties), and a slot is made up of facets (properties of a slot). One facet contains an attribute, while another contains the value, fact, rule, procedure, or pointers associated with the attribute. A rule in a facet that fires whenever the value of the slot is altered is called a demon.

Conceptually, the major advantage of an object-oriented style of programming is that it allows the user to quickly describe a subject-matter domain without having to focus on procedures. For example, the subsystems of a car and its parts, as well as the connections between parts, can be described without considering how they work or how they change in time. In object-oriented simulation the knowledge bases contain:

- various facts about each object in the system concerned
- knowledge about the object's relationship to other objects
- knowledge about the relationship between objects and consisting specifications.

Moreover, knowledge is processed about the effect of actions in the system.

Consider the example of a ship approaching a harbour. We define the objects "ship", "port authorities", "pilot", "radar-control-system", etc. Interactions are a process of information passing, upon which each object bases its behavior (or state transition). Attached to each object there is information about its possible behavior, interaction with other objects, and other relevant properties. If the object "ship" transmits a message to the object "port authorities" about its ETA, the latter will react by asking details about the depth, cargo and insurance of the ship. If an expert system establishes that the information (the object "ship" eventually sends back) is consistent with an allowance to enter the port, the object "port authorities" will grant permission to enter. In this way the port approach of a ship can be simulated and, moreover, certain expert knowledge in an expert system can be linked to the simulation model.

3.5.2 Simulation as a Tool to Test Knowledge-Based Controllers

Using intelligent systems for (real world) process control is becoming an increasingly important application domain. Compared to conventional digital control systems, there are various reasons to employ knowledge-based controllers:

- the knowledge needed to control processes can be represented in a way that is natural, explicit, adaptive and easy to modify
- symbolic processing may lead to entirely new control procedures (exploitation of symbolic data manipulation)
- intelligent systems have proven to be able to deal with complex problems, incompleteness, uncertainty and fuzziness (fuzzy control)

Often, it is possible to replace real world processes with software programs, i.e. we can simulate these processes adequately. Simulations can serve useful purposes in developing the first general structure of the integrated system (i.e., knowledge-based controller linked to the real world process concerned). Simulation would replace the often complex, expensive and fault-prone input and output interfaces and related programs between the controller and the process. It can also serve as a test-bed within which precise, well designed experiments might be run in order to check the knowledge-based controller with respect to its perfect working; this really is an example of how simulation contributes to AI developments. Hence, in the simulation domain we are now confronted with a knowledge-based system controlling a simulation model to have, for example, a desired dynamic behavior by generating "on the fly" decisions about the model's input signals and/or changes in its parameter values.

With respect to knowledge-based systems in process control there are various possibilities, such as:

- a KBS supervising a conventional controller
- a KBS able to select the appropriate algorithm from a set of different control algorithms

- a KBS watching some boundary conditions of the process (e.g., switching over from hand to automatic control and vice versa)
- a controller based on a causal model of the process [Francis, 1985a], [Francis, 1985b].

In these kinds of applications the knowledge-bases may contain process knowledge (such as the order of the process, the parameters and their values and variations, non-linearities), technical control knowledge (e.g., rules to control damping and overshoot) and heuristic knowledge based on past experiences (for example, with respect to similar processes). Because of their nature, production rules are quite obvious to represent technical control knowledge. The kind of inferences required depends on the nature of the control problem itself: backward chaining fits control actions with a precisely defined goal, while global control asks in the first instance for forward chaining.

Currently, there are just a few examples of operational realizations of knowledge-based controllers [Bristol, 1984], [Carmon, 1986], [Francis, 1985a], [Francis, 1985b], [Higham, 1985]. In contrast to static knowledge-based systems, such as MYCIN, that are merely used for consultation purposes, intelligent systems controlling real world processes and/or their simulation models should be able to operate in a (rapidly) varying environment (dynamic knowledge-based systems). In real-time applications strict conditions are put on the speed of the reasoning systems. In these real-time applications there are various aspects that can have an impact: the speed of the process, the computer used, the applied algorithms and inference mechanisms (e.g., backward chaining is faster than forward chaining), compilation versus interpretation (compilation gives faster code, interpretation provides more flexibility with respect to modifications). Currently, many research efforts are directed to increase the speed of KBS, for instance by applying parallel processing techniques [Van den Herik, 1986], [Van den Herik, 1987], [Uhr, 1987], [Wise, 1987].

3.6 A Biomental Application of Simulation and AI

The biomental information processing model described in [Gallagher, 1972] is to provide insight in the dynamical relations of the various biophysical and psychological variables as an alternative route to predicting handicapping conditions in early childhood education. Their processing model has been simulated by us, which is summarized in the following.

The model is biological, the outcome and purpose are mental variables. So the model is hybrid. The relative usefulness of diagnostic instruments in early identification of developmental difficulties in children is not explored in the majority of the published articles. Most literature is limited to examining the potential usefulness of a single instrument for a rather narrowly defined purpose.

The model as developed in [Gallagher, 1972] is comprising most of the known variables in order to deepen the insight of the user in the following two ways:

- for prediction; with help of a simulation (supported by empirical data) it is possible to predict learning behavior of a child
- for diagnostic purposes (if a child is in treatment); the model is used then as an ordinary expert system (fed with the observed data)

The functioning of a human being is partitioned in five biomental domains:

1. Reception
2. Perceptual organization
3. Cognitive processes
4. Expression
5. Control mechanism

Example: Let us say that a child is presented with a barking dog. He receives stimuli (auditory and visual) and perhaps even haptic stimuli if the dog jumps upon him. The child will process this information with his past experience of this particular dog or his associations with other dogs. He can perform the basic reasoning that the barking dog is angry and that any dog that is angry is a threat.

All of this information processing will determine the particular outcome in which the child will express himself in speech, such as "Go away, dog", or in petting the dog as a means of quieting his own reactions. The feedback on the child's response will then come back to him as to what the reaction of the dog was to his reaction. The control mechanism regulates, in part, such feedback information on the individual's own performance and determines the set of orientation of the individual. It determines how or whether the individual can focus attention on specific stimuli, or how the world is organized perceptually by the individual, or what kind of problem solving strategies the child will apply to use past associations, given a particular problem, or the manner in which he will express the results of its search and operation.

A child with a learning disability may well have inefficient control mechanisms that make the sustaining of attention impossible or can lead to improper perceptual organization of stimuli. The model also indicates that certain defects and deficiencies are more serious in influencing the individual's information processing than others. A biological handicap such as polio or speech impairment (which does not involve other handicapping conditions) would have relatively little effect on the total functioning of the individual, since perceptual organization and central processing remain relatively intact. Even the feedback mechanism would not be disastrously interfered with, since there are alternate routes by which its expressive behavior can be sent back to the child.

The effort of the modeling will be justified for those developmental problems that seem to have the most serious impact on information processing for young children. These are children with the biological handicap of hearing problems and the mental handicaps of retardation, emotional disturbance and serious learning disabilities.

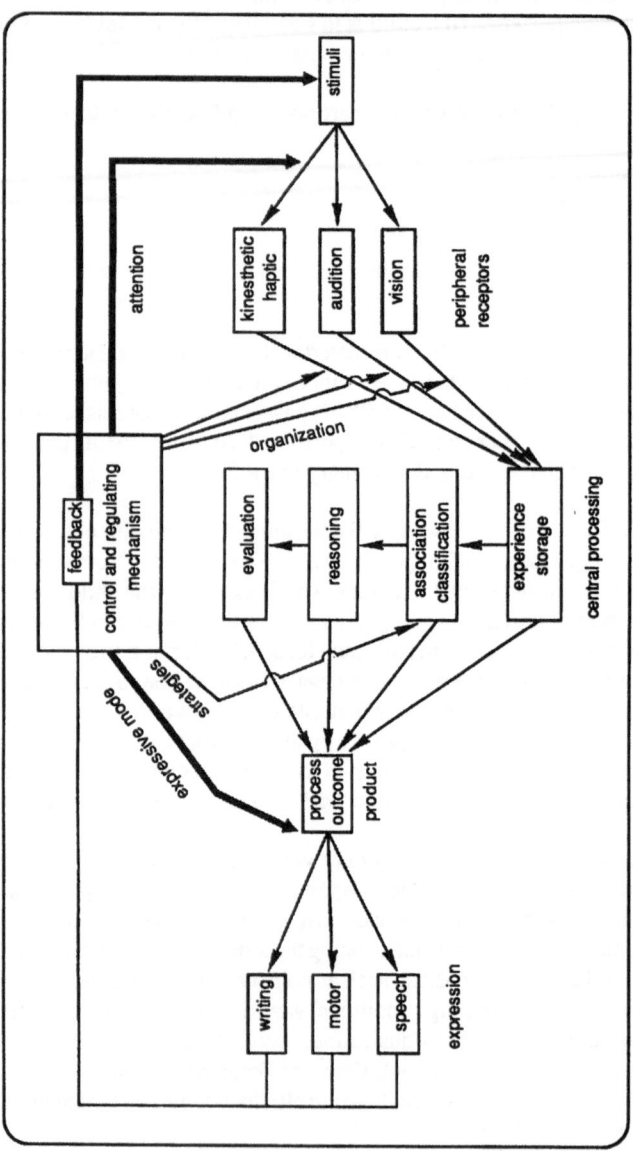

Figure 3.5. A biomental model [Gallagher, 1972]

For simulation of the model concerned (see figure 3.5) the following variables were used:

ST: stimuli
AT: attention
PR: peripheral reception
ORG: organization
ES: experience storage
AC: association classification
SR: strategies
RE: reasoning evaluation
PO: process outcome
EM: expressive mode
FU: functioning in school

The proposition derived:

1. If stimuli (ST) is low, few stimuli (ST) will arrive at the peripheral receptors (PR) and if stimuli (ST) is high, many stimuli (ST) arrive at the peripheral receptors (RP).
2. Low attention (AT) causes few stimuli (ST) for peripheral reception (PR). High attention (AT) allows many stimuli (ST) for peripheral reception (PR).
3. Ineffective organization (ORG) inhibits stimuli (ST) to the central processing system. Effective organization (ORG) yields the good possibility to pass stimuli (ST) to the central processing system.
4. Low experience storage (ES) causes low association classification (AC). High experience storage (ES) does not necessarily yield a good association classification (AC), but enhances it.
5. Low experience storage (ES) makes the process outcome (PO) lower. High experience storage (ES) makes the process outcome (PO) better.
6. Low association classification (AC) and low reasoning evaluation (RE) yield low progress outcome (PO). High association classification (AC) and high reasoning evaluation (RE) yield good process outcome (PO). Low association classification (AC) and high reasoning evaluation (RE) make the process outcome (PO) lower. High association classification (AC) and low reasoning evaluation (RE) make the process outcome (PO) higher.
7. Insufficient strategies (SR) lowers association classification (AC). High strategies (SR) raises association classification (AC).
8. Low process outcome (PO) causes bad functioning in school (FU). High process outcome (PO) causes good functioning in school (FU).
9. Bad expressive mode (EM) inhibits functioning in school (FU). Good expressive mode (EM) facilitates functioning in school (FU).
10. If functioning in school (FU) goes well, than many stimuli (ST) could be perceived. If functioning in school (FU) is bad, than less stimuli (ST) are perceived.

Three simulation runs with the expert system comprising the theory developed in
[Gallagher, 1972] were performed (see table 3.1). The model turns out to be
surprisingly insensible to variations in starting values. This was significant for
professionals (of the Utrecht University) to gain diagnostic insight.

Table 3.1.   Starting values of variables in three simulation runs

variables	starting values in three runs
stimuli ST	high, high, high
attention AT	low, medium, high
organization ORG	high, low, medium
experience storage ES	medium, medium high, unknown
association classification AC not low	low, medium high, possibly
strategies SR	high, low, medium
reasoning evaluation RE	high, medium, low
expressive mode EM	high, low, medium
functioning in school FU	unknown
process outcome PO	unknown
peripheral reception PR	unknown

3.7 Concluding Note

Biomedical and biotechnical systems belong to the most difficult systems that exist.
Although mathematical modeling and simulation have proven to be powerful tools
in understanding and dealing with these systems, techniques to enrich simulation
are certainly needed in order to enhance efficiency, flexibility and user friendliness.
In this contribution we have focussed on applying knowledge-based systems to
enrich simulation environments. On the other hand, simulation is also considered as
a tool to enrich AI-environments. All the concepts considered in this contribution
can be applied to biomedical and biotechnological systems, and can be
implemented in a PC environment.

References

Birtwistle, G., Ed.: *AI, Graphics and Simulation*, Society for Computer Simulation Int., San Diego, California, U.S.A., 1985.

Bristol, E.H.: *The Exact pattern recognition adaptive controller: a commercial success*, Proceedings Fourth Yale Workshop, 1984, pp. 50-55.

Bundy, A.: "Intelligent Front Ends", In: *State of the Art Report on Expert Systems*, Infotech Pergamon, 1984, pp. 15-24.

Carmon, A.: *Intelligent knowledge-based system for adaptive Pid-controller tuning*, Journal A, Vol. 27, Nr. 3, 1986, pp. 133-138.

Chandrasekaran, B. and S. Mittal: *Deep versus compiled knowledge approaches to diagnostic problem-solving*. Int. J. Man-Machine Studies, 19, 1983, pp. 425-436.

Dassen, W.R.M., W.P.S. Van Braam, K. Den Dulk, H.J.J. Wellens, E.D. Smith, L. Sasmor and P.P. Tarjan: *Expert systems and compiler techniques for intelligent implantable cardiac pacemakers*, Proceedings of the 1986 Summer Simulation Conference, 1986, pp. 410-414.

Deland, E.C.: *Conceptual models in physiology, where are we?*.
Proceedings of the IFIP Wg 7.1 Working Conference on Modelling and Data Analysis in Biotechnology and Medical Engineering (Brussels 1982). Eds.: G.C. Vansteenkiste and P.C.Young, North Holland Publishing Company, Amsterdam, 1983, pp. Xi-XviI.

Doukidis, G.I. and R.J. Paul: *Research into expert systems to aid simulation model formulation*. J. Op. Res. Soc. 36, 1985, pp. 319-325.

Elmaghraby, A.S. and V. Jagannathan: "An expert system for simulationists", In: Birtwistle, G., Ed.: *AI, Graphics and Simulation*, Society for Computer Simulation Int., San Diego, California, U.S.A., 1985., pp. 106-109.

Fjellheim, R.A.: "A knowledge based interface to process simulation", In: Kerckhoffs, E.J.H., G.C. Vansteenkiste and B.P. Zeigler, Eds.: *AI Applied to Simulation*, Simulation Series, Vol. 18, No. 1, Simulation Councils, Inc. (Society for Computer Simulation Int.), San Diego, California, Usa, February 1986., pp. 97-102.

Ford, D.R. and B.J. Schroer: *An expert manufacturing simulation system*. Simulation, 48, 1987, pp. 193-200.

Fox, M., N. Sathi, V. Baskaran and J. Bouer: *Simulation Crafttm: An expert system for discrete event simulation*. Proceedings of the Eastern Simulation Conference, Norfolk (U.S.A), March 1986.

Francis, J.C. and R.R. Leitch: "Artifact: A real-time shell for intelligent feedback control", In: *Research and Development in Expert Systems*, Ed.: M.A. Bramer, Cambridge University Press, 1985, pp. 151-162.

Francis, J.C. and R.R. Leith: *Intelligent knowledge based process control*. Proceedings Ieee International Conference Control 85, 1985, pp. 483-488.

Gallagher, J. and R.H. Bradley: *Early identification of developmental difficulties*, Early Childhood Education 71th year book, Nsse, 1972.

Haddock, J.: *An expert system framework based on a simulation generator*. Simulation, 48, 1987, pp. 45-53.

Harmon, P. and D. King: *Expert Systems -- Artificial Intelligence in Business*. John Wiley & Sons, Inc., New York, 1985.

Van Den Herik, H.J., E.J.H. Kerckhoffs and H. Koppelaar: "Simulation of the parallel knowledge-based system Hydra". *Proceedings of the 1986 Summer Computer Simulation Conference*, Eds.: R. Crosbie and P. Luker, (Reno, Nevada, Usa, July 28-30, 1986), Simulation Councils. Inc. (Society for Computer Simulation Int.), San Diego, California, U.S.A, 1986, pp. 972-977.

Van Den Herik, H.J., A.G. Hofland, J. Henseler and C.R.J. Verhoest: "Parallel processes in the knowlege-based system Hydra". *Proceedings of Imacs International Symposium on AI, Expert Systems and Languages in Modelling and Simulation*, (Barcelona, Spain, June 1987), North Holland Publishing Company, 1987.

Higham, E.H.: "A selftuning controller based on expert systems and artificial intelligence". *Proceedings IEE International Conference Control 85*, 1985, pp. 110-115.

Hill, T.R. and S.D. Roberts: *A prototype knowledge-based simulation support system*. Simulation, 48, 1987, pp. 152-161.

Hockney, R.W. and C.R. Jesshope: *Parallel Computers*, Adam Hilger, Ltd., Bristol, UK, 1981.

Holmes, W.M., Ed.: *Artificial Intelligence and Simulation*. Simulation Council, Inc. (Society for Computer Simulation Int.), San Diego, California, Usa, 1985.

Johnson, T.: *The Commercial Application of Expert Systems Technology.* Ovum Ltd., London, 1984.

Karplus, W.J., Ed.: *Peripheral Array Processors.* Simulation Series, Vol. 14 No. 2, Simulation Council, Inc. (Society for Computer Simulation Int.), San Diego, California, U.S.A., 1984.

Karplus, W.J., Ed.: *Multiprocessors and Array Processors.* Simulation Series. Vol. 18, No. 2, Simulation Council, Inc.(Society for Computer Simulation Int.), San Diego, California, U.S.A, 1987.

Kerckhoffs, E.J.H. and S.W. Brok: *The Delft Parallel Processor DPP81: Properties and utilization in simulation and related fields.* Systems Analysis, Modelling and Simulation (Journal of Mathematical Modelling and Simulation in Systems Analysis). Akademieverlag Berlin, Vol. 2, 1985, pp. 175-208.

Kerckhoffs, E.J.H.: *Parallel Processing and Advanced Environments in Continuous Systems Simulation.* PhD-Thesis in Computer Science, University of Ghent, Ghent, Belgium, June 1986.

Kerckhoffs, E.J.H. and G.C. Vansteenkiste: *The impact of advanced information processing on simulation : An illustrative review.* Simulation, Vol. 46, Nr. 1, January 1986, pp. 17-26.

Kerckhoffs, E.J.H., G.C. Vansteenkiste and B.P. Zeigler: *AI Applied to Simulation,* Simulation Series, Vol. 18, No.1, Simulation Councils, Inc. (Society for Computer Simulation Int.), San Diego, California, U.S.A, February 1986.

Klahr, P. and W. Faught: "Knowledge-based simulation". *Proceedings National Conference on Artificial Intelligence,* Stanford, California, U.S.A, August 1980.

Klahr, P.: "Expressibility in Ross: An object-oriented simulation system". In: *AI Applied to Simulation,* Kerckhoffs, E.J.H., G.C. Vansteenkiste, B.P. Zeigler, Eds., Simulation Series, Vol. 18, No.1, Simulation Councils, Inc. (Society for Computer Simulation Int.), San Diego, California, Usa, February 1986., pp. 136-139.

Lehmann, A., B. Knodler, E. Kwee and H. Szczerbicka: "Dialog-oriented and knowledge-based modeling in a typical PC environment". In: *AI Applied to Simulation,* Kerckhoffs, E.J.H., G.C. Vansteenkiste, and B.P. Zeigler, Eds., Simulation Series, Vol. 18, No. 1, Simulation Councils, Inc. (Society for Computer Simulation Int.), San Diego, California, Usa, February 1986., pp. 91-96.

Lehnert, W.G., M.G. Dyer, P.N. Johnson, C.J. Yang and S. Harley: *Boris -- An experiment in in-depth understanding of narratives*. Artificial Intelligence, 20, 1983, pp. 15-62.

Levine, A.P.: "An expert system for computer performance modeling: design issue". *Proceedings of the Cmg-Conference*, 1984, pp. 227-233.

Luker, P.A. and H.H. Adelsberger: *Intelligent Simulation Environments*. Simulation Series, Vol. 17, No. 1, Simulation Councils, Inc. (Society for Computer Simulation Int.), San Diego, California, U.S.A, January 1986.
Luker, P.A. and G. Birtwistle, Eds.: *Simulation and AI*. Simulation Series, Vol. 18 No. 3, Simulation Councils, Inc.(Society for Computer Simulation Int.), San Diego, California, U.S.A, January 1987.

Mellichamp, J.M. and A.F. Wahab: *An expert system for Fms design*. Simulation, 48, 1987, pp. 201-208.

Middleton, S. and R. Zanconato: "Blobs: An object-oriented language for simulation and reasoning". In: *AI Applied to Simulation*, Kerckhoffs, E.J.H., G.C. Vansteenkiste and B.P. Zeigler, Eds., Simulation Series, Vol. 18, No.1, Simulation Councils, Inc. (Society for Computer Simulation Int.), San Diego, California, U.S.A, February 1986., pp.130-135.

Moser, J.G.: *Integration of artificial intelligence and simulation in a comprehensive decision-support system*, Simulation, 47, 1986, pp. 223-229.

Muetzelfeldt, R., A. Bundy, M. Uschold and D. Robertson: "Eco --An intelligent front end for ecological modelling", In: *AI Applied to Simulation*, Kerckhoffs, E.J.H., G.C. Vansteenkiste and B.P. Zeigler, Eds., Simulation Series, Vol. 18, No.1, Simulation Councils, Inc. (Society for Computer Simulation Int.), San Diego, California, U.S.A, February 1986., pp. 67-70.

O'Keefe, R.M.: "Advisory systems in simulation". In: *AI Applied to Simulation*, Kerckhoffs, E.J.H., G.C. Vansteenkiste and B.P. Zeigler, Eds., Simulation Series, Vol. 18, No.1, Simulation Councils, Inc. (Society for Computer Simulation Int.), San Diego, California, U.S.A, February 1986., pp. 73-78.

Oren, T.I. and B.P. Zeigler: *Concepts for advanced simulation methodologies*, Simulation 32, 1979, pp.69-82.

Reddy, R.: *Epistemology of knowledge-based simulation*. Simulation, 48, 1987, pp. 162-166.

Spriet, J.A. and G.C. Vansteenkiste: *Computer-Aided Modelling and Simulation : International Lecture Notes in Computer Science*. Academic Press, London 1982.

Standridge, C.R.: *Using the Simulation Data Language (Sdl)*. Simulation, Vol. 37 No. 3, 1981, pp. 73-81.

Standridge, C.R. and A.A.B. Pritsker: "Using data base capabilities in simulation". In: *Progress in Modelling and Simulation*, Ed.: F.E. Cellier, Academic Press, 1982, pp. 347-365.

Stefik, M.: *Planning and meta-planning*. (Molgen: Part 2), Artificial Intelligence, 16, 1981, pp. 141-170.
Stewart, S.D.: *Expert systems invades military*. Simulation, 46, 1986, pp. 69-70.

Swartout, W.R.: *Xplain: A system for creating and explaining expert consulting programs*. Artificial Intelligence, 21,1983, pp. 285-325.

Uhr, L.: *Multi-Computer Architectures for Artificial Intelligence: Towards Fast, Robust, Parallel Systems*. John Wiley & Sons; New York, 1987.

Vansteenkiste, G.C. and E.J.H. Kerckhoffs: "Information base support in simulation of biological systems". In: *Proceedings of the 1984 Uksc Conference on Computer Simulation* (Bath/England, 12-14 September 1984), Ed.: D.J. Murray-Smith, Butterworths, London 1984, pp.198-218.

Wise, M.J.: *Prolog Multiprocessors*. Prentice-Hall 1987.

Zeigler, B.P.: *Theory of Modeling and Simulation*. Robert E. Krieger Publishing Company, Malabar, Florida, 1976.

Zeigler, B.P.: *Multifaceted Modeling and Discrete Event Simulation*. Academic Press, London, 1984.

4
Simulation: A General Design Tool

Paul P.J. van den Bosch

Abstract

Simulation is an accepted and very powerful tool for the analysis and design of dynamical systems. This chapter will introduce an example of a block-oriented, interpretative simulation program with its many additional facilities. It represents about the maximum result achievable with an interactive, interpretative, block-oriented simulation program. Still, some limitations can be recognized. In order to circumvent these limitations new developments have been started. Some of these future developments will be described, such as the use of different submodels based on block schemes, equations, bond-graphs or even higher-order programming languages.

4.1 Introduction

The roots of many simulation programs reach to times when digital computers were equipped with 8 to 16 kB and later even with 64 kB (PDP11) and when discussions were held to decide whether an analog or a digital computer was to be purchased to satisfy simulation needs.

Nowadays, simulation programs are widely accepted as a universal tool for analyzing dynamical systems. Problems associated with simulating continuous models with digital computers (numerical integration, algebraic loops) have almost all been conquered. The availability of cheap and yet powerful digital computers has realized an impressive penetration of these tools to almost any level of engineering. At the same time computer users have come in contact with highly sophisticated software for Personal Computers (PC), workstations or superminis. They pose almost the same requirements on interaction, user-friendliness, graphic facilities, etc. as offered by programs such as LOTUS 123 and WordPerfect. Based on this perception, new simulation programs have to satisfy more demanding requirements than earlier simulation programs.

In this chapter we will describe BIOPSI. BIOPSI is the name of the general-purpose simulation program PSI as it is used for studying bio-physical and bio-chemical phenomena. PSI has been developed at the Delft University of Technology in the past years and has emerged in many engineering applications. We will give brief statements on its facilities, its usages, and its weak and strong points, and we will compare it with other existing simulation facilities. In the last section we will formulate a framework for a new simulation environment, called PHI,which will be developed in the next two years to satisfy university and industry needs in the near future.

In studying the dynamical behavior of systems with simulation tools we can distinguish the following situations:

- *Model building*. Based on the physical laws of the system, a model must be derived to describe the relations among the many inputs, states and outputs of the system. Generally, this model is based on a description with differential and/or difference equations. Sometimes a description with transfer functions or state models is allowed or even a description via bond graphs;
- *Data acquisition*. Data must be collected from real-world experiments in order to obtain a set of input and output signals that are representative of the system behavior;
- *Parameter estimation*. Based on these input and output signals, researchers must obtain estimates of the values of the parameters of the model, that approximate the system dynamics as well as possible;
- *Design*. Designs can be performed with the aid of trial-and-error. In simulation programs a design is mainly based on optimization of a user-defined criterion. Another approach utilizes a linear model and linear design methods. A linear model can be made available via appropriate linearization techniques.
- *Implementation*. If a control structure has been derived, it is useful if it can be tested immediately in real-world experiments.

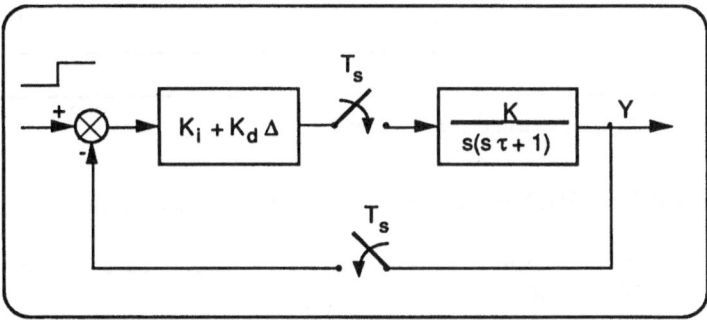

Figure 4.1. Control system

Many programs have been created that can be used for one or more of these applications. In the next sections a number of these programs will be discussed with an indication as to what extent they satisfy the needs of the users. As an example a control system as depicted in figure 4.1 has been selected for illustrating the way a program represents its simulation model.

4.2 Block-oriented Programs

Based on the approach of CSMP 1130, several block-oriented programs have been realized. Sets of (non-)linear, first-order differential equations can easily be solved. These programs are highly interactive and easy to use. In general, they fit the needs of naive users but lack the flexibility and power to satisfy the requirements of complex simulations. By a proper design the flexibility and capabilities can be extended to a high level. In this chapter we will discuss PSI [Van denBosch, 1985] as an example of such an approach. The philosophy of these programs is that any simulation model has to be constructed from basic elements like an integrator, gain, transfer function, table-lookup facilities, non-linearities, etc. Each element has just one output and several inputs. As a consequence of the availability of just one output,the name of a block can represent the value of its output. The structure of the control system of figure 4.1 is elucidated in figure 4.2.

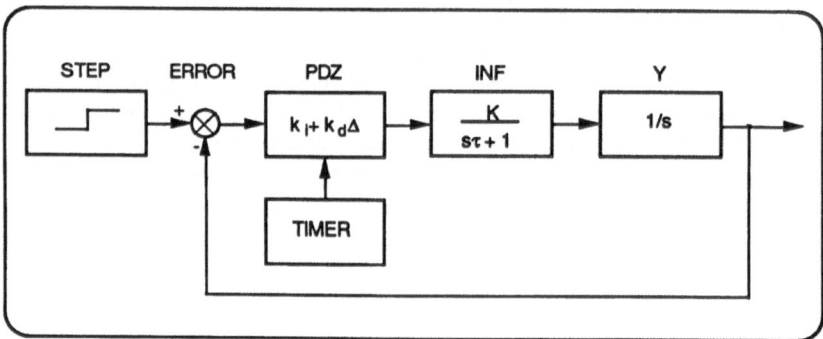

Figure 4.2.   Control system implemented in PSI

The description in PSI becomes sorted according to the block names:

Block	name	Type	Input1	Input2	Par1	Par2	Par3
	ERROR	SUB	STEP	Y			
	INF	INF	PDZ		0	1	1
	PDZ	PDZ	TIMER	ERROR	1	1	1
	STEP	CON			1		
	TIMER	TIM	STEP		.1		
	Y	INT	INF		0	1	

Because the structure and parameters of each block are stored in tables, calculation of the required outputs is achieved by interpretation and performed by calculating all blocks according to some calculation sequence. A modification in either the structure or the parameters does not require any recompilation or linking. Consequently, these programs can be highly interactive and can respond immediately to any modification required by the user. As a drawback of such an approach, the calculation speed is limited. Moreover, a user is not able to add additional block types or commands to the program, which limits its flexibility.

PSI has been extended with several facilities to circumvent, at least partially, some of the drawbacks of interpretative, block-oriented programs.
First of all, about 60 powerful block types are offered, including several different types of integrators (limited, mode-controllable, resettable), controllers (continuous and discrete PI and PD), delays, a pulse-width modulator, etc. Many (about 100) commands are available to satisfy existing needs, such as optimization, interactive table-lookup facilities, storage and manipulation of variables and multi-run facilities. Moreover, fixed and variable step size integration methods, iterative solution of algebraic loops and initial, dynamic and terminal facilities are available.

Within the structure of PSI, a user can write user-supplied blocks in Fortran, which then become, after compilation and linking, standard block types for PSI. Of course, these block types have to satisfy the definition of a block: a maximum of 3 parameters, 3 inputs and only one output is allowed. In PSI no distinction is made between a user-defined and a regular block type. For example, the following user-defined block type can easily be inserted in PSI:

```
Subroutine XX1(i1,i2,i3,p1,p2,p3,y,s,is)
Integer is
Real i1,i2,i3,p1,p2,p3,y,s
c
y = (p1*i1 + p2*i2)/(p3*i3 + s)
s = y
Return
End
```

A user can build his simulation model in PSI. Next, he can save his model and translate it into Fortran code, via PSICOM, the PSI-compiler. Because this code is well documented, he is able to modify it and add any Fortran subroutine to satisfy his requirements. After compilation of this dedicated Fortran program and linking it with the remaining parts of PSI, a dedicated version of PSI has been created. All commands and facilities of PSI are still available, except the facility to modify the structure of the simulation model. This structure is stored in Fortran code, which cannot be changed interactively. On the other hand any function can be realized and the calculation speed is improved since the calculations are based on a compiled code instead of being based on interpretation of data stored in tables. For example, the Fortran code obtained with PSICOM of the model of figure 4.2 becomes:

```
          Subroutine DERIV
C
          Include:'common.inc'
C
C----------------------------------------------
C   1 Y        INT - Integrator
C   Input(s):  2 INF
C
          DYDT(1)=PAR2(1)*C(2)
C
C----------------------------------------------
C   4 ERROR  SUB - Subtractor
C   Input(s):  3 STEP   1 Y
C
          C(4)=C(3)-C(1)
C
C----------------------------------------------
C   6 TIMER    TIM - Timer
C   Input(s):  3 STEP
C
              If (TEST5.EQ.1) Then
                If (C(3).GT.0.) Then
                    BLKDTR(6)=BLKDTR(6)+DT
                    If (BLKDTR(6).LT.(PAR1(6)-0.09*DT)) Then
                        C(6)=0.
                    Else
                        BLKDTR(6)=0.
                        C(6)=1.
                    Endif
                Else
                    BLKDTR(6)=PAR1(6)
                    C(6)=0.
                Endif
              Endif

C----------------------------------------------
C   5 PDZ      PDZ - Discrete PD controller
C   Input(s):  6 TIMER  4 ERROR
C
              If ((TEST5.EQ.1).AND.(C(6).GT.0)) Then
                C(5)=PAR2(5)*C(4)+PAR3(5)*(C(4)-BLKDTR(5))
                BLKDTR(5)=C(4)
              Endif
C
C----------------------------------------------
C   2 INF    INF - Integrator used as 1st order system
C   Input(s):  5 PDZ
C
          DYDT(2)=(PAR2(2)*C(5)-C(2))/PAR3(2)
          Return
          End
```

A user can define his simulation model in a kind of language that makes the model more easily surveyable and improves the readability and documentation of his model. This language allows submodels with local variables. Indeed a simulation model with about 1000 blocks, which we have made to simulate a model of the Dutch economy, lacks readability if it is shown as 20 pages of 50 lines with blocks ordered according to some sorting sequence. This language is translated via PSITXT, which translates the model described in the PSI language into an ordinary PSI model. For example, the following text can be interpreted by PSITXT and will yield the PSI model of figure 4.2:

```
SYSTEM  REG SYST
* The model is described via two submodels
SUBSYS REG                              * the discrete controller
    STEP=CON()(1)
    ERROR=SUB(STEP,SYST#Y)              * SYST#Y is Y in submodel SYST
    PDZ=PDZ(TIMER,ERROR)(1,1,1)
    TIMER=TIM(STEP)(.1)
ENDSUB
SUBSYS SYST                             * the second-order model
    INF=INF(REG#PDZ)(0,1,1)
    Y=INT(INF)(0,1)
ENDSUB
ENDSYS
```

Owing to these supplementary facilities PSI yields an attractive performance. PSI is easy to use for novices due to its very fast interaction and its similarity with ordinary block diagrams. Still, PSI also yields the expert many added facilities. Consequently PSI is used by several thousands of users both in industry and educational institutes, mainly on IBM PC or PS/2 systems for modeling, parameter estimation and design.

Characteristic applications of PSI deal with a highly complex non-linear model of a continuous process. This model can be analyzed in order to obtain additional insight or it can be controlled by a relatively simple controller. Models with up to hundred integrators and about 1000 to 2000 blocks have been implemented for studying the dynamic behavior of, for example, (bio)-physical phenomena, power systems, satellites, ships, submarines, planes, (bio)-chemical processes, climate systems, economic models, etc.

Generally, a process contains many non-linearities and many variables representing physical quantities. These variables have to be accessible for study. Therefore, high-order transfer functions or state models do not occur very often in describing a physical process. In contrast, controllers are based on a polynomial or state description and require adequate models.

However, one limitation is quite fundamental and restricts the applicability of PSI. Each block has a maximum of just one output. Therefore, PSI can be used very well for SISO (single-input, single-output) models, but using vectors and matrices is almost impossible. Applications such as least-squares estimators, predictive controllers and minimum-variance controllers cannot be realized in PSI.

We presume that PSI, with its many supplementary simulation facilities, represents about the maximum result achievable with an interactive, interpretative, block-oriented simulation program. Another approach must be selected for MIMO (multiple-input, multiple-output) applications.

4.3. Comparison with Existing Simulation Programs

ACSL and SIMNON [Elmqvist, 1977; Astrom, 1985] are other general-purpose simulation programs. In contrast with PSI their model description is based on equations rather than on blocks. This allows more complicated expressions to be used. From that point of view these types of simulation programs are more powerful. On the other hand, these programs are compiler-based. Any modification of the model's structure requires translation of the model into Fortran, compilation of this Fortran program and linking it together with the required software tools. Consequently, these simulation facilities yield less interaction and are suited for the experienced user.

In SIMNON the control system of figure 4.1 is inserted via:

```
CONTINUOUS SYSTEM SYST
" SYST describes the second-order model
     Input u
     Output y
     State x1, x2
     Der dx1, dx2
     dx1 = x2
     dx2 = u - x2
     y = x2
     . . .
     END
     DISCRETE SYSTEM REG
" REG represents the discrete PD controller
     Input r,y
     Output u
     State s
     New ns
     e = r - y
     v = kp*e + kd*(e - s)
     ns = e
     . . .
     END
     CONNECTING SYSTEM CON
" CON defines the connections among the blocks
     r[REG] = 1
     y[REG] = y[SYST]
     u[SYST] = u[REG]
     END
```

Recently, MATRIX$_x$ [Shah, Floyd and Lehman, 1985] has become available not only for linear analysis and controller design but also for simulation of (non-)linear models. It offers many attractive simulation features. MATRIX$_x$ is quite expensive and more suited for the experienced user.

4.4. Outlook

As a consequence of the still increasing power of digital computers, graphics devices and software, new innovative approaches can be applied in order to create an attractive analysis and design environment based on simulation. Appearing on the market are many new computer architectures that offer parallel execution, for example transputers, multi-processor systems, broad-instruction set architectures, etc. We still expect that general-purpose hardware with general-purpose operating systems will continue to dominate the market in performance per price unit. This indicates that the acceptation of general-purpose simulation software increases provided that it is available for a general-purpose computer.

Among many improvements in version 7 of PSI, the following are most striking for a user:

- The use of a cursor-controlled editor for modifying or extending the current simulation model enhances the user interface considerably compared with using separate commands.
- The possibility of replacing the names of inputs by algebraic expressions reduces the required number of blocks and improves the readability of a model. Especially with many algebraic relations, which for example occur in describing the nonlinear interaction in satellites or robots, expressions reduce the number of required block considerably. For example, the following single block with an expression instead of inputs:

   $X = INT(2*Y^4+ABS(Z)/(SIN(T)+1))$

   replaces the 9 blocks necessary to realize the required expression:

   ```
   Y3 = MUL(Y,Y,Y)
   Y4 = MUL(Y3,Y)
   ABSZ = ABS(Z)
   ONE = CON
   SINT = SIN(T)
   DENUM = ADD(SINT,ONE)
   DIVIDE = DIV(ABSZ,DENUM)
   SUM = SUM(Y4,DIVIDE)
   X = INT(SUM)
   ```

Based on the needs as formulated in the first section, we expect an integration of the attractive facilities from existing simulation programs into a new simulation environment, which we have called PHI. This environment PHI will be based on a multi-tasking operating system that allows the parallel execution of tasks. For example, printing or plotting, editing or calculating can be executed easily in parallel. Associated with multi-tasking, windowing techniques will improve the user interface by increasing the user's ability to control his simulation.

A second requirement will be the presence of a hierarchy in the simulation model. The distinction of the simulation model into several levels and/or submodels allows the user to combine several different model descriptions into one overall simulation model. Moreover, it improves the readability. A major calculation-oriented argument for distinct submodels is the availability of multi-rate sampling. Each submodel can be assigned its own "optimal" sampling rate and/or numerical integration method. In particular, implicit stiff methods favor a low dimension of a submodel. If submodels are allowed, a submodel can be modeled according to, for example,

-    a block-oriented simulation program like PSI
-    an equation-oriented simulation program like ACSL or SIMNON
-    a Fortran, Pascal or C procedure
-    a linear transfer function or state model like MATRIX$_x$
-    a bond graph-oriented model description

Each user is able to select a model description most suited for his needs. Moreover, a mixture of these different submodels is allowed.

These models describe a time-driven, parallel system. So, the independent variable of the simulation, in general the time, increases with fixed or variable steps to its final value. However, many applications of simulation concern event-driven or state-driven models, such as production systems, or power electronic convertors equipped with electronic switches. These models require that the value of the time becomes dependent on the point in time in which an event occurs. Up to now, these models have been excluded in programs such as PSI, SIMNON and MATRIX$_x$. With these programs it is rather difficult or even impossible to simulate, for example, a state-driven model of an electronic switch that becomes active if the value of some voltage passes through some threshold. SIMULA is a program especially suited for event-driven models, although time-driven models can be incorporated. As systems become more and more complex, the boundaries among time-driven, event-driven or state-driven models vanish. Consequently, there will be a need to create a simulation environment in which these separate models can be mixed without sacrificing calculation speed.

Appropriate general-purpose operating systems for the simulation environment PHI can be OS/2, UNIX and VMS. Only VMS can combine multi-tasking and real-time. The other operating systems have to be extended with communication with a real-time target computer. Then, in one environment, the user can execute all his requirements on data acquisition, modeling, parameter estimation, design and implementation.

The IBM PC and its many successors yield an unbeatable performance compared with their relatively low price. With their high penetration in society, they yield an attractive hardware environment for simulation-programs.

Although Fortran is used to create many simulation programs, including PSI, the selection of another structured language is obvious.Whether the language selection in writing PHI will be Ada, C, Modula-2 or Pascal has not yet been decided.

Research is going on to determine the possible contribution of expert systems in a simulation environment. We expect a modest contribution in improving the user interface and as an advisor in selecting appropriate values for the integration interval and integration method, in selecting a non-linear optimization method and an optimization criterion in order to satisfy the design requirements. Moreover, an expert system can be utilized in avoiding algebraic loops and in improving calculation speed by an appropriate calculation sequence that avoids all superfluous calculations.

References

Astrom, K.J.: "Computer-Aided Tools for Control System Design". In: *Computer-Aided Control Systems Engineering*, Eds.: M. Jamshidi, and C.J. Herget, North-Holland, Amsterdam, 1985, pp. 3-40.

Bosch, P.P.J. van den: "Interactive Computer-Aided Control System Analysis and Design". In: *Computer-Aided Control Systems Engineering*, Eds.: M. Jamshidi and C.J. Herget, North-Holland, Amsterdam, 1985, pp. 229-242.

Elmqvist, H.: *SIMNON - An Interactive Simulation Program for Nonlinear Systems*. Proceedings of Simulation 1977, Montreux, 1977.

Jamshidi, M. and C.J. Herget, Eds.: *Computer-Aided ControlSystems Engineering*. North-Holland, Amsterdam, 1985.

Shah, S.C, M.A. Floyd and L.L. Lehman: "MATRIX$_x$: Control Design and Model Building CAE Capability". In: *Computer-Aided Control Systems Engineering*, Eds.: M. Jamshidi, and C.J. Herget, North-Holland, Amsterdam, 1985, pp. 181-208.

5
Experience in Teaching with the Help of Models

Zdenek Wünsch

At the Institute of Physiology,[1] computer models began to be used in the education of Medical Faculty students in 1970. From the outset, this undertaking was considered to be an experiment, and it continues to be an experiment to date. It could not be otherwise, because computer technology, scientific knowledge and its horizons, and the attitudes of students and teachers are in a state of constant change; new possibilities arise and ways for their adequate utilization must be sought. Our experience with and ideas on instruction involving models should accordingly be looked upon as a case history; they are, of necessity, subjectively tinted and influenced by time and local specifics no less than any other individual experience. It is thus not off the point to begin with a brief backward look and then to characterize our teaching program and its aims.

The idea of using computers and computer models in teaching physiology emerged with the nascent concepts of the role of biocybernetics in medicine, several years ahead of our first applications. It appeared simultaneously with the valuation of the importance of cybernetics and other system sciences, prognostications about computer technology and its availability to users, and the presumed parallel growth of the cognitive and the practical value of computer simulations. At that time (i.e. in the mid-sixties) we were not in possession of computers suitable for practical teaching at a medical school. However, it was natural to try to begin right away and prepare for the future use of these methodological and technical contrivances for this particular type of work, verifying at the same time the adequacy of one's ideas about the useful content and extent of the new topics that would have to be incorporated into the already rather extensive curricula for medical students. Under these circumstances, it seemed a first-rate task to influence our students' way of thinking: students often have (not only at our Faculty) an inclination to diversely motivated negativism towards mathematic formulations and some abstract concepts, and they are not even satisfactorily equipped with concepts that would enable

---

[1] Institute of Physiology, Medical Faculty, Charles University Prague, CSSR.

them to accept and utilize at least the basic knowledge and methods of system disciplines in contemporary biological sciences.

We accordingly started by gradually including biocybernetic topics in the teaching of physiology, seeing that, in the main, its area comprises the functioning of the human organism and its subsystems. As a teaching subject, physiology represents a synthesis of the basic theoretical parts of medical studies on which the disciplines of later years can draw. It was our opinion - and experience upholds it - that physiology provides an adequate context for acquainting students with the basic general notions and methodological possibilities of the study of dynamic systems, including control and information systems, and with their applications; computer models and simulation techniques rank among these methodological possibilities. The role of control systems in the organization of physiological processes is one of the principal reasons why this general system theme and its applications have been (and will probably continue to be) presented under the heading of biocybernetics in our course of instruction.

To begin with, biocybernetics and modeling were included in the teaching of physiology in the form of a lecture course. Students could acquire only rather superficial information from these lectures, but the lecturer had an opportunity to check the limits of acceptable comprehensibility of his unaccustomed subject; within the scope of the course, the centre of gravity was restricted to the elementary aspects of control systems and some simple, appropriate examples of applications to physiology.[2]

The first computer models used in practical exercises were analog models. They enabled the direct graphic output of simulation task solutions and, therefore, the actual performance of simulation experiments within the limited time of the practical exercises. A number of different models were produced, covering both the general properties of dynamic, and especially of control, systems and some simple (simplified) physiological systems (e.g. compartment models of transport kinetics for some substances in the organism, models of partial aspects of the endocrine and the circulatory systems, or a model of the myotatic reflex arc and some of its elements). In addition, use was made of single-purpose electronic models (e.g. the Hodgkin-Huxley model), the adaptive linear neuron ADALINE (a model of pattern recognition), and so on. In general, the models were used in two ways: for demonstrating of the basic properties of dynamic systems (static and dynamic characteristics, responses to different input signals, stability, etc.) and for simulation experiments.

Analog models have some generally recognized and useful properties (e.g. fast solution of sets of differential equations), and because of these we still use them for solving some problems. However, they also have various disadvantages; their lower precision is not so much a shortcoming from the point of view of teaching as is the fact that, for various reasons, it is not efficient to allow a student to practice on an

---

[2] From 1968 onwards students had at their disposal a text, *Minimum of Medical Cybernetics,* by the author of this chapter, where the subject was expounded in greater detail and more systematically than was feasible in the lectures.

analog machine on his own. Leaving the pursuit of experiments with models to the teacher, the student does not have direct, personal contact with the model and easily remains in the position of a more or less passive onlooker. Problems in which students must take an active part and possibly compete have proved optimal. An example may be a problem in which the student performs the function of an operator who tries to compensate the effect of a random course of a disturbance variable; the results of his regulative interventions are quantitatively evaluated by an analog computer and may be compared with the results of other students. (The role of operator enables the student to form an opinion of the role of control on the basis of his/her personal experience; it also furnishes an opportunity for explaining different physiological and methodological aspects and applicabilities of related control systems.)

During the "analog era" we had opportunity to realize that:

1. Many interesting models can be devised, but not all of them can be practically employed in teaching (for want of time, organization and other reasons).
2. The maxim "less is sometimes more" can also apply to the complexity of the model (especially at the beginning of the course).
3. Active participation in simulation experiments has a positive impact on the student's attitude towards the subject and on the knowledge acquired.
4. The understanding of experiments with physiological-system models can be facilitated by familiarity with general properties of dynamic systems, and vice versa.
5. Students are capable of comprehending all sorts of things when they must (e.g. when they must sit for interim and final tests).
6. It is useful and desirable for the whole subject under discussion (basic theoretical knowledge of system and modeling as well as practical simulation experiments) to be regarded as an integral part of the whole discipline, in this case physiology, by all who are involved in its teaching.
7. The student should know that the knowledge and experience gained will be useful and necessary in further study and possibly in his/her future work.

Similar conclusions can also be drawn from our experience with models implemented on digital computers.

The introduction of digital computers into instruction has meant a qualitative change, because, disregarding their other advantages, even a technically illiterate person can communicate with a user-friendly implemented digital model without any particular training. Our first positive experience had already been acquired with the desk-top calculator HP 9830A equipped with an extended operational memory, a floppy disk, and a thermal printer. A little later we could implement our tasks on an SM 4-20 minicomputer provided with a network of alphanumerical terminal units.[3]

---

[3] The system is similar to the PDP 11/34 of DEC.

In using this technology, one gets block diagrams of a system's structure and various curves delineated in the form of pseudographs; pseudographs representation cannot compete with graphic display as regards the aesthetic effect and resolution power, but simplicity has its benefits too. At present we also use 8-bit and 16-bit (IBM PC compatible) microcomputers with graphics; in practical exercises they serve not only for modeling, but also for obligatory statistical evaluation of experiments, biosignal processing or knowledge testing.

Similarly to analog models, digital models are used mainly in two ways: for demonstrating properties of dynamic systems and for implementing models with which students experiment. The students may experiment with the models either voluntarily or within the framework of compulsory exercises. We consider it important that with digital, in contrast to analog models, active student-model interaction has predominance over demonstration. The possibilities are, in this respect, relatively unlimited as regards topic variety, complexity of models, and sophisticated conditions of experimental situations. One can certainly foresee, and all the other chapters of this book also point that way, that future developments will optimize and further extend these possibilities in all directions; it would undoubtedly be useful if some models, or lessons plus the models, became verified and generally available standards, part of a library from which one could make selections according to one's needs.

We never considered it our aim to replace traditional forms of instruction by computer teaching. However, there are topics and tasks that, for various reasons, can be better presented or handled only with the aid of computer models. Such tasks involve relatively simple demonstrations and experiments on one hand (e.g. the kinetics of a substance in a system of two compartments or the dependence of the course of a regulated variable on the transfer of the regulator), and on the other hand interactions with models of rather complex subsystems of the organism. An example of the latter variant is Coleman's model HUMAN,[4] a version of which, modified for our purposes, we also use. A developmental prospect for this category of models are models that qualify as trainers for future doctors.

The utility of a models task depends not only on the adequacy of the topic and the model, the qualities of its implementation, and so forth, but also on the scenario of the task as a whole. Here, as in other experiments, the student should be aware of the limits to the conditions within which sensible results can be expected. To be able to experiment with a model, he/she need not be capable of programming but should at least understand the basics, that is, the simple principles of model construction and of solving the given problem (e.g. aware that the validity of the model has limits).

Computer models can influence teaching simultaneously in several directions that are not mutually independent. The application of a model should not be an end in itself, but should - at least in our opinion - be called for by a concrete subject within a particular discipline of medical study (e.g. physiology in our case) by the

---

[4] Coleman, T.G., J.E. Randall, *Human, a comprehensive model for body function.*
University of Mississipi Medical Center, Jackson, Mississipi, 1983.

requirements of the teaching process, and by the estimate of the profile of knowledge and capabilities that will be required of the students in their continuing study and later by the need to keep abreast of progress in science. Within the framework of physiology, for example, the employment of models should:

a. mediate a better understanding of the functional interrelations and properties of the organism's systems,
b. enhance the object-lesson qualities of instruction,
c. enhance the efficiency of instruction,
d. extend the area of physiological teaching topics amenable to practical instruction,
e. facilitate acquaintance with the necessary minimum of basic concepts, knowledge, and methods in the disciplines of dynamic systems.

Teaching with the aid of models has or can have an impact on the students' knowledge and their manner of scientific thinking; it also influences the methods and organization of the teaching process. All of this naturally implies that the teaching staff is influenced as well. The result is a rather complicated interplay of relations that, depending on still further factors, undergoes changes in the course of time. This is also one of the overall reasons why we have always looked on model-involving education as an experiment: Whether models should be used or not is not the question; but one can always ask how a model could be employed the most efficiently under the given conditions. The answer to such a question does not depend solely, in our experience, on the properties and manner of implementation of the model itself.

References

Coleman, T.G. and J.E. Randall: Human, a comprehensive model for body function. University of Missisipi Medical Center, Jackson, Mississippi, U.S.A, 1983.

6
How to Use This Book

Jos E.C.M. Aarts and Ben J. Jansen

This book is about modeling and simulation in the fields of physiology and biomedical engineering. The computer experiments accompanying each model are intended to explore these models.

In the first part of this book several views about modeling and simulation are presented. In the second part, specific models and computer experiments are treated. The last part of the book deals with guidelines for teachers and the use of the simulation language BIOPSI.

The demonstration/simulation character of the computer experiments is of interest to the student and the teacher. The model experiments can augment lectures and will fit into any teaching setting. The exercises require the students to show initiative; they can change model parameters and run the programs again. Manipulating the material in an active way enhances the student's learning process. The teacher is referred to the third part, 'Guide for teachers and experienced users' in particular.

The presentation of each model follows an identical strategy. The learning objectives for a particular model are described with references to relevant background information. In describing a model, only those mathematical relations are given which are indispensable for the explanation of the model. Special attention is devoted to the relevant parameters and their physiologically meaningful ranges. Also the limitations of the models are outlined since every model is only a limited description of reality. This approximation is not only essential to present the basics of a physiological phenomenon but also to reduce the complexity of its description. In the case of the BIOPSI models, sufficient specification is given to be able to understand the model structure and parameters, intended as a guideline for further exploration. In most cases, an outlook is given for the specific case.

The computer simulations in this book allow the user to change relevant parameters within given ranges and to study their effect on the physiological system.

In order to reap the desired benefits from this book, some considerations should be taken into account.

-   The reader is supposed to be acquainted with the basic physiology relevant to each model. The required physiological background can be found in standard textbooks of physiology (see references).

- The models have been realized as executable programs or in the simulation language BIOPSI. No previous knowledge of the language is required in order to be able to experiment with the models.
  The experiments are selfinstructive, starting with a 'default demo'. Special help-screens have been designed for several experiments with additional information and/or exercise suggestions.
- A more intimate knowledge of the models is required if the reader wants to expand them to include new phenomena.

The models included pertain to the following fields:

Electrophysiology

The Excitable Membrane: The Hodgkin-Huxley Model

A 'classic' in modeling, presented here as the first example of the use of BIOPSI and its facilities in physiology and therefore extensively documented.

The Specific Conduction System of the Heart

A simple animation showing the depolarization and repolarization fronts traveling through the heart muscle and the relation between the electrical heart vector and the ECG leads. Based on Durrer's experiments and real-time signals.

Electrodes for Bioelectric Signals

The transfer characteristics of bioelectrodes in relation to the amplifier input stage and filtering are demonstrated for a variety of real-time and simulated signals. Thus, a better insight into signal distortion presented as time-course and spectra is given.

Circulation

The Heart as a Pump

The model 'CARDIO' describes the circulation and its regulation with the compartments: heart, vascular system, intra- and extracellular fluid volumes, kidney and nerve reflex system. Pathological cases and drug administration are included.

The Baroreflex-Controlled Circulation

The baroreflex is one of the systems controlling blood pressure stabilization and offers an interesting application of control theory in physiology, demonstrated in a fairly simple circulation model.

Pumping and Wall Mechanics of the Left Ventricle

The cardiac pump function is analyzed as a system with blood pressure and cardiac output as output parameters, determined by heart rate, left atrial pressure (preload), peripheral resistamce (afterload), and contractility. This allows open-loop experiments, including hemorrhagic shock and myocardial ischemia.

Heart Rate Regulation During Physical Load

Sport physiology studies: the human body under 'extreme' physical load situations. A model is presented describing heart rate regulation as one of the mechanisms in circulatory adaptation, also of importance in cardiology.

Catheter-Manometer Systems

Catheter-manometer systems are still widely used for blood-pressure measurement. The hydraulic transmission line introduces signal distortion, to be explained fundamentally by the traveling wave phenomenon. A treatment in terms of transmission-line theory shows aspects which cannot to be demonstrated by modeling with the lumped-circuit models frequently used.

Respiration

Respiration Regulation

The respitory system is modeled as a 'metabolic regulator', including a setpoint and a comparator. The dependency of the gas tensions on the ventilation are analyzed under steady-state conditions by modeling the controller as well as the 'plant'. Artificial ventilation and depression of the system by drugs are included.

The Bain Model for the Capnogram

The 'BAIN' circuit, used in capnography, is simulated with respect to the CO_2 flows, the pressures and the volumes in the system. Several parameters of the patient-ventilator system may be varied.

Clearance

Renal Function and Blood Pressure Stabilization

A renal function model is presented as an example of long-term regulation, based on the regulation of blood volume via the pressure-controlled urinary output, which increases with an elevated arterial blood pressure. Normal as well as pathological cases (Goldblatt hypertension) are included.

Urodynamics of the Lower Urinary Tract

The biomechanics of the lower urinary tract in the storage phase and the expulsion phase are modeled in BIOPSI, based on basic research. Some clinical procedures (cystometry, stoptest) can be demonstrated with the model, thus allowing clinicians to test a hypothesis for their clinical findings.

Compartmental Analysis

Fluid Volumes

The program 'FLUIDS' describes control mechanisms of body fluid volumes and electrolytes as well as repiratory control mechanisms. Thirst, fluid loss, exaggerated drinking, carbon dioxyde inhalation, severe physical exercise etc. may be simulated. Infusing fluids of different compositions and giving a diuretic represent clinical interventions to be studied.

Pharmacokinetics

The concept of compartments in the administration of pharmaca can be modeled to various degrees of sophistication, linear as well as non-linear. Sensitivity analysis and optimization are important points here.

Optimal Experiment Design in Pharmacokinetics

A model is presented which allows the definition of the optimal time points for concentration measurements and drug administration in relation to the cost and the accuracy of measurement. Parameter estimation is the central technique used in pharmacokinetic experiment design.

Cerebrospinal Fluid Circulation

The compensatory capacities of the cerebrospinal fluid space are modeled, yielding insight into the rate of CSF formation and absorption and into the storage capacities of the system.

Physiological Control Systems

Endocrine Control Systems

Blood Glucose Regulation

The student is provided with some simple example control systems derived from the complex endocrine system maintaining a constant glucose concentration in the blood. Thus, the concept of negative feedback in homeostasis is introduced in the 'discovery' of endocrine control systems .

Regulation of Gastric Acidity

Even with a relatively simple model of the very complex regulation of gastric acidity in humans, many aspects of the regulation can be analyzed and understood. The influence of different parts of the system on regulation under various conditions can be examined.

Neuronal Control Systems

Thermoregulation

Simulations can be performed to gain insight into the basic processes of thermoregulation. Shivering, sweating and heat loss through the skin are included. External effects taken into account are clothing, air velocity and humidity. The variety of parameters allow the construction of a large number of 'cases' to be studied.

Muscle Control

The sliding filament model, together with the activation-contraction coupling via Ca^{++} ions and the spindle and tendon transducers constitute the fundamentals for the simulation of muscle control. The model will be used in a small-signal linear approximation at a certain rest-length.

Thus, a scattered but relevant survey of several fields in physiology and biomedical engineering is presented, which can be useful in a variety of courses or in 'investigative learning'.

References

Guyton, A.C.: *Human Physiology and Mechanisms of Disease.*
3rd ed., Philadelphia, W.B. Saunders Company, 1982.

Mountcastle V.B. Ed.: *Medical Physiology.*, (2 volumes),
14th ed., St. Louis, The C.V. Mosby Company, 1980.

7
The Excitable Membrane:
The Hodgkin–Huxley Model

Rogier P. van Wijk van Brievingh and Ignacio A. García Alves

7.1 Aim of Instruction

These simulations include the Nernst potential for the relevant ions and the Goldman potential as the mechanism for calculating the resting potential [Plonsey, 1969].
The Hodgkin-Huxley equations themselves are elaborated with numerical values according to [Rendall, 1980] and [Talbot and Gessner, 1973].

The parameters to be controlled by the user are:

- extracellular K^+ and Na^+ concentrations
- two block-shaped current stimuli with adjustable amplitude, duration and interval
- temperature

The phenomena to be presented as functions of time are:

- activation parameters
- ion conductances
- individual ion currents
- the transmembrane potential
- the threshold potential

7.2 Theory and Reference to Standard Textbooks

7.2.1 Membranes and Ions

Potential gradients are known to exist across the membrane of excitable cells. In resting cells the potential difference is in the order of 100 mV with the inside negative. The strength and duration of a current stimulus determine whether an activation threshold is reached. During stimulation below the threshold, the membrane behavior may be described as passive and linear. When the transmembrane potential is raised above the threshold value, a brief transition in membrane properties occurs, bringing about an ionic current flow through the internal and external media. This response, during which the transmembrane potential rapidly changes from negative to positive and recovers more slowly, is called the *action potential*. In nerve cells this 'all-or-nothing' phenomenon is propagated along the length of the cell. In muscle fibers it constitutes the trigger mechanism for contraction.

The electric potential across the cell membrane is determined by the concentration gradients of ions and membrane permeability.

Besides Na^+, K^+, and Cl^- from salts, proteinates and other organic compounds may be ionized, represented by A^+ and A .

Tables of ion concentrations are to be found in many textbooks [Plonsey, 1969; Noble, 1975; Guyton, 1986].

Typical values are:

Table 7.1. Intracellular and extracellular ion concentrations

Ion [mmol/l]	Intracellular [mmol/l]	Extracellular
Na^+	12	145
K^+	155	4
A^+	0	5
Cl^-	4	120
A^-163	34	

A membrane that is permeable for only one ion will have an equilibrium potential across it as the net result of the diffusion gradient and the electrical force acting on the ions. This potential is called the *Nernst potential* [Plonsey, 1969].

$$V_{NERNST} = V_e - V_i = - \frac{RT}{z_i F} \ln \left\{ \frac{(C_i)_e}{(C_i)_i} \right\} \qquad (7.1a)$$

with: R = universal gas constant [8314.41 J.K^{-1}.mol^{-1}]
      T = temperature [K]
      z_i = valence of ith ion
      F = Faraday's constant [9.648 . 10^4 C.mol^{-1}]
      C_i = concentration of ith ion [mmol.l^{-1}]

In the case of i types of ions for which the membrane is permeable, only one steady state potential can be present -the *Donnan equilibrium potential*-, dictating the concentration gradients for all ion types.

$$V_{DON} = -\frac{RT}{z_i F} \ln \left\{ \frac{(C_i)_e}{(C_i)_i} \right\} \text{ for } all \text{ i} \qquad (7.1b)$$

For Na$^+$ and Cl$^-$ it follows:

$$V_{DON} = \frac{RT}{F} \ln \left\{ \frac{(C_{Na})_i}{(C_{Na})_e} \right\} = \frac{RT}{F} \left\{ \frac{(C_{cl})_e}{(C_{cl})_i} \right\} \qquad (7.1c)$$

If a membrane of uniform thickness with a homogeneous electrical field is assumed, the *Goldman quasi-static state potential* is found:

$$V_{GOLD} = V_e - V_i = \frac{RT}{F} \ln \left\{ \frac{P_K(C_K)_i + P_{Na}(C_{Na})_i + P_{Cl}(C_{Cl})_e}{P_K(C_K)_e + P_{Na}(C_{Na})_e + P_{Cl}(C_{Cl})_i} \right\} \qquad (7.2a)$$

with indices: e = extracellular
                  i = intracellular

The ratio of the permeabilities of the relevant ions has to be known in order to calculate V_{Gold}. [Plonsey , 1969] gives the following values for the resting state:

$$P_K : P_{Na} : P_{Cl} = 1:0.04 :0.45 \qquad (7.2b)$$

K$^+$: P_K/P_K =   1;   deviation from equilibrium -7.5 mV
Na$^+$: P_{Na}/P_K =   0.04;   deviation from equilibrium 156.4 mV
Cl$^-$: P_{Cl}/P_K =   0.45;   in equilibrium.

Generation of the action potential is governed by the change in time of the permeabilities. If the cell is excited, the sodium permeability is enhanced, which results in a local influx of sodium ions and a reversal in sign of the transmembrane potential. This is followed by a relatively greater outflow of potassium current and a recovery to the resting condition.

A dynamic description of an action potential was formulated by [Hodgkin and Huxley, 1952] through the use of a membrane model that includes as parameters the sodium and potassium conductances, their respective Nernst potentials and the membrane capacitance.

7.2.2   The Hodgkin-Huxley Equations

The basis of the HH-equations is the *ion-gating mechanism*, e.g. the existence of differently controlled gates for each type of ion. In the resting state, very few of the ionic channels are found to be conducting. As the transmembrane potential is varied in a depolarizing direction, the fraction of open channels increases. The gating mechanism is therefore voltage dependent. Since the response to an instantaneous voltage change appears not to be instantaneous, the gating mechanism is also time dependent. The HH-scheme for describing the gating mechanism assumes that each channel is controlled by a charged gate, which may be fully open (state α) or fully blocked (state β). A first-order reaction is assumed between these two states. If the fraction of gates in the β state is y, the fraction in the α state will be (1-y). If the opening rate coefficient is β_y and the closing rate coefficient is α_y the rate of opening will be given by $\alpha_y(1-y)$ and the rate of closing will be $\beta_y y$.

Thus, the net rate of change dy/dt in the fraction of open channels will be given by:

$$dy/dt = \alpha_y(1-y) - \beta_y y \qquad (7.3a)$$

with the steady-state value:

$$y_\infty = \alpha_y/(\alpha_y + \beta_y) \qquad (7.3b)$$

When the potential is changed step-wise from one value to another, the value of y will follow an exponential time course when α_y and β_y are constants (rate coefficients).

The time-constant for this process equals $1/(\alpha_y + \beta\beta_y)$.

The current carried by the channels will be given by the maximum current multiplied by the fraction of channels conducting.

When the membrane potential is not equal to the Nernst potential, there will be a net flow of ions determined by the difference between the membrane potential and the equilibrium potential. Together with the membrane capacitance, this is expressed in the electrical analog (Fig. 7.2).

$R_{Na} = 1/g_{Na}$; $R_K = 1/g_K$; $R_l = 1/g_l$ vary with time and membrane potential; the other components are constant.

The capacitive current equals dV/dt and the ion currents are given by Ohm's law, with g_i the conductances. Thus, the total transmembrane current I (not counting the stimulus current) is given by equation 7.4a.

From an examination of data acquired from voltage-clamp experiments, Hodgkin and Huxley evaluated g_{Na} and g_K as functions of voltage and time. For potassium they postulated four identical gates (activation variable n) for each K^+ channel, and these must all be in the open state in order to conduct.

Sodium required an extra mechanism to describe the fact that the increase in Na^+ current during depolarization is not maintained. Following its relatively fast onset it decays exponentially: the inactivation process. In the kinetic scheme this may be

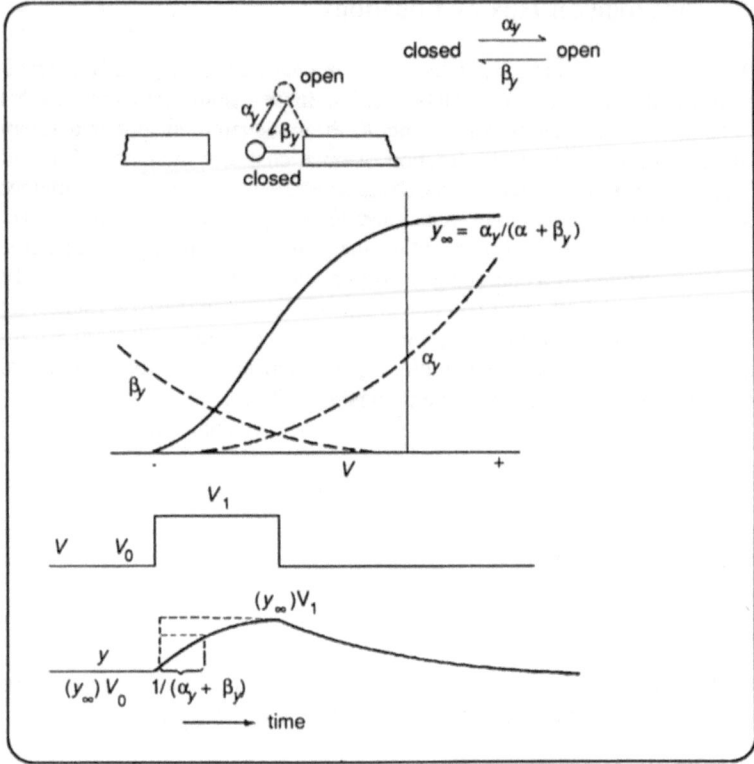

Figure 7.1. (after [Noble, 1975])

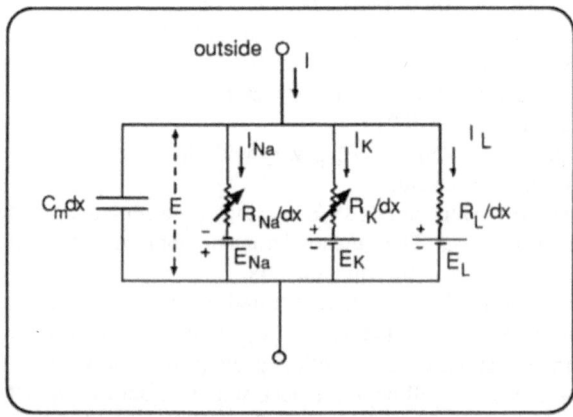

Figure 7.2. The equivalent circuit proposed by Hodgkin and Huxley for the axon membrane (Squid)

described by supposing that one of the gates moves more slowly than the others and moves in the opposite direction, i.e. it closes on depolarization of the membrane. Three activation gate variables (m) and one inactivation gate variable (h) were found to give the best fit to experimental data. The variables act on the maximum value of the respective conductance $\bar{g}_i$ as factors to the power of their number (eq. 7.4a).

The activation parameters n, m and h are calculated according to equation 7.3 from the first-order differential equations 7.4b through 7.4d. The rate coefficients α and β for these three equations are given in equations 4e through 4j, based on curve-fitting to experimental data. These were obtained from experiments carried out on the giant axon of a squid at 6.3 °C; the temperature correction factor φ acting on the time derivatives of the activation parameters, is given in equation 7.4k.

The Hodgkin and Huxley equations:

$$I = C\dot{V} + \bar{g}_{Na}\, m^3\, h(V-V_{Na}) + \bar{g}_K\, n^4(V - V_K) + \bar{g}_L(V-V_L) \tag{7.4a}$$

$$\dot{m} = \phi\, [(1-m)\alpha_m(V) - m\beta_m(V)] \tag{7.4b}$$

$$\dot{n} = \phi\, [(1-n)\alpha_n(V) - n\beta_n(V)] \tag{7.4c}$$

$$\dot{h} = \phi\, [(1-h)\alpha_h(V) - h\beta_h(V)] \tag{7.4d}$$

$$\alpha_m = 0.1\,(V + 25) \left[\exp\left(\frac{V+25}{10} \right) - 1 \right]^{-1} \tag{7.4e}$$

$$\alpha_n = 0.1\,(V + 10) \left[\exp\left(\frac{V+10}{10} \right) - 1 \right]^{-1} \tag{7.4f}$$

$$\alpha_h = 0.07\,[\exp\,(V/20)] \tag{7.4g}$$

$$\beta_m = 4\,[\exp\,(V/18)] \tag{7.4h}$$

$$\beta_n = 0.125\,[\exp\,(V/80)] \tag{7.4i}$$

$$\beta_h = \left[\exp\left(\frac{V+30}{10} \right) + 1 \right]^{-1} \tag{7.4j}$$

$$\phi = 3^{(T-6.3)/10} \tag{7.4k}$$

Legend:

n = (K+)-activation variable [-]
m = (Na+)-activation variable [-]
h = (Na+)-inactivation variable [-]
α_i = ion-transfer factor, for ion i [1/ms]
β_i = ion-transfer factor, for ion i [1/ms]
V_i = V_{NERNST} - $V_{GOLDMAN}$ for ion i , V_{OUT} - V_{IN} [mV]
V = transmembrane potential - $V_{GOLDMAN}$, V_{OUT} - V_{IN} [mV]
c = membrane capacity [$\mu F/cm^2$]
$\bar{g}_i$ = maximal conductivity for ion i [1/Ω]
φ = temperature factor [-]
I = transmembrane current, inward positive [μA]
T = temperature [°C]

7.2.3   The Stimulation Threshold

Fitzhugh [1960] observed that the activation variables n and h vary timewise but slowly and that their sum is almost constant, (see exercise 7.5.2.2.3), whereas m and V vary rapidly but asynchronously. This offers the opportunity to reduce the four-dimensional state-space S(V,m,n,h) to a more intelligible V-m plane if n and h were regarded as constant.

Thus, with $\dot{n} = \dot{h} = 0$ and letting I = 0 after the stimulus has ceased to exist, the *isoclines* are obtained from equations 7.4c and 7.4d:

$$\frac{dm}{dV} = \frac{\dot{m}}{\dot{V}} = \frac{C\phi\,[(m-1)\alpha_m(V) + m\beta_m(V)]}{\bar{g}_{Na}\,m^3 h(V+V_{Na}) + \bar{g}_K\,n^4(V+V_K) + \bar{g}_L(V+V_L)} \qquad (7.5)$$

From equation 7.4b follows the $\dot{m} = 0$ nullcline:

$$m(V) = \frac{\alpha_m}{\alpha_m + \beta_m} = \frac{0.1(V+25)\left[\exp\left(\dfrac{V+25}{10}\right) - 1\right]^{-1}}{0.1(V+25)\left[\exp\left(\dfrac{V+25}{10}\right) - 1\right]^{-1} + 4\,\exp(V/18)} \qquad (7.6)$$

With I = 0 the $\dot{V} = 0$ nullcline is found:

$$m(V) = \left[\frac{-\,\bar{g}_K\,n^4(V-V_K) - \bar{g}_L\,(V-V_L)}{\bar{g}_{Na}\,h(V-V_{Na})}\right]^{1/3} \qquad (7.7)$$

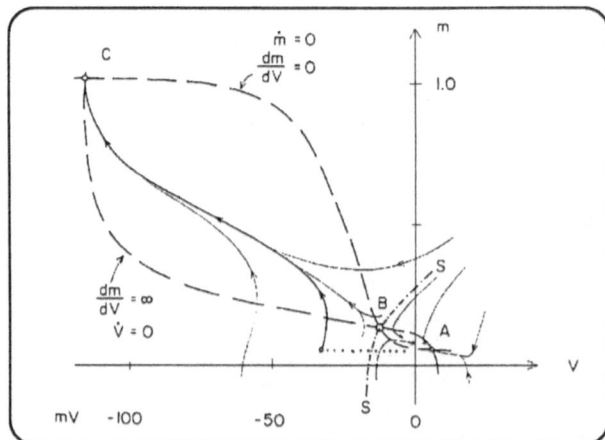

Figure 7.3.   (after [Talbot and Gessner, 1975])

In this phase plane of the reduced m(V) system, A, B and C are singular points, of which A represents the resting state, C the stable (as long as n = h = 0) depolarization state, and B is an unstable point through which passes the *threshold separatrix* S-S between the stable fields around A and C. In single-pulse operation of the HH-equations, a negative pulse carries the phase point from A beyond S-S to a trajectory along which it moves to stable depolarization at C. As time progresses, the values of n and h change, raising nullcline V = 0 so that intersections C and B coalesce and vanish, leaving the potential to return by a path through the upper half-plane to its rest point at A [Talbot and Gessner, 1975].

Thus, the reduced system may be used to determine the threshold B as the point m(V) for which both nullclines intersect with the minimal negative value of V.

7.3   Aspects to Be Considered

The classic work of Hodgkin and Huxley [1952] has been thoroughly described and verified. The kinetics of the membrane conductances for the relevant ions can account for many phenomena encountered in nerve physiology. Many of these cannot be measured in vivo, but simulation allows them to be presented in graphical form, thus giving insight into the processes involved. Especially the specific conductances, governed by activation parameters via rate coefficients, and the individual ionic currents can be studied effectively. The work of Fitzhugh [1961] provides a justifiable simplification of the state space, allowing for a criterion to establish the threshold potential, as described by Talbot and Gessner [1973] within the framework of the stability analysis of the equations involved. Parameter values are kept within ranges consistent with physiological experiments.

7.4   Model Description

7.4.1   Foundations and Assumptions

The model is described by equations 7.4a through 7.4k.
The threshold voltage V_d is found as the intersection of the nullclines (equations 7.6 and 7.7) with the smallest negative value of V.

7.4.2   Structure

The model is composed of the parts specified below, which communicate through the parameters mentioned. The conventions for voltage polarities and current directions are chosen according to [Hodgkin and Huxley, 1952]: inward currents are counted positive and voltages are referred to the intracellular potential.

- "Temperature" submodel
  Function   : Calculation of the temperature dependencies

- "CNa_ex & CK_ex" submodel
  Function   : Choice of the extracellular concentrations of Na^+ and K^+

- "Vr-Goldman" submodel
  Function   : Calculation of the resting potential of the cell

- "α_i & β_i " submodel
  Function   : Calculation of the ion rate coefficients

- "Stimuli" submodel
  Function   : Generation of one or two stimulus currents

- "N, M & H" submodel
  Function   : Calculation of the activation variables

- "GNa & GK" submodel
  Function   : Calculation of the ion permeabilities of the cell

- "INa, IK & IL" submodel
  Function   : Calculation of the ion currents

- "HH-potential" submodel
  Function   : Calculation of the transmembrane potential

- "Isoclines" submodel
  Function   : Calculation of the isoclines, for determining the threshold value

These submodels are functionally independent of one another; the interaction between the blocks takes place via signals that constitute a well-defined physical quantity as described in the literature.
Thus it is feasible to rearrange parts of the model for the purpose of different simulations.

The specification of the submodels is as follows:
On top of the blocks the name of the block is indicated in *italics*; the function is given inside the block.
The parameters of the blocks are indicated as p1, p2 or p3 according to the BIOPSI convention.

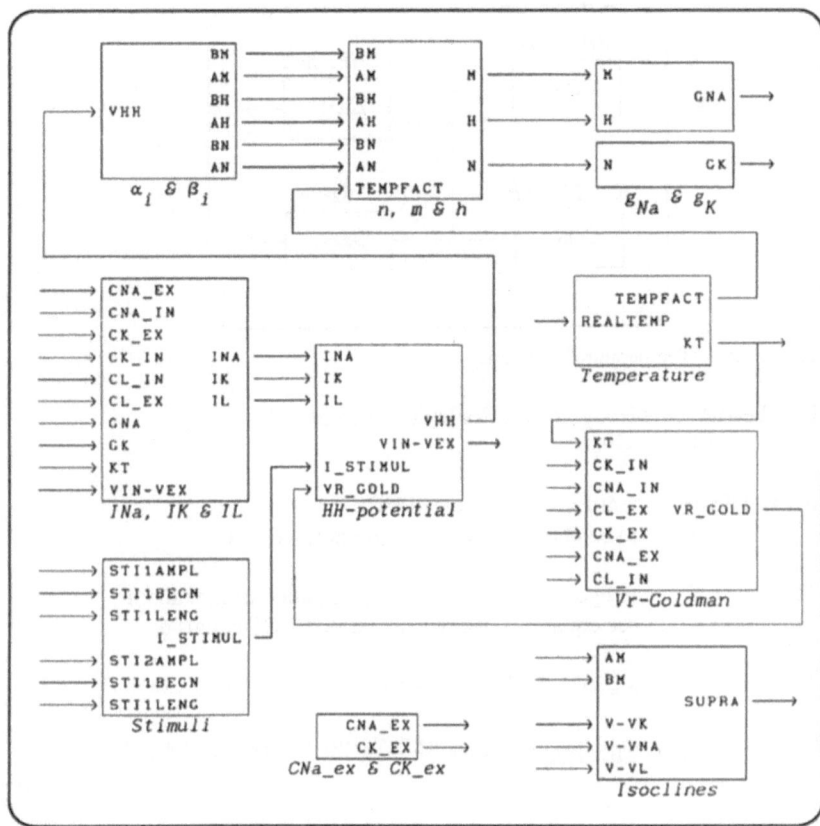

Figure 7.4.   Survey of model

7.4.2.1   Temperature

This submodel calculates the temperature correction factors for the time derivatives of the activation parameters and for the Nernst equations.

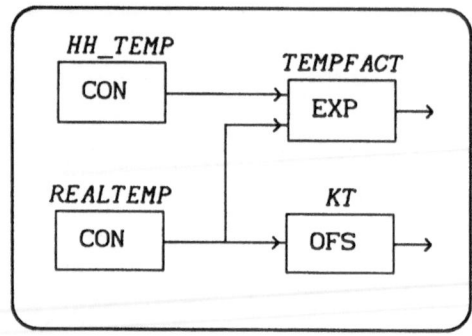

Figure 7.5. "Temperature" submodel

HH_TEMP   : Hodgkin and Huxley temperature
         p1 : -6,3 °C

REALTEMP : temperature of the cell [°C]
         p1 : to be chosen by the user (default = 6.3 °C)

TEMPFACT : temperature factor for the calculation of n,m and h
equation   : see equation 7.4k, where TEMPFACT equals ϕ

$$\phi = 3^{(T-6.3)/10} \text{ , so}$$

$$TEMPFACT = 10^{((REALTEMP + HH\_TEMP) * 10\log 3)/10}$$

p1 : 1
p2 : (log 3)/10 = 4.771E-2
p3 : 2

KT : (gas constant/Faraday's constant) * Kelvin temp
p1 : (gas constant/Faraday's constant) * 10^3
       [k = R/F = 8.617E-2 J/K]
p2 : factor to convert degrees Celsius to Kelvin [273.15]

7.4.2.2 CNa_ex & CK_ex

Separate submodels are used for the extracellular concentrations of the (Na$^+$) and
(K$^+$) ions separate submodels are used, allowing flexible simulations.
The extracellular concentrations may be made to vary linearly between chosen
minimal and maximal values with an adjustable slope. The intracellular
concentrations are kept constant.

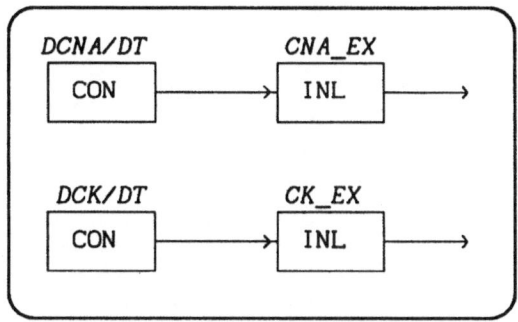

Figure 7.6. "CNa_ex and CK_ex" submodel

DCNA/DT : slope of the extracellular concentration of Na$^+$ [mmol/l.ms]
    p1 : to be chosen by the user (default = 0)

CNA_EX : extracellular concentration of Na$^+$ [mmol/l]
equation : CNA_EX = p1 + DCNA/DT * t
    p1 : starting value of $(CNa^+)_{ex}$ [145 mmol/l]
    p2 : minimum value of $(CNa^+)_{ex}$ (default = 1)
    p3 : maximum value of $(CNa^+)_{ex}$ (default = 300)

DCK/DT : slope of the extracellular concentration of K$^+$ [mmol/l.ms]
    p1 : to be chosen by the user (default = 0)

CK_EX : extracellular concentration of K$^+$ [mmol/l]
equation : CK_EX = p1 + DCK/DT * t
    p1 : starting value of $(CK^+)_{ex}$ [4 mmol/l]
    p2 : minimum value of $(CK^+)_{ex}$ (default = 1)
    p3 : maximum value of $(CK^+)_{ex}$ (default = 300)

7.4.2.3 Vr-Goldman

This submodel calculates the resting potential of the nerve-cell from the intracellular and extracellular concentrations of Na$^+$, K$^+$ and L$^-$ as chosen by the user. The resting potential is calculated with Goldman's equation [equation 7.3]. L$^-$ represents the ions causing the membrane's 'leaking' current.
The default concentrations are chosen as the values in Table 7.1.

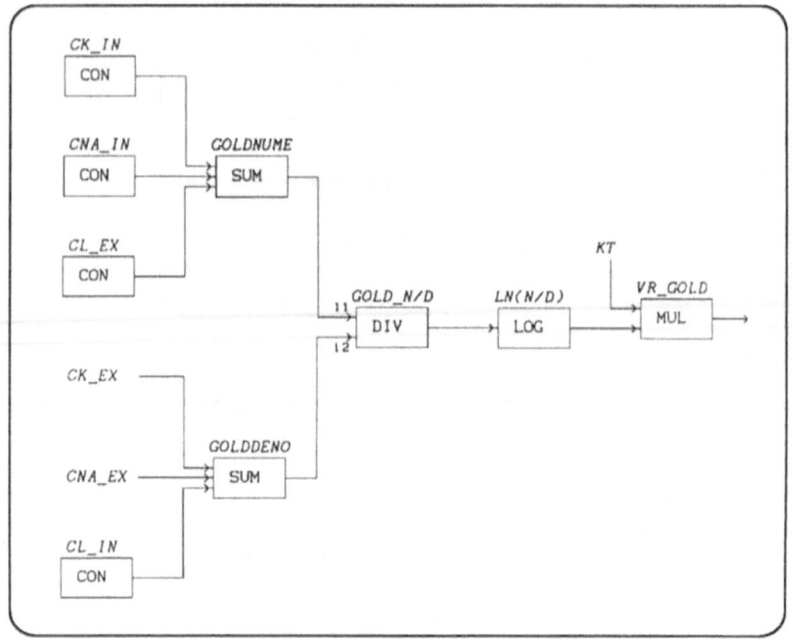

Figure 7.7. "Vr-Goldman" submodel

Ci_EX   : extracellular concentration of ion i [mmol/l]
     p1   : see Table 7.1

Ci_IN   : intracellular concentration of ion i [mmol/l]
     p1   : see Table 7.1

GOLDNUME : numerator of Goldman's equation.
     p1   : $P_K/P_K = 1$
     p2   : $P_{Na}/P_K = 0.04$
     p3   : $P_L/P_K = 0.45$

GOLDDENO : denominator of Goldman's equation
     p1   : $P_K/P_K = 1$
     p2   : $P_{Na}/P_K = 0.04$
     p3   : $P_L/P_K = 0.45$

GOLD_N/D : (Numerator/Denominator) of Goldman's equation

LN(N/D)   : ln(Numerator/Denominator) of Goldman's equation
   p1   : 1
   p2   : 1
   p3   : 1

VR_GOLD   : resting-potential
equation   : equation 7.2 => KT * LN(N/D)

7.4.2.4 α_i & β_i

This submodel calculates the rate coefficients α_i and β_i, which are used in the calculation of n, m and h from V_{HH} (equations 7.4e through 7.4j). V_{HH} [Hodgkin and Huxley, 1952] represents the deviation of the transmembrane potential from the resting potential (V_{out} - V_{in} - $V_{Goldman}$).

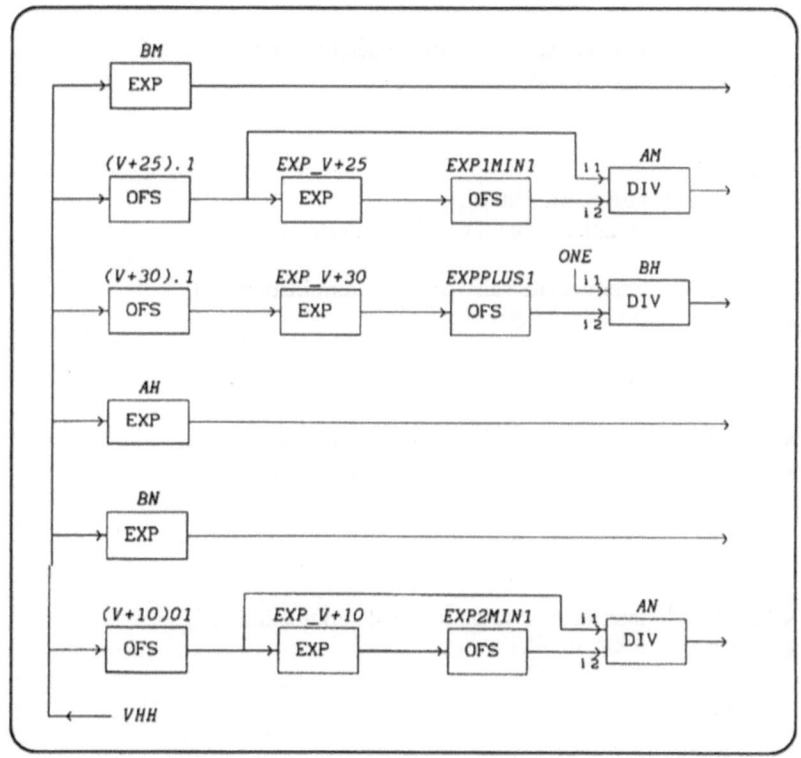

Figure 7.8.   "αi & βi " submodel

BM   : rate coefficient β_m
  equation   : equation 7.4h
      p1   : 4
      p2   : 1/18 = 5.56E-2
      p3   : 1

(V+25).1   : intermediate value for the determination of α_m
  equation   : $0.1\ (V_{HH} + 25)$
      p1   : 0.1
      p2   : 25

EXP_V+25   : intermediate value for the determination of α_m
  equation   : $\exp [0.1\ (V_{HH} + 25)] = \exp [(V+25).1]$
      p1   : 1
      p2   : 1
      p3   : 1

EXP1MIN1   : intermediate value for the determination of α_m
  equation   : EXP_V+25 - 1
      p1   : 1
      p2   : -1

AM   : rate coefficient α_m
  equation   : equation 7.4e = (V+25).1 / EXP1MIN1

(V+30).1   : intermediate value for the determination of β_h
  equation   : $0.1\ (V_{HH} + 30)$
      p1   : 0.1
      p2   : 30

EXP_V+30   : intermediate value for the determination of β_h
  equation   : $\exp [0.1\ (V_{HH} + 30)] = \exp [(V+30).1]$
      p1   : 1
      p2   : 1
      p3   : 1

EXPPLUS1   : intermediate value for the determination of β_h
  equation   : EXP_V+30 + 1
      p1   : 1
      p2   : 1

ONE   : constant with value 1
     p1   : 1

BH   : rate coefficient β_h
  equation   : equation 7.4j = ONE / EXPPLUS1

AH   : rate coefficient α_h
  equation   : equation 7.4g
     p1   : 0.07
     p2   : 1/20 = 0.05
     p3   : 1

BN   : rate coefficient β_n
  equation   : equation 7.4i
     p1   : 0.125
     p2   : 1/80 = 1.25E-2
     p3   : 1

(V+10)01:   intermediate value for the determination of α_n
  equation   : 0.01 (V_{HH} + 10)
     p1   : 0.01
     p2   : 10

EXP_V+10   : intermediate value for the determination of α_n
  equation   : exp [0.1 (V_{HH} + 10)] = exp [10 * *(V+10)01*]
     p1   : 1
     p2   : 10
     p3   : 1

EXP2MIN1   : intermediate value for the determination of α_n
  equation   : EXP_V+10 - 1
     p1   : 1
     p2   : -1

AN   : rate coefficient α_n
  equation   : equation 7.4f = (V+10)01 / EXP2MIN1

7.4.2.5 Stimuli

In this submodel the stimulus currents and timing are generated.
A stimulus consists of two block-pulses, of which the starting moments, the durations and the interval may be chosen by the user. Inward currents are counted positive.

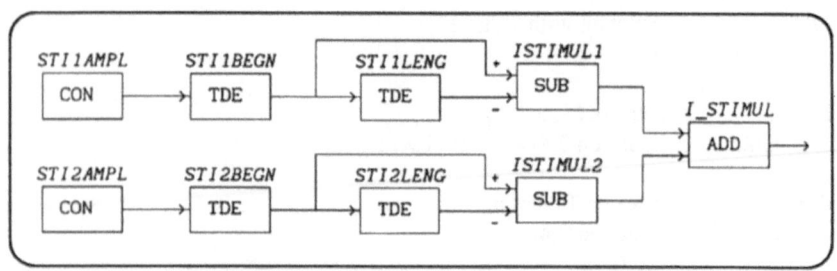

Figure 7.9. "Stimuli" submodel

STI1AMPL   :   amplitude of pulse #1
        p1   :   amplitude (user-adjustable) [$\mu A/cm^2$]

STI1BEGN   :   start of pulse #1
        p1   :   0 (initial condition of TDE)
        p2   :   start (user-adjustable) [ms]
        p3   :   integration time = 0.04 [ms]

STI1LENG   :   duration of pulse #1
        p1   :   0 (initial condition of TDE)
        p2   :   duration (user adjustable) [ms]
        p3   :   integration time = 0.04 [ms]

STI2AMPL   :   amplitude of pulse #2
        p1   :   amplitude (user-adjustable) [$\mu A/cm^2$]

STI2BEGN   :   start of pulse #2
        p1   :   0 (initial condition of TDE)
        p2   :   start (user-adjustable) [ms]
        p3   :   integration time = 0.04 [ms]

STI1LENG   :   duration of pulse #2
        p1   :   0 (initial condition of TDE)
        p2   :   duration (user-adjustable) [ms]
        p3   :   integration time = 0.04 [ms]

ISTIMUL1   :   amplitude of pulse #1
    equation :   STI1BEGN - STI1LENG

ISTIMUL2   :   amplitude of pulse #2
    equation :   STI2BEGN - STI2LENG

I_STIMUL   :   sum of amplitudes of pulse #1 and pulse #2
    equation :   ISTIMUL1 + ISTIMUL2

7.4.2.6 N, M & H

N (K$^+$-activation), M (Na$^+$-activation) and H (Na$^+$-inactivation) are calculated according to equations 7.4b through 7.4d. These equations are expressed in factors, reducing the number of blocks as well as calculation speed (from α_i and β_i to only two blocks of N, M or H). The starting values of N, M and H are according to [Randall, 1980].

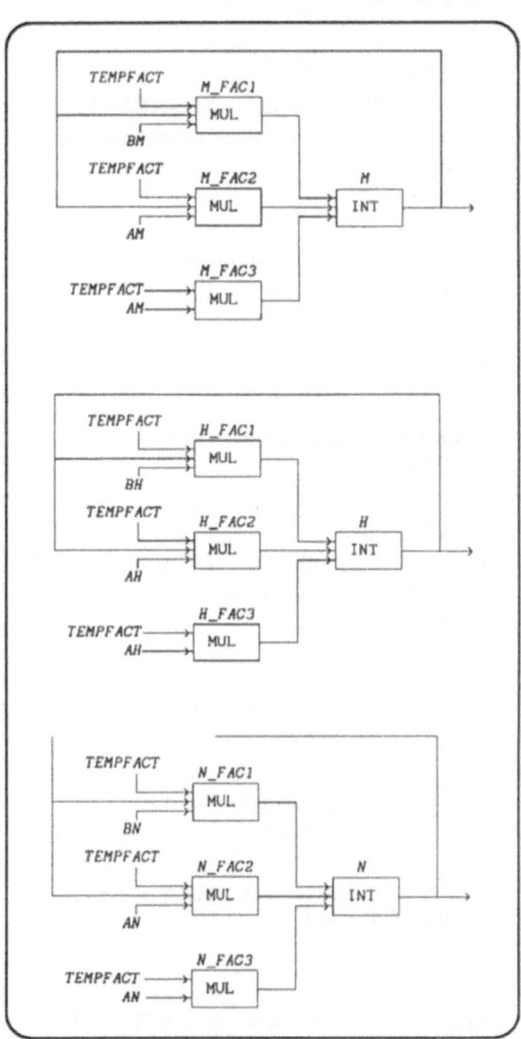

Figure 7.10. "N, M & H" submodel

M_FAC1   :   Factor #1 for the calculation of m
equation :   TEMPFACT * M * BM

M_FAC2   :   Factor #2 for the calculation of m
equation :   TEMPFACT * M * AM

M_FAC3   :   Factor #3 for the calculation of m
equation :   TEMPFACT * AM

M   :   m (Na^+-activation)
equation :   equation 7.4b => dm/dt = - M_FAC1 - M_FAC2 + M_FAC3
p1   :   starting value of m = 0.04391 [Randall, 1980]
p2   :   -1(-M_FAC1)
p3   :   -1(-M_FAC2)

H_FAC1   :   Factor #1 for the calculation of h
equation :   TEMPFACT * H * BH

H_FAC2   :   Factor #2 for the calculation of h
equation :   TEMPFACT * H * AH

H_FAC3   :   Factor #3 for the calculation of h
equation :   TEMPFACT * AH

H   :   h (Na^+-inactivation)
equation :   equation 7.4b => dh/dt = - H_FAC1 - H_FAC2 + H_FAC3
p1   :   starting value of h = 0.29407 [Randall, 1980]
p2   :   -1(-H_FAC1)
p3   :   -1(-H_FAC2)

N_FAC1   :   Factor #1 for the calculation of n
equation :   TEMPFACT * N * BN

N_FAC2   :   Factor #2 for the calculation of n
equation :   TEMPFACT * N * AN

N_FAC3   :   Factor #3 for the calculation of n
equation :   TEMPFACT * AN

N   :   n (K^+-activation)
equation :   equation 7.4b => dn/dt = - N_FAC1 - N_FAC2 + N_FAC3
p1   :   starting value of n = 0.64906 [Randall, 1980]
p2   :   -1(-N_FAC1)
p3   :   -1(-N_FAC2)

7.4.2.7 GK & GNa

GK (K^+-conductivity) and GNA (Na^+-conductivity) are calculated from n, m and h with the aid of two constants, $\bar{g}_K$ and $\bar{g}_{Na}$. See equation 7.4a.

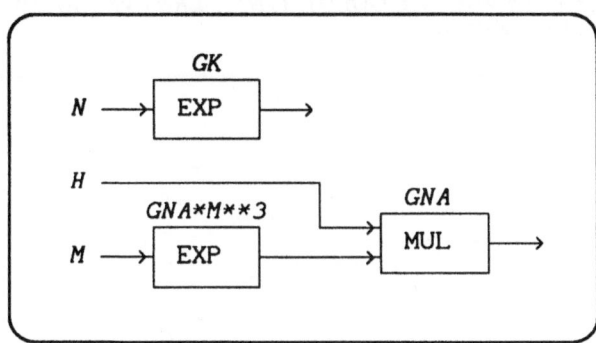

Figure 7.11. "GNa & GK" submodel

GK   : Conductivity for ion K^+ [mmol/cm^2].
  equation : $\bar{g}_K * n^4$
       p1 : value of $\bar{g}_K = 36.00$ mmol/cm [Randall, 1980].
       p2 : 4
       p3 : 3

GNA*M**3 : Intermediate value for the determination of g_{Na}
  equation : $\bar{g}_{Na} * m^3$
       p1 : value of $\bar{g}_{Na} = 120.00$ mmol/cm^2 [Randall, 1980].
       p2 : 3
       p3 : 3

GNA   : Conductivity for ion Na^+ [mmol/cm^2].
  equation : GNA * M ** 3 * h.

7.4.2.8 I_{Na}, I_K & I_L

The currents I_{Na}, I_K and I_L are determined with equation 7.4a, where the potentials V_{Na}, V_K and V_L are calculated with the aid of the Nernst-equation [equation 7.1a]. Inward currents are counted positive. All potentials are defined as $V_{out} - V_{in}$, according to [Hodgkin and Huxley 1952], except for the transmembrane potential itself, which follows the usual conventions.

In equation 7.4a, V stands for V_{HH} and V_{Na} while V_K and V_L stand for the deviation of the Nernst potential of the resting potential (e.g. $V_{Na\text{-Nernst}} - V_{r\text{-Goldman}}$). V_{HH} is defined as $V_{out} - V_{in} - V_{r\text{-Goldman}}$.

Thus, $V - V_{Na}$ is equivalent to:

$V_{HH} - V_{Na} = (-VIN\text{-}VEX - VR\_GOLD) - (VNA\_NERN - VR\_GOLD) =$
$- VIN\text{-}VEX - VNA\_NERN.$

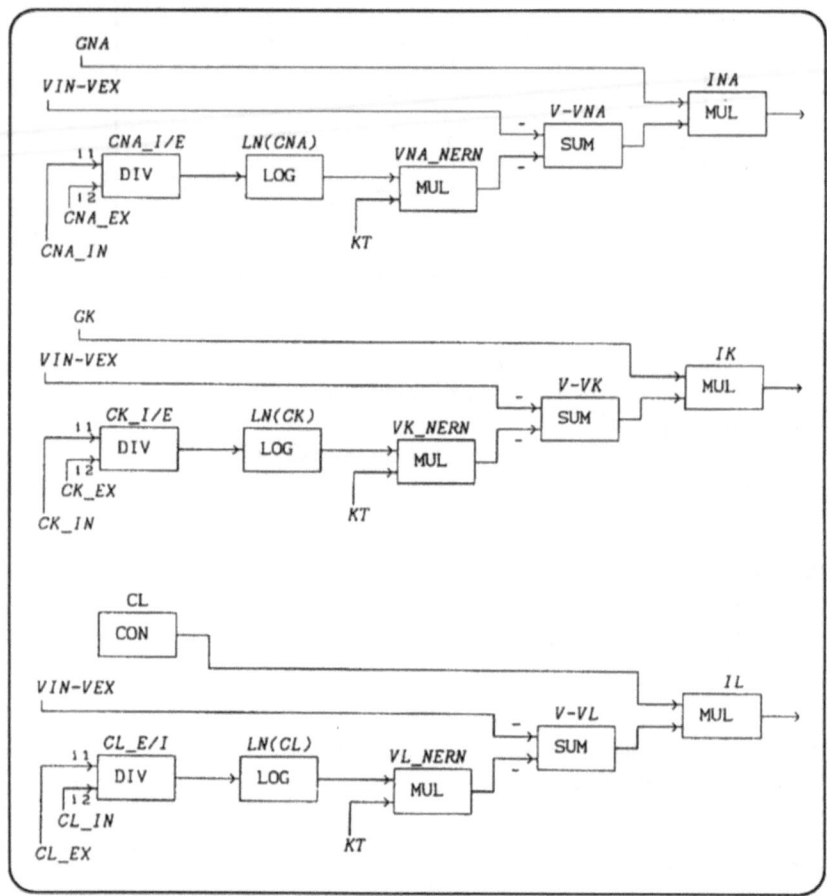

Figure 7.12. "INA, IK & IL" submodel

CNA_I/E :   Intermediate value for the determination of V_{Na}-Nernst.
   equation : $(C_{Na})_i / (C_{Na})_e$

LN(CNA):   LN(CNA_I/E)
      p1 : 1
      p2 : 1
      p3 : 1

VNA_NERN: Nernst potential for (Na^+)-ions
   equation : KT * LN(CNA)

V-VNA   :   Deviation of VNa-Nernst from V_{HH}
   equation : - VIN-VEX - VNA_NERN
       p1 : -1
       p2 : -1

INA   : (Na^+)-current, positive inward $[\mu A/cm^2]$
   equation : GNA * V-VNA

CK_I/E   :   Intermediate value for determinating
            V_{Na}-Nernst
   equation : $(C_K)_i/(C_K)_e$
LN(CK   :   LN(CK_I/E)
      p1 : 1
      p2 : 1
      p3 : 1

VK_NERN :   Nernst potential for (K^+)-ions
   equation : KT * LN(CK)

V-VK   :   Deviation of VK-Nernst from V_{HH}
   equation : - VIN-VEX - VK_NERN
       p1 : -1
       p2 : -1

IK   : (K^+)-current, positive inward $[\mu A/cm^2]$
   equation : GK * V-VK

CL_E/I   :   Intermediate value for determinating V_L-Nernst
   equation : $(C_L)_e/(C_L)_i$

GL   :   Conductivity of (L^-)-ions $[1/\Omega]$
      p1 : 0.3 mmol/cm^2

LN(CL)   :   LN(CL_E/I)
       p1   :   1
       p2   :   1
       p3   :   1

VL_NERN   :   Nernst potential for (L⁻)-ions
   equation   :   KT * LN(CL)

V-VL   :   Deviation of VL-Nernst from V_{HH}
   equation   :   - VIN-VEX - VL_NERN
       p1   :   -1
       p2   :   -1

IL   :   (L⁻)-current, positive inward [$\mu A/cm^2$]
   equation   :   GL * V-VL

7.4.2.9 HH-potential

V_{HH} (the deviation of the transmembrane-potential relative to the resting-potential) is determined with the aid of the currents I_{Na}, I_K and I_L. V_{in} - V_{out} is calculated as:
V_{in} - V_{out}= -(V_{HH} - $V_{r-Goldman}$)

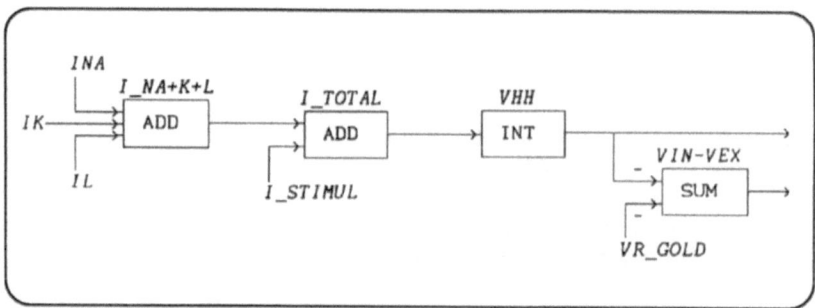

Figure 7.13. "HH-potential" submodel

I_NA+K+L   :   The sum of ion currents [$\mu A/cm^2$]
   equation   : I_{Na} + I_K + I_L

I_TOTAL   :   The sum of ion currents and stimulus current [$\mu A/cm^2$]
   equation   :   I_NA+K+L + I_STIMUL

VHH   :   "Hodgkin and Huxley" potential [mV]
    equation :   equation 4a, $V_{out} - V_{in} - V_r$-Goldman = $- \partial I\_TOTAL/\partial t$
        p1   :   starting value (0)
        p2   :   Membrane capacity per area (-1 cm^{-2}),
                    counted negative to obtain positive sign for VHH

VIN-VEX   :   transmembrane potential
    equation : $V_{in} - V_{out}$ = $-$ VHH - VR_GOLD [mV]
        p1   : -1 (-VHH)
        p2   : -1 (-VR_GOLD)

7.4.2.10   Isoclines

With the aid of equations 7.5, 7.6 and 7.7 the isoclines can be determined, relating V and m. The two isoclines (equations 6 and 7) intersect at three points, where ISO1**3 equals ISO2_V=0.

The threshold of the membrane is found at point B in Fig. 7.3. In order to facilitate the calculation, the third power of the isoclines is used. The block SUPRA detects the intersection points: for point A, SUPRA = 1; between A and B SUPRA = 0; between B and C SUPRA = 1 and beyond C SUPRA = 0.

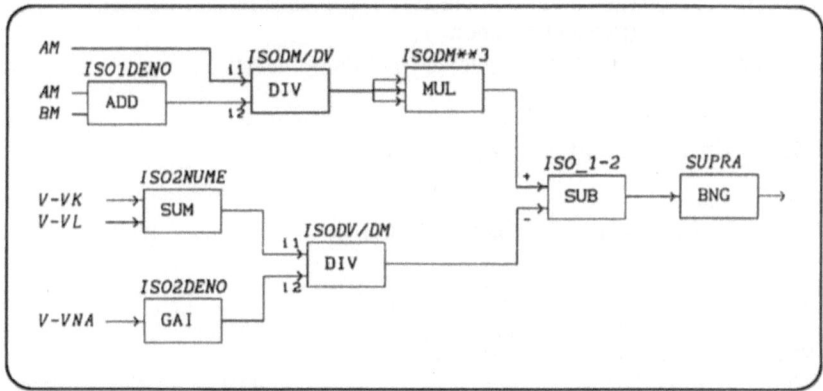

Figure 7.14.   "Isoclines" submodel

ISO1DENO :   Denominator of the Isocline where dm/dV = 0
equation   :   denominator of equation 7.6, $\alpha_m + \beta_m$

ISODM/DV :   Isocline where dm/dV = 0
equation   :   equation 7.6, => $\alpha_m/(\alpha_m + \beta_m)$

ISODM**3   :   (ISO_M=0)**3
equation   :   equation 7.6 to the third power

ISO2NUME   :   Numerator of the Isocline where dV/dm = 0
equation   :   equation 7.7
    p1   : $-\bar{g}_K.n^4$ (n = 0.3180 [steady-state], $\bar{g}_K$ = 36)
    p2   : $-\bar{g}_L$ (= 0.3)

ISO2DENO   :   Denominator of the Isocline where dV/dm = 0
equation   :   equation 7.7p1
           : $\bar{g}_{Na}.h$ (h = 0.5960 [steady-state], $\bar{g}_{Na}$ = 120)

ISODV/DM   :   Isocline where dV/dm = 0
equation   :   equation 7.7, to the third power

ISO_1-2   :   Difference between Isocline dm/dV and Isocline dV/dm
equation   :   ISODM**3 - ISODV/DM

SUPRA   :   Indicates if VHH passes the threshold value
equation   :   Equals 1, if VHH lies before A or between B and C
               Equals 0, if VHH lies between A and B or after C
    p1   :   0 (minimum stimulus)
    p2   :   1 (maximum stimulus)
    p3   :   0 (criterion for generation of the stimulus)

7.4.3   Structure and Parameter Listing

Block	Type	Input1	Input2	Input3	Par1	Par2	Par3
ISO1DENO	ADD	AM	BM				
I_NA+K+L	ADD	INA	IK	IL			
I_STIMUL	ADD	ISTIMUL1	ISTIMUL2				
I_TOTAL	ADD	I_NA+K+L	I_STIMUL				
SUPRA	BNG	ISO1-2			.0000	1.000	.0000
CK_IN	CON				155.0		
CL_EX	CON				120.0		
CL_IN	CON				4.000		
CNA_IN	CON				12.00		
DCK/DT	CON				200.0		
DCNA/DT	CON				200.0		
GL	CON				.3000		
HH_TEMP	CON				-6.300		
ONE	CON				1.000		
REALTEMP	CON				6.300		
STI1AMPL	CON				.0000		
STI2AMPL	CON				.0000		

AM	DIV	(V+25).1	EXP1MIN1				
AN	DIV	(V+10)01	EXP2MIN1				
BH	DIV	ONE	EXPPLUS1				
CK_I/E	DIV	CK_IN	CK_EX				
CL_E/I	DIV	CL_EX	CL_IN				
CNA_I/E	DIV	CNA_IN	CNA_EX				
GOLD_N/D	DIV	GOLDNUME	GOLDDENO				
ISODM/DV	DIV	AM	ISO1DENO				
ISODV/DM	DIV	ISO2NUME	ISO2DENO				
AH	EXP	VHH			7.0000E-02	5.0000E-02	1.000
BM	EXP	VHH			4.000	5.5600E-02	1.000
BN	EXP	VHH			.1250	1.2500E-02	1.000
EXP_V+10	EXP	(V+10)01			1.000	10.00	1.000
EXP_V+25	EXP	(V+25).1			1.000	1.000	1.000
EXP_V+30	EXP	(V+30).1			1.000	1.000	1.000
GK	EXP N				36.00	4.000	3.000
GNA*M**3	EXP	M			120.0	3.000	3.000
TEMPFACT	EXP	HH_TEMP	REALTEMP		1.000	4.7710E-02	2.000
ISO2DENO	GAI	V-VNA			71.52		
CK_EX	INL	DCK/DT			4.000	1.000	1000.
CNA_EX	INL	DCNA/DT			145.0	1.000	1000.
H	INT	H_FAC1	H_FAC2	H_FAC3	.5960	-1.000	-1.000
M	INT	M_FAC1	M_FAC2	M_FAC3	5.3000E-02	-1.000	-1.000
N	INT	N_FAC1	N_FAC2	N_FAC3	.3180	-1.000	-1.000
VHH	INT	I_TOTAL			.0000	-1.000	
LN(CK)	LOG	CK_I/E			1.000	1.000	1.000
LN(CL)	LOG	CL_E/I			1.000	1.000	1.000
LN(CNA)	LOG	CNA_I/E			1.000	1.000	1.000
LN(N/D)	LOG	GOLD_N/D			1.000	1.000	1.000
GNA	MUL	GNA*M**3	H				
H_FAC1	MUL	TEMPFACT	H	BH			
H_FAC2	MUL	TEMPFACT	H	AH			
H_FAC3	MUL	TEMPFACT	AH				
IK	MUL	GK	V-VK				
IL	MUL	GL	V-VL				
INA	MUL	GNA	V-VNA				
ISODM**3	MUL	ISODM/DV	ISODM/DV	ISODM/DV			
M_FAC1	MUL	TEMPFACT	M	BM			
M_FAC2	MUL	TEMPFACT	M	AM			
M_FAC3	MUL	TEMPFACT	AM				
N_FAC1	MUL	TEMPFACT	N	BN			
N_FAC2	MUL	TEMPFACT	N	AN			
N_FAC3	MUL	TEMPFACT	AN				
VK_NERN	MUL	LN(CK)	KT				
VL_NERN	MUL	LN(CL)	KT				
VNA_NERN	MUL	LN(CNA)	KT				
VR_GOLD	MUL	KT	LN(N/D)				
(V+10)01	OFS	VHH			1.0000E-02	10.00	
(V+25).1	OFS	VHH			.1000	25.00	
(V+30).1	OFS	VHH			.1000	30.00	
EXP1MIN1	OFS	EXP_V+25			1.000	-1.000	
EXP2MIN1	OFS	EXP_V+10			1.000	-1.000	
EXPPLUS1	OFS	EXP_V+30			1.000	1.000	

KT	OFS	REALTEMP			8.6170E-02	273.1	
GOLDDENO	SUM	CK_EX	CNA_EX	CL_IN	1.000	4.0000E-02	.4500
GOLDNUME	SUM	CK_IN	CNA_IN	CL_EX	1.000	4.0000E-02	.4500
ISO1-2	SUB	ISODM**3	ISODV/DM				
ISO2NUME	SUM	V-VK	V-VL		-.3681	-.3000	
ISTIMUL1	SUB	STI1BEGN	STI1LENG				
ISTIMUL2	SUB	STI2BEGN	STI2LENG				
V-VK	SUM	VIN-VEX	VK_NERN		-1.000	-1.000	
V-VL	SUM	VIN-VEX	VL_NERN		-1.000	-1.000	
V-VNA	SUM	VIN-VEX	VNA_NERN		-1.000	-1.000	
VIN-VEX	SUM	VHH	VR_GOLD		-1.000	-1.000	
STI1BEGN	TDE	STI1AMPL			.0000	1.000	4.0000E-02
STI1LENG	TDE	STI1BEGN			.0000	1.000	4.0000E-02
STI2BEGN	TDE	STI2AMPL			.0000	1.000	4.0000E-02
STI2LENG	TDE	STI2BEGN			.0000	1.000	4.0000E-02

7.5 Experiments

7.5.1 Demonstrations

Demonstrations are simulation runs with default values of the parameters, and they are not adjustable by the user.

7.5.1.1 The Default Demo

The default demo consists of a set of runs of all exercises.

7.5.1.2 Ion Concentrations and the Nernst Potential

The Nernst potentials are calculated in the submodel "I_{Na}, I_K & I_L". The external concentrations of Na^+ and K^+ are changed in the submodel "CNa_ex & CK_ex".

Contents of the corresponding AUTOEX.PSI file are:

```
MR NERVE                                    { Read the model             }
Y                                           { Confirm the reading        }
AINFLUENCE OF HIGHER CNA ON VNA_NERN {        Give it a title            }
P                                           { Change parameters          }
DCNA/DT,200        ***USER***               { Increase CNa 200 mmol/l/s  }
                                            { Exit change parameters     }
O                                           { Change output parameters   }
CNA_EX,VNA_NERN                             { New output parameters      }
DS1 0,300                                   { Adjust scaling of CK_EX    }
```

```
DS2 -50,100                              {  Adjust scaling of VK_NERN   }
DTT 1                                    {  Simulate during 1 ms        }
R                                        {  Start the simulation-run    }
WAIT                                     {  Wait for a keystroke        }
AINFLUENCE OF HIGHER CK ON VK_NERN       {  Give it a new title         }
P                                        {  Change parameters           }
DCK/DT,200        ***USER***             {  Increase CK 200 mmol/l/s    }
                                         {  Exit change parameters      }
O                                        {  Change output parameters    }
CK_EX,VK_NERN                            {  New output parameters       }
DS1 0,300                                {  Adjust scaling of CK_EX      }
DS2 -50,100                              {  Adjust scaling of VK_NERN   }
DTT 1                                    {  Simulate during 1 ms        }
R                                        {  Start the simulation run     }
WAIT                                     {  Wait for a keystroke        }
EXIT                                     {  Leave the BIOPSI session     }
```

7.5.1.3   Ion Concentrations and the Goldman Potential

The Goldman resting-potential is calculated in submodel "Vr_Goldman". The external concentrations of Na^+ and K^+ can be changed in the submodel "CNa_ex & CK_ex".

Contents of the corresponding AUTOEX.PSI file are:

```
MR NERVE                                      {  Read the model              }
Y                                             {  Confirm the reading         }
AINFLUENCE OF HIGHER CNA ON VR_GOLD           {  Give it a title             }
P                                             {  Change parameters           }
DCNA/DT,200      ***USER***                   {  Increase CNa 200 mmol/l/s   }
                                              {  Exit change parameters      }
O                                             {  Change output parameters    }
CNA_EX,VR_GOLD                                {  New output parameters       }
DS1 0,500                                     {  Adjust scaling of CNA_EX    }
DS2 -50,100                                   {  Adjust scaling of VR_GOLD   }
DTT 1                                         {  Simulate during 1 ms        }
R                                             {  Start the simulation run     }
WAIT                                          {  Wait for a keystroke        }
AINFLUENCE OF HIGHER CK ON VR_GOLD            {  Give it a new title         }
P                                             {  Change parameters           }
DCK/DT,200       ***USER***                   {  Increase CK 200 mmol/s      }
                                              {  Exit change parameters      }
O                                             {  Change output parameters    }
CK_EX,VR_GOLD                                 {  New output parameters       }
DS1 0,300                                     {  Adjust scaling of CK_EX     }
DS2 -50,100                                   {  Adjust scaling of VR_GOLD   }
DTT 1                                         {  Simulate during 1 ms        }
R                                             {  Start the simulation-run    }
WAIT                                          {  Wait for a keystroke        }
EXIT                                          {  Leave the BIOPSI session     }
```

7.5.2   Exercises

The exercises allow the interactive choice of one parameter each, within minimum and maximum values, checked by the menu program.

7.5.2.1   Stimulation and the Action Potential

Subliminal and Supraliminal Stimulation

The stimulation parameters are chosen via the submodel "Stimuli". I_STIMUL as well as $V_{in} - V_{ex}$, from the submodel "HH_potential", are presented.

Contents of the corresponding AUTOEX.PSI file are:

```
MR NERVE                          {   Read the model                  }
Y                                 {   Confirm the reading             }
ASUBTHRESHOLD STIMULATION         {   Give it a title                 }
P                                 {   Change parameters               }
STI1AMPL,25                       {   Stimulusamplitude 25 µA         }
STI1BEGN,,0.5,                    {   Stimulusbegin, 0.5 ms           }
STI1LENG,,1,                      {   Stimuluslength, 1 ms            }
                                  {   Exit change parameters          }
O                                 {   Change output parameters        }
I_STIMUL,VIN-VEX                  {   New output parameters           }
DS1 0,100                         {   Adjust scaling of I_STIMUL      }
DS2 -90,70                        {   Adjust scaling of VIN-VEX       }
DTT 4                             {   Simulate during 4 ms            }
R                                 {   Start the simulation-run        }
WAIT                              {   Wait for a keystroke            }
ASUPRATHRESHOLD STIMULATION       {   Give it a new title             }
P                                 {   Change parameters               }
STI1AMPL,100   *** USER ***       {   Stimulus amplitude 100 µA       }
STI1BEGN,,0.5,                    {   Stimulus begin, 0.5 ms          }
STI1LENG,,1,                      {   Stimulus length, 1 ms           }
                                  {   Exit change parameters          }
O                                 {   Change output parameters        }
I_STIMUL,VIN-VEX                  {   Same output parameters          }
DS1 0,100                         {   Adjust scaling of I_STIMUL      }
DS2 -90,70                        {   Adjust scaling of VIN-VEX       }
DTT 4                             {   Simulate during 4 ms            }
R                                 {   Start the simulation run        }
WAIT                              {   Wait for a keystroke            }
EXIT                              {   Leave the BIOPSI session        }
```

7.5.2.1.2   Refractory Period

The stimulation parameters are chosen via the submodel "Stimuli".
I_STIMUL as well as $V_{in} - V_{ex}$ from the submodel "HH_potential" are presented.
Contents of the corresponding AUTOEX.PSI file are:

MR NERVE		{	Read the model	}
Y		{	Confirm the reading	}
ASTIMULATION WITH TWIN-PULSES		{	Give it a title	}
P		{	Change parameters	}
STI1AMPL,100	*** USER ***	{	Stimulus amplitude 100 µA	}
STI1BEGN,,0.5,	*** USER ***	{	Stimulus begin, 0.5 ms	}
STI1LENG,,1,	*** USER ***	{	Stimulus length, 1 ms	}
STI2AMPL,200	*** USER ***	{	Stimulus amplitude 200 µA	}
STI2BEGN,,5,	*** USER ***	{	Stimulus begin, 5 ms	}
STI2LENG,,2,	*** USER ***	{	Stimulus length, 2 ms	}
		{	Exit change parameters	}
O		{	Change output parameters	}
I_STIMUL,VIN-VEX		{	New output parameters	}
DS1 0,250		{	Adjust scaling of I_STIMUL	}
DS2 -90,70		{	Adjust scaling of VIN-VEX	}
DTT 10		{	Simulate during 10 ms	}
R		{	Start the simulation run	}
WAIT		{	Wait for a keystroke	}
EXIT		{	Leave the BIOPSI session	}

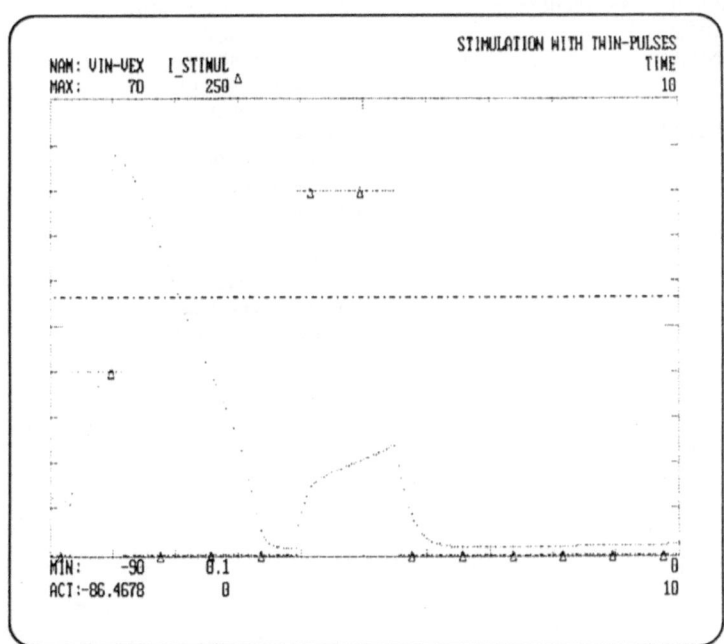

Figure 7.15. Simulation with twin-pulses

7.5.2.2   Special Hodgkin-Huxley-Model Features

So that combinations of different parameter values may be used for the next simulation exercise, the values chosen for STI1AMPL may be kept for the next run in a new model file if the last two lines in the AUTOEX.PSI file are changed into:

WAIT	{ Wait for a keystroke	}
MW NERVE	{ Write the changed model	}
Y	{ Confirm the reading	}
Y	{ Confirm duplication	}
EXIT	{ Leave the BIOPSI session	}

Relation Between N, M, H and V_{in} - V_{ex}

The parameters N, M and H from the submodel "n, m & h" as well as V_{in} - V_{ex} are presented.

Contents of the corresponding AUTOEX.PSI file are:

MR NERVE	{ Read the model	}
Y	{ Confirm the reading	}
ARELATION BETWEEN N M H AND VIN-VEX	{ Give it a title	}
P	{ Change parameters	}
STI1AMPL,100 *** USER ***	{ Stimulus amplitude 100 μA	}
STI1BEGN,,0.5,	{ Stimulus begin, 0.5 ms	}
STI1LENG,,1,	{ Stimulus length, 1 ms	}
	{ Exit change parameters	}
O	{ Change output parameters	}
N,M,H,VIN-VEX	{ New output parameters	}
DS1 0,1	{ Adjust scaling of N	}
DS2 0,1	{ Adjust scaling of M	}
DS3 0,1	{ Adjust scaling of H	}
DS4 -90,70	{ Adjust scaling of VIN-VEX	}
DTT 4	{ Simulate during 4 ms	}
R	{ Start the simulation run	}
WAIT	{ Wait for a keystroke	}
EXIT	{ Leave the BIOPSI session	}

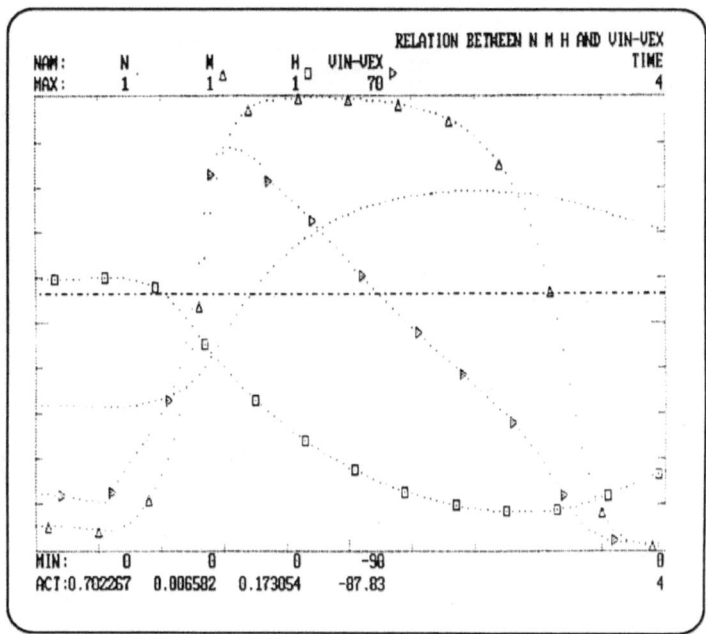

Figure 7.16

7.5.2.2.2 Ion Channel Conductances

The parameters GNA, GK and GL from the submodel "GNa & GK" as well as V_{in} - V_{ex} are presented.
Contents of the corresponding AUTOEX.PSI file are:

MR NERVE	{ Read the model	}
Y	{ Confirm the reading	}
ASTIMULATION AND IONIC CONDUCTANCES	{ Give it a title	}
P	{ Change parameters	}
STI1AMPL,100 *** USER ***	{ Stimulus amplitude 100 μA	}
STI1BEGN,,0.5,	{ Stimulus begin, 0.5 ms	}
STI1LENG,,1,	{ Stimulus length, 1 ms	}
	{ Exit change parameters	}
O	{ Change output parameters	}
GNA,GK,GL,VIN-VEX	{ New output parameters	}
DS1 0,50	{ Adjust scaling of GNA	}
DS2 0,50	{ Adjust scaling of GK	}
DS3 0,50	{ Adjust scaling of GL	}
DS4 -90,70	{ Adjust scaling of VIN-VEX	}
DTT 4	{ Simulate during 4 ms	}
R	{ Start the simulation run	}
WAIT	{ Wait for a keystroke	}
EXIT	{ Leave the BIOPSI session	}

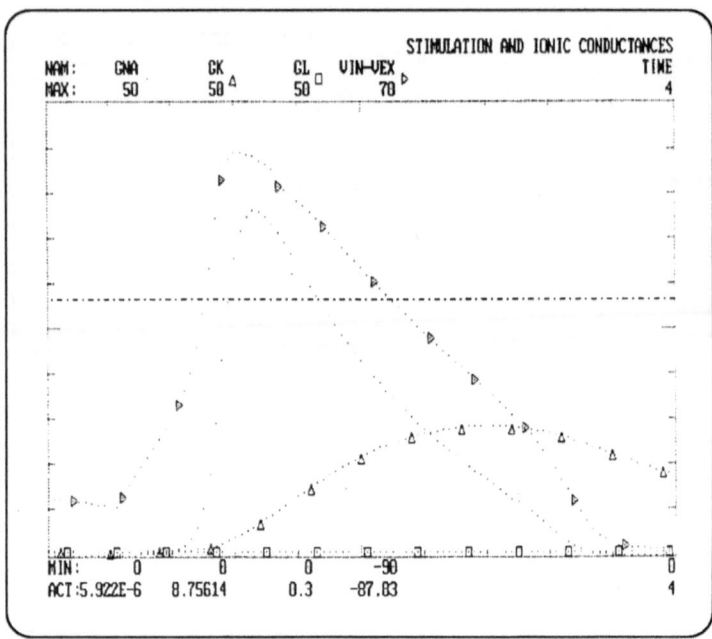

Figure 7.17. Simulation and ionic conductances

7.5.2.2.3 Ionic Currents

The parameters INA, IK and IL from the submodel "INa, IK & IL" as well as V_{in} - V_{ex} are presented. Contents of the corresponding AUTOEX.PSI file are:

```
MR NERVE                              { Read the model                  }
Y                                     { Confirm the reading             }
ASTIMULATION AND IONIC CURRENTS { Give it a title                 }
P                                     { Change parameters               }
STI1AMPL,100    *** USER ***          { Stimulus amplitude 100 µA       }
STI1BEGN,,0.5,                        { Stimulus begin, 0.5 ms          }
STI1LENG,,1,                          { Stimulus length, 1 ms           }
                                      { Exit change parameters          }
O                                     { Change output parameters        }
INA,IK,IL,VIN-VEX                     { New output parameters           }
DS1    -1200,1200                     { Adjust scaling of INA           }
DS2    -1200,1200                     { Adjust scaling of IK            }
DS3    -1200,1200                     { Adjust scaling of IL            }
DS4    -90,90                         { Adjust scaling of VIN-VEX       }
DTT    4                              { Simulate during 4 ms            }
R                                     { Start the simulation run        }
WAIT                                  { Wait for a keystroke            }
EXIT                                  { Leave the BIOPSI session        }
```

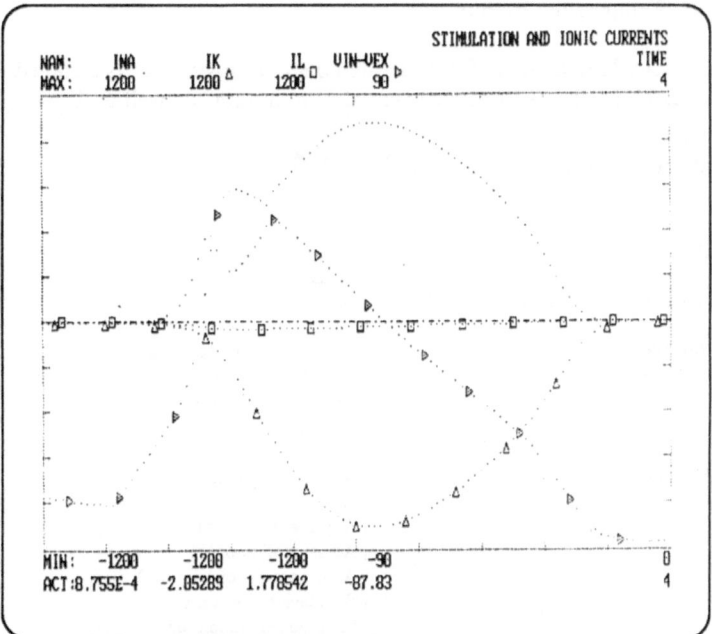

Figure 7.18. Simulation and ionic currents

7.5.2.2.4 Stimulation Threshold

The parameter SUPRA from the submodel "Isoclines" as well as V_{in} - V_{ex} from the submodel "HH_potential" are presented. Contents of the corresponding AUTOEX.PSI file are:

```
MR NERVE                              {  Read the model                    }
Y                                     {  Confirm the reading               }
ASUPRA INDICATES WHEN DM/DV > DV/DM   {  Give it a title                   }
P                                     {  Change parameters                 }
STI1AMPL,100    *** USER ***          {  Stimulus amplitude 100 µA         }
STI1BEGN,,0.5,                        {  Stimulus begin, 0.5 ms            }
STI1LENG,,1,                          {  Stimulus length, 1 ms             }
                                      {  Exit change parameters            }
O                                     {  Change output parameters          }
SUPRA,VIN-VEX                         {  New output parameters             }
DS1     0,1                           {  Adjust scaling of SUPRA           }
DS2     -90,70                        {  Adjust scaling of VIN-VEX         }
DTT     4                             {  Simulate during 4 ms              }
R                                     {  Start the simulation run          }
WAIT                                  {  Wait for a keystroke              }
EXIT                                  {  Leave the BIOPSI session          }
```

7.5.2.2.5   Temperature Effects

The parameter REALTEMP from the submodel "Temperature" is changed. "The parameters N, M and H from the submodel "n, m & h" as well as V_{in} - V_{ex} are presented.
Contents of the corresponding AUTOEX.PSI file are:

```
MR NERVE                                { Read the model                     }
Y                                       { Confirm the reading                }
ATEMPERATURE EFFECTS, T = 6.3 DEG       { Give it a title                    }
P                                       { Change parameters                  }
STI1AMPL,100                            { Stimulus amplitude 100 µA          }
STI1BEGN,,0.5,                          { Stimulus begin, 0.5 ms             }
STI1LENG,,1,                            { Stimulus length, 1 ms              }
                                        { Exit change parameters             }
O                                       { Change output parameters           }
N,M,H,VIN-VEX                           { New output parameters              }
DS1      0,1                            { Adjust scaling of N                }
DS2      0,1                            { Adjust scaling of M                }
DS3      0,1                            { Adjust scaling of H                }
DS4      -90,70                         { Adjust scaling of VIN-VEX          }
DTT      4                             { Simulate during 4 ms               }
R                                       { Start the simulation run           }
WAIT                                    { Wait for a keystroke               }
ATEMPERATURE EFFECTS, T = T CHOSEN      { Give it a title                    }
P                                       { Change parameters                  }
REALTEMP,20   *** USER ***              { Change temperature to 20°          }
                                        { Exit change parameters             }
R                                       { Start the simulation run           }
WAIT                                    { Wait for a keystroke               }
EXIT                                    { Leave the BIOPSI session           }
```

References

Fitzhugh, R.: "Thresholds and plateaus in the Hodgkin-Huxley nerve equations". *J. Gen. Physiol.* 43, 1960, 867-896.

Fitzhugh, R.: "Impulses and physiological states in theoretical models of nerve membrane". *Biophys. J.* 1, 1962, 445-466.

Guyton, A.C.: *Textbook of medical physiology*.
W.B. Saunders Company, 1986.

Goldman, L. and C.L. Schauf: "Quantitative description of sodium and potassium currents and computed action potentials in Myxicola giant axons". *J. Gen. Physiol.* 61, 1973, 361-384.

Hodgkin, A.L. and A.F. Huxley: "A quantitative description of membrane current and its application to conduction and excitation in nerves". *J. Physiol.* 117, 1952, 500-544.

Noble, D.: *The initiation of the heartbeat*.
Clarendon Press, Oxford, 1975.

Plonsey, R.: *Bioelectric Phenomena*.
McGraw-Hill Book Company, Case Western Reserve University, 1969.

Randall, J.E.: *Microcomputers and Physiological Simulation*.
Addison-Wesley Publishing Company, Massachussetts, 1980.

Talbot, S.A. and U. Gessner: *Systems Physiology*.
John Wiley & Sons, Basel, 1973.

8
The Specific Conduction System of the Heart

Rogier P. van Wijk van Brievingh

8.1 Aim of Instruction

The program consists of two parts:

- An animation[1] of the time-space course of the depolarization and repolarization fronts in perspective view
- A presentation of the frontal vector loops together with the corresponding ECGs in standard and Goldberger projections.

The programs in this section are simple demonstrations, offering a dynamic illustration of the process of electrical activation of the heart and the ECG.
The position of the electrical heart axis within the thorax may be altered interactively, effecting the ECG. The activation process remains the same.

8.2 Theory and Reference to Standard Textbooks

All textbooks on cardiac physiology describe the electrical activation system.
The structures shown in the model are:

- SA-node
- Atrial muscle
- AV-node
- His bundle
- Ventricular muscle

The animations are based on illustrations in Netter's famous book of medical illustrations [1978] as well as on the experiments of [Durrer,1970] on the human heart.

---

[1] Thanks are due to Ronald Krom, research assistant, for programming the animation.

8.3   Features of the Model

The animation is performed in time steps of 10 ms and shows a frontal vector loop and a "normal" ECG. The animation speed may be slowed or stopped to momentarily inspect the situation.

The ECG's presented in the second part can be chosen from several signal files. The ECG leads are assumed to be the projections of the electrical heart vector on the equilateral Einthoven triangle in the frontal plane only. Standard I, II and III leads as well as aVL, aVR and aVF leads are included and shown in slow-motion together with the loops in the frontal plane. In order to illustrate the projection equations, a circular loop is included also, showing the phase relations between the sinusoidal projections. Axis deviations to the right and left determine the initial phase.

8.3.1   Animation

The model reads data files which contain the standard leads obtained from volunteers.[2] Data files from patients may also be included. The "background" is a schematic view of the atria and the ventricles with the AV-node and the His bundle according to [Netter,1978]. The depolarization and repolarization fronts are superimposed on this structure in a different color. The electrical heart vector and the ECG are shown in separate boxes.

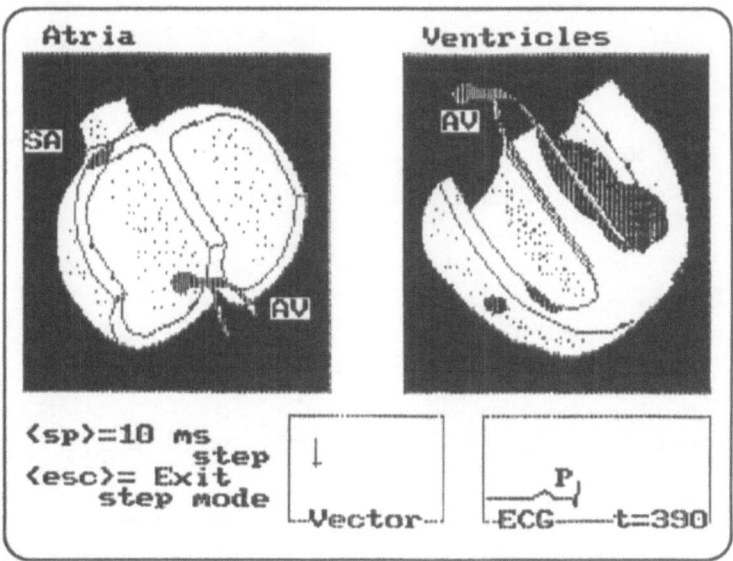

Figure 8.1

---

[2] Thanks are due to Dr. R.M. Heethaar, Department of Medical Physics, Utrecht University, the Netherlands

8.3.2   Projections

The Einthoven triangle or the Goldberger "star" (the lines of gravity) is presented with the loop in the center and the projections in separate boxes. After viewing these, the corresponding ECG's are shown separately. The orientation of the electrical heart axis may be changed within physiological limits.

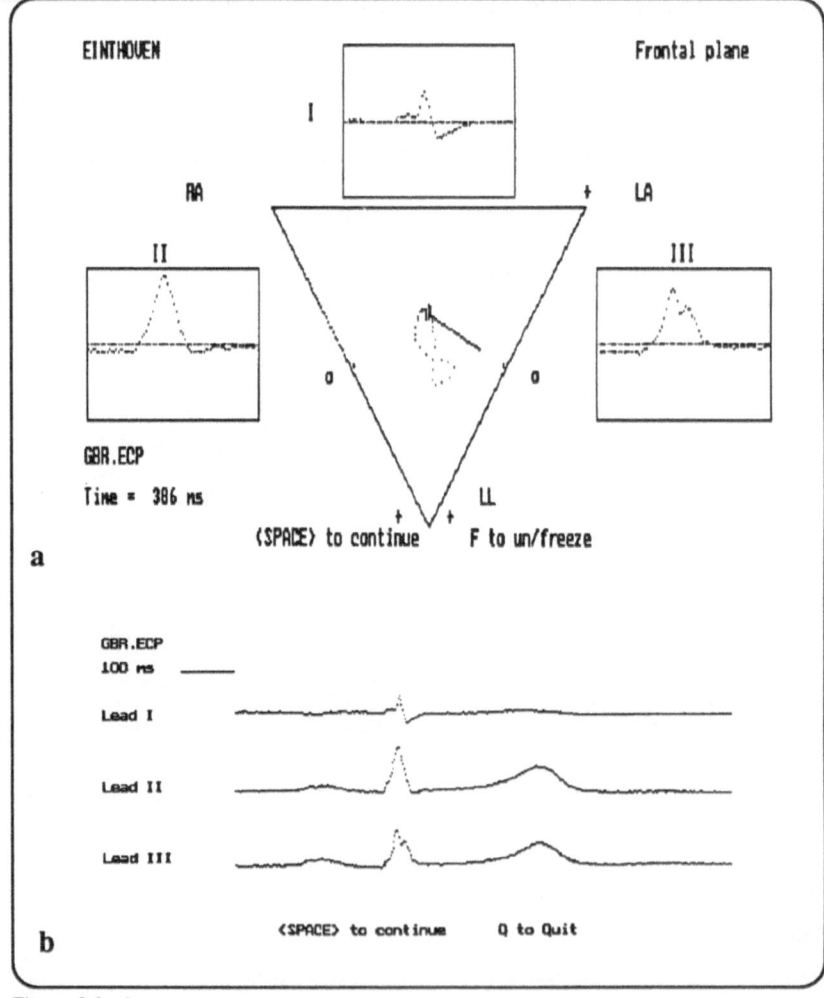

Figure 8.2 a,b

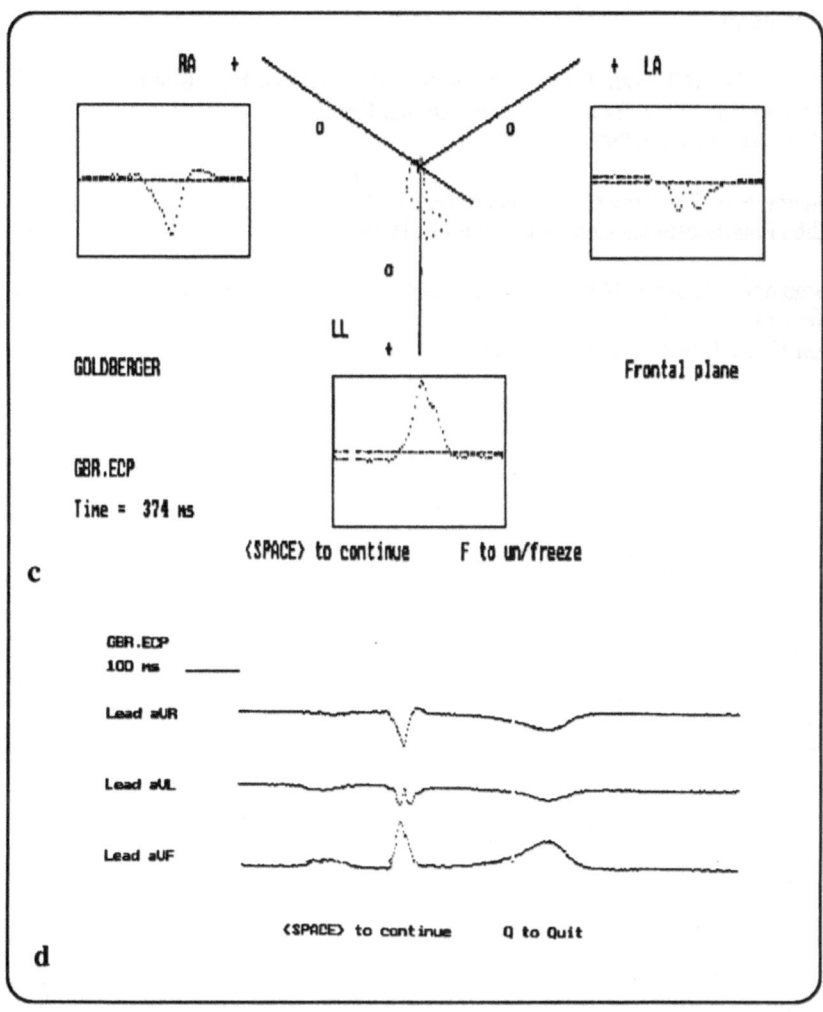

Figure 8.2 c,d

8.4 Conclusions

Illustrations in textbooks are static and require long descriptions to convey the message of a dynamic process. Computer graphics based on physiological data allow the dynamic presentation of the processes to be illustrated and can thus help us gain insight.

3D computer graphics yield an even closer approximation of reality, but require more computer power [Solomon,1973].

References

Durrer, D., R.T. van Dam, G.E. Freud, M.J. Jansen, F.L. Meijler and R.C. Arzbaecher: "Tidal Excitation of the Human Heart:.
Circ., Vol XLI, pp. 899-912, 1970.

Netter, F.H: *The Ciba Collection of Medical Illustrations*, Vol. 5 The Heart.
Ciba Pharmeceutical Company, Summit, 1978.

Solomon, J.C and R.H Selvester: "Simulation of measured activation sequence in the human heart".
Am.Heart J. Vol 85, No 4, pp.518-524, 1973.

9
Electrodes for Bioelectric Signals

Rogier P. van Wijk van Brievingh and Antoine J.C. de Reus

9.1   Aim of Instruction

Bioelectric electrodes are transducers that convert ionic currents within the body into electron currents in the metal leads to the amplifier input stage. This transducer system comprises a metal/electrolyte interface with an electrical double layer while the electrode/amplifier system exhibits a frequency transfer with interesting properties. Most often, the distortion of the biosignal by this transfer is acceptable; however, in order to justify the conclusions drawn from the signal in the diagnostic process at least some elementary insight is required.

Especially when the measurement results are interpreted in terms of a frequency spectrum, the frequency transfer of the measurement system has to be taken into account [Bekey, 1982]. Whereas the electrode properties are dictated by the laws of electrochemistry, the amplifier and filter characteristics can be manipulated extensively. It is the responsibility of the biomedical engineer to assure the best possible total system for each type of measurement. The system also includes the choice of the electrode material and size, as well as proper skin treatment and electrode placement. The above-mentioned aspects, except the last one, are included in this simulation and can be manipulated by the user.

9.2   Theory and Reference to Standard Textbooks

9.2.1   The Metal/Electrolyte Interface

When a metallic electrode is placed in an electrolytic solution, metal ions will enter

---

[1]Thanks are due to J.M. van Ouwerkerk and F.T.C. Koenis, research assistants, for their programming contributions

into the solution and ions will combine with the electrode material. The net result is a charge distribution at the electrode-electrolyte interface, depending on the species of metal and the type of electrolyte. The electrode properties are dictated by this ionic distribution. The electrode acquires a *contact potential* as the result of the charge distribution, which may be orders of magnitude greater than the biosignal to be measured. This unavoidable phenomenon must be minimized by proper choice of the electrode material and the electrolyte used [Geddes, 1972]. Furthermore, the double layer changes when electrical current passes through it, an effect called *polarization. Stability* of the double layer can be achieved by coating the metal surface with one of its non-soluble salts, which introduces an equilibrium reaction with the electrolytic solution. The double layer is considered to be composed of a *compact layer* in which charges of opposite sign keep one another closely bonded and a *diffusion layer* in the electrolyte with a diminishing charge density depending on its distance from the interface.

Ag/AgCl with KCl is the classic combination of a non-polarizable electrode. At the interface between two solutions with different concentrations and ionic mobilities, a *liquid-junction potential* develops. The use of solutions with equal ionic mobilities favors minimization of junction potentials. Potassium chloride appears to be the substance of choice for a *salt bridge*, to couple ionic systems. The relevant properties of the Ag/AgCl/KCl system are summarized in Table 9.1:

Table 9.1   Aspects of Ag/AgCl transducer design

Component	Metal	Coating	Salt Bridge	Skin/Body
Function	Output current	Stability	Coupling	Input current
Electrochemical reaction	$Ag = Ag^+ + e^-$	$AgCl = Ag^+ + Cl^-$	$KCl = K^+ + Cl^-$	Varying $[Cl^-]$
Relevant property	Solid	Solid	Equal mobilities	
		Majority charge carriers		

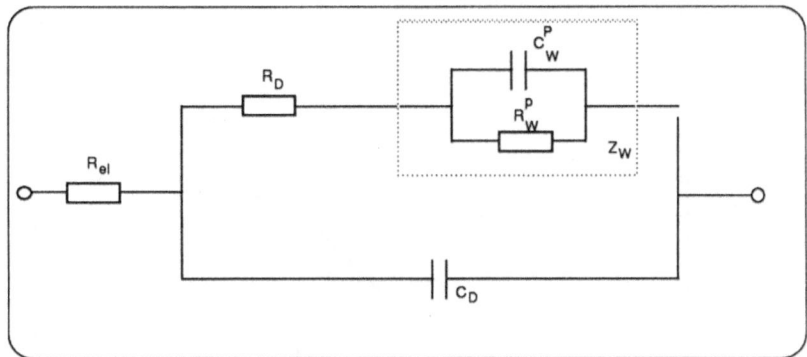

Figure 9.1   Electric substitution network

R_{el} Resistance of the electrolyte "bulk" in the body
C_D Capacitance of the compact part of the double layer
R_D Transport Resistance, a function of the current passing
Z_w "Warburg Impedance", describing the diffusion layer
[Note that this circuit allows the passing of DC current]

9.2.2   Electrode Impedance

The network shown in Fig. 9.1 can be used as an approximate model for the electric properties of a bioelectric electrode in contact with the object to be measured.
The Warburg concept for the electrode-electrolyte interface is an approximation of Fricke's law, which states that the series-equivalent capacitance of such an interface varies with frequency as:

$$C = K.f^{-\alpha} \tag{9.1}$$

whereas the phase angle:

$$\sigma = \pi/2.(1-\alpha) \tag{9.2}$$

depends on α also.
To a first approximation, $\alpha = 0.5$, allowing for description by a parallel RC-network with $R^P.C^P = 1/\omega$, with $\omega = 2\pi f$ [Geddes, 1972].
More detailed measurements [van Oosterom, 1978] showed that in

$$R = K/\omega^m \tag{9.3}$$

$$C = 1/K.\tan(m.2\pi/2).\omega^{1-m}) \tag{9.4}$$

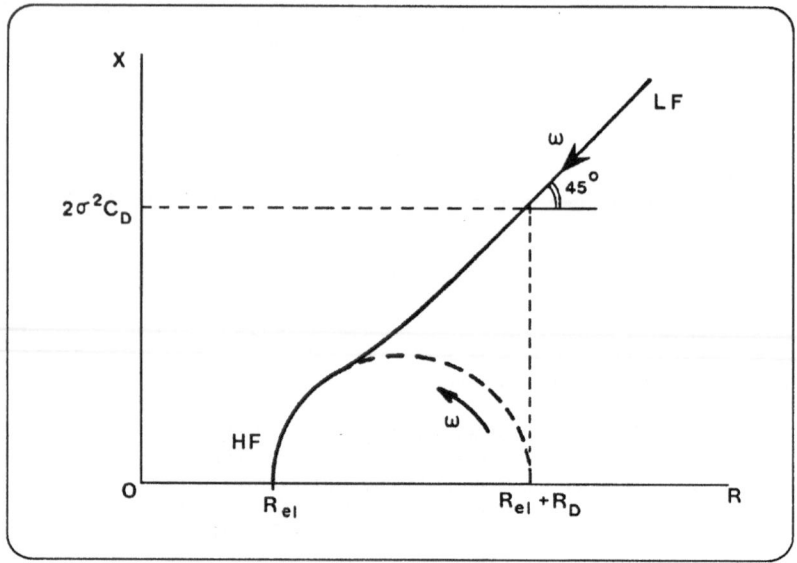

Figure 9.2   Polar diagram of electrode impedance

C_D Transport Capacity
$\sigma = (R.T/z^2F^2).1/C_o\sqrt{2D}$, a factor depending on the
ion equilibrium concentration C_o and on temperature.

the parameter m has a value of about 0.75 for platinum electrodes. K is dependent on the surface condition and inversely proportional to the electrode area.

The frequency behavior of the network with the Warburg approximation shown in Fig. 9.1 can be demonstrated by a polar diagram, where $X = 1/\omega C$ is plotted against R with ω as a parameter along the curve.

From this diagram it becomes evident that the diffusion process is dominant in the low-frequency range, whereas the compact layer determines the high-frequency behavior.

9.2.3   Electric Properties of the Skin

Using surface electrodes for bioelectric measurements, the skin must also be taken into account. The dry epidermis has a far higher electric resistance than the interstitial fluid, through which the volume conductance takes place. Sweating lowers this resistance as sweat contains electrolytes [Geddes, 1972].

A model for skin impedance is given in [Miller, 1974] and it can be approximated by a parallel CR network. The epidermis determines low-frequency impedance, whereas the dermis determines high-frequency impedance. Skin impedance decreases with increasing frequency. Skin preparation can decrease impedance at low frequencies but has little effect at higher frequencies.

The skin resistance is lowered by a factor of about 60 by applying conductive paste; abrading gives a factor of about 250. At the deeper layers of the epidermis, a liquid-junction potential is also present. Since the skin acts as an ion-sensitive membrane, allowing the passage of K^+ better than Cl^- between the surface of the skin and the underlying tissue, a *potential difference* develops, which may be between 50 and 60 mV. Movement of the skin changes the interface, notably the diffusion layer, causing *movement artifacts* that are difficult to quantify and may be considerably greater than the signal generated by skin deformation [Ödman, 1980].

9.2.4   Amplifier Input Stage

The *amplifier* serves to amplify the electrode signal in the presence of interference from the environment. Therefore, the first stage usually has a differential input, which allows rejection of common-mode signals.

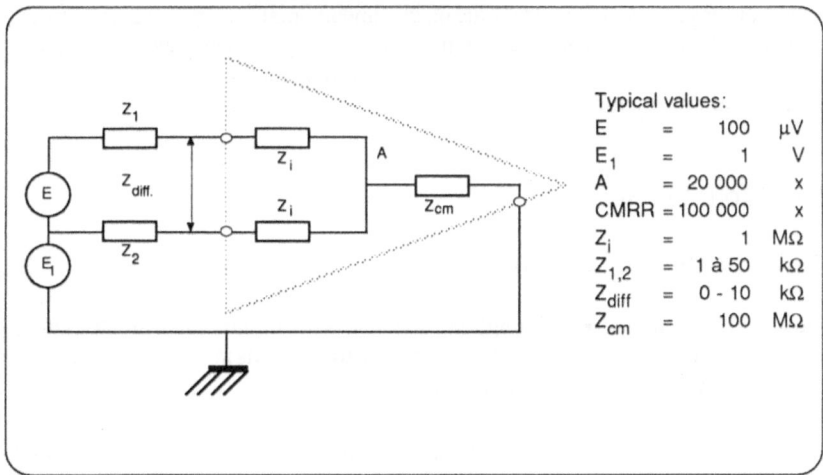

Figure 9.3   The amplifier input stage configuration

Four types of signals reach the next amplifier stages:
- the desired signal, E
- an undesired component due to asymmetry in the source impedances Z_1 and Z_2
- an undesired component due to asymmetry in the amplifier input impedances Z_i and Z_{diff}
- an undesired component from the common-mode signal E_1, due to the finite impedance $(Z_1 + Z_i)//(Z_2 + Z_i) + Z_{cm}$ through which E_1 draws current. Note that in general $E_1 \gg E$.

The common-mode rejection ratio (CMRR) is defined as the ratio of the signal values for E_1 and E that give equal output voltages. The CMRR is in principle frequency dependent due to ever-present parasitic capacitances. The design value of Z_i is a compromise: the high value desired for the CMRR cannot be applied because the electrode double layer has to stabilize, drawing current from the input circuit. Too high an input impedance would require a long stabilization time and thus cause drift.

9.3   Limitations of the Model

As already mentioned under 9.1, the aspect of electrode position in relation to the relevant source of bioelectricity and the ever-present undesired sources (e.g. myographic interference in the ECG during exercise or conversely ECG components present in the EMG of thoracic muscle) cannot be included readily in an electrical network model. The same applies to movement artifacts [Ödman, 1980]. Although thermal and 1/f noise and mains interference might have been included, the authors have chosen to concentrate on the fundamental concepts of the electrode properties themselves. There are many electrode types, materials and gels the relevant parameters of which are not always specified, thus, only a limited number of cases could be included.

9.4   Model Description

9.4.1   Structure of the Model

The model is computational throughout, and makes use of numerical methods with complex numbers, notably the Fast Fourier Transform and its inverse. The electrode model shown in Fig. 9.1. is expressed as:

$$Z_e = R_b + K/\omega^m.[1 - j.\tan(m.\pi/2)] \qquad (9.5)$$

The simulation results are presented as functions of time or frequency for the various signals and as polar plots for impedances.

9.4.2   Parameters and Signals

The values for model parameters were derived from tables and graphs in the literature and thus reflect measurement results from renowned authors. For details refer to section 9.5. Also, bioelectric signals from volunteers were recorded and digitized to serve as inputs for the models.

As an example, measurement results from [During ,1962] are given in Fig. 9.4.

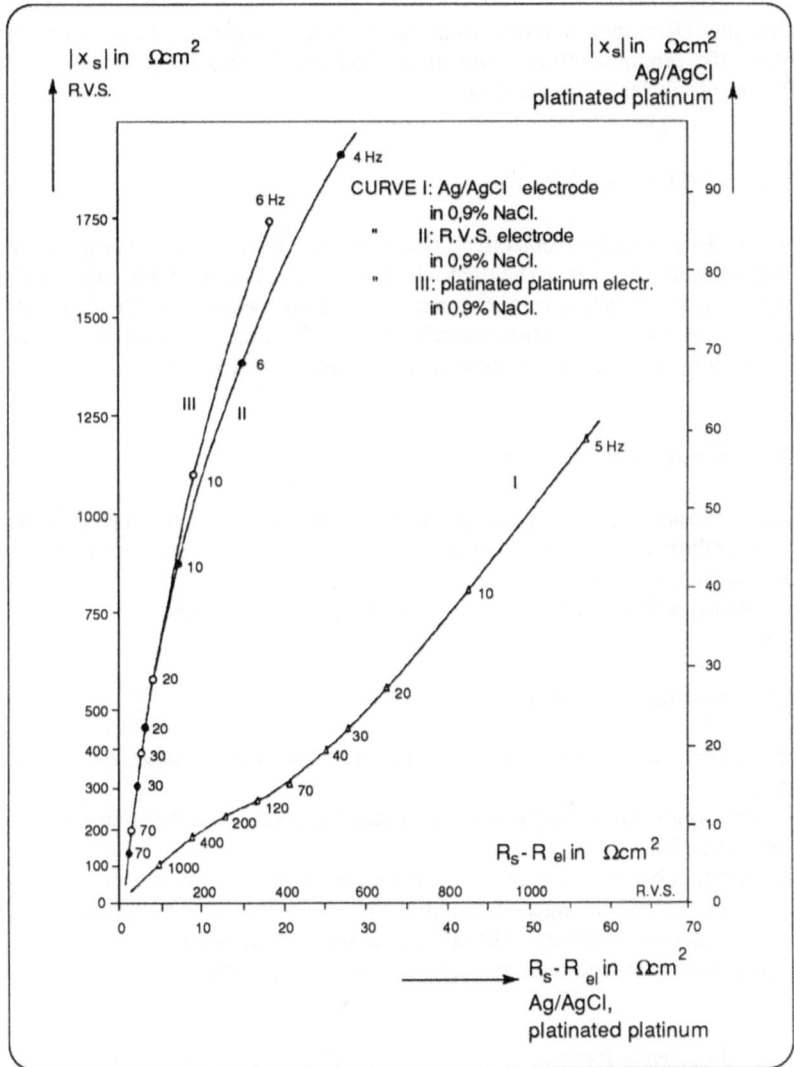

Figure 9.4   Polar diagrams of metal electrodes
  Curve   I: Ag/AgCl
          II: Stainless Steel
         III: Platinated Platinum
  all immersed in 0.9% NaCl.

9.5   How to Use the Program

To simulate electrode properties in relation to the electronic part of the measurement system, the program offers many illustration possibilities related to the main concepts of bioelectric measurements.

9.5.1   The Default Demo

The default demo has two objectives. It explains which selections can be made, and which relevant values are to be used with the options chosen. In this way it offers insight into the possibilities of the simulations to be performed. The demo also shows the system's transfer function with the default values. It is a good idea to run the default demo before attempting to choose options and values.

9.5.2   Simulation Parameters

The user can adjust the many simulation parameters from the menu at will. Default values for the parameters are set at start up, so the program is ready for simulations from the main menu level.
The defaults will be restored by selecting the default demo again.

9.5.2.1   Number of Data Points

The number of data points can be selected from four predefined values: 64, 128, 256 and 512.
If the user is more interested in speed than in accuracy and detail, the minimum value is recommended.
The larger numbers of data points result in increased accuracy and more detail in the plots, at the expense of longer computation time. Thus, the user can make a trade off between speed and accuracy. The default number of data points is 64.
Keeping this value until more detail is required is suggested.

9.5.2.2   Frequency Range

With this option the bandwidth, which determines the frequency components present, can be selected. Plots of the amplifier and electrode impedance, the transfer function and the system's transfer function including CMRR influence are affected by this option.

It is possible to choose from:

$$
\begin{array}{llll}
0.01 & - & 10 & \text{Hz} \\
0.1 & - & 100 & \text{Hz} \\
1 & - & 1 & \text{kHz} \\
10 & - & 10 & \text{kHz} \\
& \text{auto-adjust} & &
\end{array}
$$

Auto-adjust automatically adapts the frequency range to the frequency content of the signal selected. For signals read from disk, the frequency range is defined according to the sample frequency of these signals.
Default is the automatic adjustment of the frequency range.

9.5.3 Electrodes

The program has many electrodes from which to select.
The metal/electrolyte interface can be chosen from a wide range of materials. Also, the area of the electrode can be adjusted and the skin impedance is variable. The model for the electrodes has been discussed in section 9.2. and shown in Fig. 9.1. R_b represents the bulk resistance of the electrolyte in the body, including the patient's skin impedance, which is assumed to be purely resistive.
The other impedances in the model represent the electrical properties of the metal/electrolyte interface. In the program the serial representation for the generalized Warburg impedance has been taken. R and C have values that depend on frequency and on the type of metal/electrolyte layer [Geddes, 1972].

9.5.3.1 Materials

For the equation from section 9.4.1 we used parameters presented by [Geddes, 1972] and [Van Oosterom, 1978]). The presented parameters were calculated for various bandwidths.
Almost all values for the parameters have been calculated from measurements with unspecified temperature and thus have an uncertainty margin in the frequency range for which the measurements are valid.
For these two reasons, we extrapolated the curves for other frequencies and ignored the temperature in the calculations.
It appeared that these assumptions did not violate the expectations one might have when investigating the electrode impedance in a somewhat qualitative way for educational purposes.

9.5.3.2 Area

The capacitance and the resistance in the model also depend on the electrode area. For the sake of simplicity, the geometry of the electrode has been neglected and only the area has been made variable. The program allows the user to choose from the (wide) range of 0.0001 cm^2 (micro-electrodes) up to 2.0 cm^2 (surface electrodes).

9.5.3.3 Skin Impedance

In the measuring situation, the impedance of the patient's skin is an uncertain factor. However, it can be influenced by rubbing the skin or using electrode gel. The program allows the choice of values from extremely dry skin to skin treated with gel. A value for this resistance in an *in vitro* situation is also included.

9.5.4 Amplifiers

The program offers the choice of five amplifiers. The models consist of a resistance and a capacitance in parallel, which simulate the amplifier's input stage. The gain of all amplifiers is 1 (= 0.0 dB).

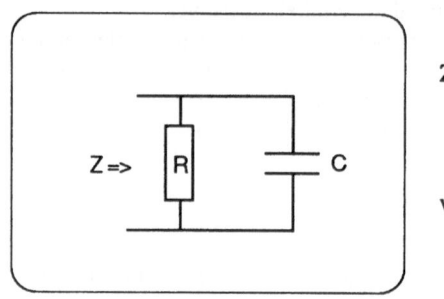

$$Z = R/(j\omega RC+1)$$

$$= (R/(\omega^2 R^2 C^2 +1)) \cdot (1 - j\omega RC)$$

with: $j^2 = -1$ and
$\omega = 2\pi f$
the angular frequency in [rad/s]

Figure 9.5

R represents the input resistance [Ω] and C the input capacitance [F]. An indication of the application is given in the menu.

The parameter options are:

100	kΩ	//	18	pF	(Default)
1	MΩ	//	22	pF	
10	MΩ	//	40	pF	
100	MΩ	//	1	nF	
200	MΩ	//	500	pF	

9.5.5   Filters

A low-pass filter can be applied to the amplifier's output.
It has one pole, so it a has a smooth first-order character (-6 dB/octave). The transfer function of the selected electrode-amplifier combination is modified according to this filter.
The filter's cut-off frequency can be selected from five values, all taken from amplifiers used in practice:

    15 Hz
    30 Hz
    41 Hz
    70 Hz
    300 Hz

Default is "filter off".

9.5.6   Signals

The user can choose between simulating a signal or reading it from disk. An EEG signal, a "white noise" spectrum and some test signals, can be simulated. An ECG, EEG, EMG or any other test signal supplied by the user can be read from disk.
Default is the "white noise" spectrum, with a fixed maximum frequency, representing a Sinc-function ($\sin(a \cdot t)/(a \cdot t)$) in the time domain.
If an error occurs when reading or simulating a signal, this "white noise" signal will be substituted with a corresponding message.

9.5.6.1   Digital Signal Generator

The program has a built-in digital signal generator with many options. Besides the signals mentioned above, several test signals may be generated through suitable algorithms.

Simulated EEG

An EEG can be simulated, using Kemp's model of a person in the drowsiness sleep stage [Kemp, 1987]. This model consists of a Gaussian white noise generator followed by a low-pass filter.
The steps in the simulation of the EEG are:
-   generate samples of the noise signal in the time domain
-   transform this signal to the complex frequency domain
-   apply the low-pass filter (cut-off frequency 1.0 Hz)
-   perform an inverse transform to the time domain

As the number of data points is changed, the duration of the EEG-signal is adapted proportionally, thus the sample frequency has a fixed value.

"White Noise"

A signal with a uniform frequency spectrum has an unlimited bandwidth [Papoulis, 1981]. As a consequence, the corresponding signal contains infinite energy. Such a signal is a mathematical abstraction of the real world. In practice, a signal with a flat power spectrum in a frequency range which is wide compared to the bandwidth of the system to be investigated is considered to be a "white noise" signal. With the aid of the Fourier transform, a "uniform" spectrum can be transformed to the time domain. The corresponding time signal is a Sinc-function, $\sin(a \cdot t)/(a \cdot t)$, in with a, a constant, and t representing time).

The program generates a spectrum as well as a signal with parameters suitable to the system being investigated.

Square Wave

The square wave generated has a basic frequency of 2 Hz.
This leads to odd harmonic frequency components in the frequency domain (2, 6, 10, 14, ...). These components decay proportional to $1/f$ [Papoulis, 1981].

Triangular Wave

A triangular wave has frequency components that decay proportional to $1/f^2$. The basic frequency is also 2 Hz.

Sine Wave

The sine wave signal is most often used to investigate linear systems. Then the output is also sinusoidal with a phase shift and an amplitude change representing the system's properties.
The program generates a 2 Hz sine wave.

Synthetic Sine Waves

A signal composed of five sine waves with frequencies of 2, 4, 6, 8 and 10 Hz of the same amplitude is also available for demonstration purposes.

Discrete Sine Wave

A sine wave is provided that is discrete in amplitude. Five voltage levels are taken (-1, -0.5, 0, 0.5 and 1), derived from a normal sine wave, rounded to the given levels.

9.5.6.2 Signal Files

Signals can also be read from disk. These files should have the .SIG-format. Refer to the appendix for details on the files supplied with the package and to section 9.6.1 for the format.

Signals the user might want to include should comply with the following characteristics:
- a small or no DC-component
- a sample frequency chosen with care, resulting in no
  sub-sampling
- start and end values about equal

A large DC-component could hide the other frequency components in the spectrum since automatic scaling of the plots is performed.

When the third constraint is not fulfilled, high-frequency components arise in the spectrum. In other words, the signal should be approximately periodic. This is expected in the FFT (Fast Fourier Transform) routines used.

9.5.7   Graphical Presentation

With the program it is possible to show all useful combinations of (bio-)signal, electrode, amplifier and low-pass filter in plots made by the universal graphics program included in the package.

9.5.7.1 Signals

Signals can be presented as a function of time or as their power spectrum in the frequency domain. The input signal as well as the chosen output signal from, the system configuration can be selected for presentation.

9.5.7.2 Impedances

Electrode and amplifier impedances can be judged from impedance plots, which demonstrate their frequency dependency.

The electrode impedance can be presented in a polar diagram with the real part (resistance) on the abscissa and the imaginary part (reactance) of the complex impedance on the ordinate. The electrode model is described in section 9.5.3. The amplifier input impedance can be shown as a Bode plot. The logarithmic abscissa presents the frequency and the ordinate represents the normalized impedance. The amplifier model is given in section 9.5.4.

9.5.7.3   The System with Common Mode Rejection

Two more plots are provided: one demonstrates the influence of the Common Mode Rejection on the transfer function, using an appropriate amplifier model in combination with the electrode selected; the other shows the transfer function, using the amplifier model of section 9.5.4.

Transfer Function

The influence of the total system as it is used in the calculations for the output signal can be shown in terms of its transfer function in the chosen frequency range. The selected electrode and amplifier parameters are used to compute this function. When the low-pass filter is switched on, its influence is included and the ordinate is normalized. This allows a comparison with the many system configurations that can be selected by the user.

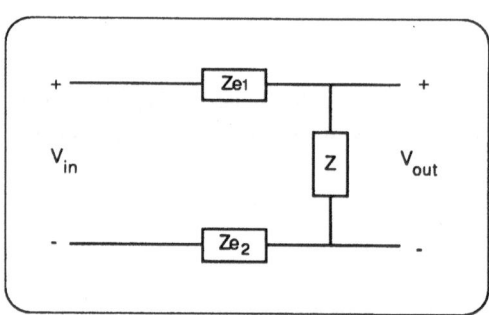

The transfer function $H(j\omega)$ equals:

$$H(j\omega) = \frac{Z}{Ze_1 + Ze_2}$$

Ze_1 and Ze_2 represent the electrode impedances. In the program they are assumed to be identical. Z represents the amplifier's input impedance.

Figure 9.6

Transfer Function with CMRR influence

With this option, three curves are calculated and shown in one plot. The transfer function accounting for the electrode and the electronics is given according to [Bekey, 1983]. In this simulation the common mode influence is represented by two common mode (CM) impedances.

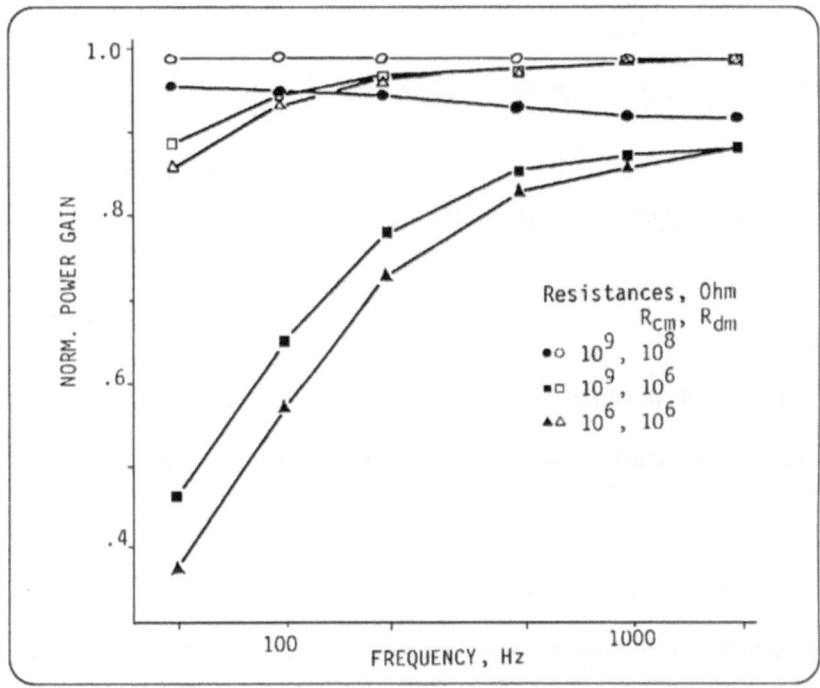

Figure 9.7 Normalized power gain vs. frequency for surface electrodes. Empty shapes - 50 μm diamter electrodes, filled shapes - 25 μm electrodes. ●○ - balanced input circuit, ▫▪ - unbalanced input circuit, ▲▲ - unbalanced with no input guard, ▼▼ - unbalanced, no input guard with low common mode resistance (1 MΩ) (Bekey, 1983, by permission).

To simplify the model, the authors assumed both the common mode as well as the differential mode impedances to be pure resistive. This results in a reduction of the computation time, but even with this (optimistic) assumption, the results are striking. Compare this with Fig. 9.7 [Bekey, 1983].

Note that the low-pass filter is not used in this model, as it would veil the preponderant issues.

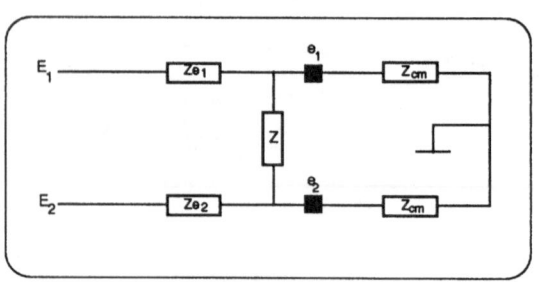

$$H(j\omega) = \frac{e_1 - e_2}{E_1 - E_2}$$

with:

E_1, E_2 the voltages on the patient's skin, e_1, e_2 the voltages to be amplified

Z_{cm} = CM-impedance
Z = DM-impedance

Figure 9.8

With this model, the DM (differential mode) transfer $H(j\omega)$ is computed for varying values of Z_{cm}.
The user is asked if help is required with this demonstration. It is suggested that you read this help page first.

9.6   How to Do it Yourself

The user can run the program stand-alone by typing from the DOS command line:

ELECTRO [UniDir [Letter]]

With:

UniDir   the directory for locating UNIGRAPH.EXE, the universal graphics program. Default is the current directory.

Letter   a letter from ELECTRO's main menu. When you specify this letter only the corresponding menu item will be executed.

If only one command line option is given ELECTRO assumes that this is the Unigraph directory.

When using the program for the first time, it is suggested to run the demonstration. The demo gives some insight into the many selections that can be made. From the main menu, press "D" or use the cursor keys and <ENTER> to run the default demo.

9.6.1   Format of the Signal Files

The .SIG-files contain signals that can be used as the input signal for the electrode-amplifier-filter system.
The user can include his own signal files, as described below. Also review section 9.5.6.2 for constraints on these signals.

The *.SIG-file is an ASCII (text) file, composed as follows:

Table 9.2

Line #	Contents	Format	Purpose/Limitations
1	Signal Type	Integer	Identifiers: 1=ECG, 2=EEG, 3=EMG, other=Test Signal Comments are allowed, following the identifier
2	F_sample	Real	The sample frequency in Hertz (Hz)
3	Value	Real	The sampled values as floating point numbers, one value per line. If there are insufficient samples compared with the number of data points, a suitable number of zeros will be inserted (zero padding).

9.6.2 Experiment Files

An aspect of the program's educational philosophy is the large number of parameter combinations that can be chosen.
In order to facilitate the reproduction of interesting combinations, the option of experiment files is offered.

9.6.2.1 Reading and Saving Experiments

An 'experiment' comprises the settings of all parameters selected from the menus. It is possible to save these settings and to read experiments from disk files. The automatically generated extension for experiment files is ".PRE" (preset), required for the read procedure. Pressing <ALT><S> from the main menu saves the current parameter values. The user is then prompted for the filename. Reading is achieved by pressing <ALT><R> from the main menu. A window shows the files from which a selection may be made. It is also possible to generate experiment files with the general editor. In that case, however, a correct structure cannot be guaranteed. Errors in a .PRE file may result in an abortation of the reading.

9.6.2.2 Format of the .PRE files

Experiment files are in ASCII-format, composed as follows:

Table 9.3

Line #	Contents	Format	Effect
1	Name	string, 255 characters, may be empty	Allows to include a descriptive title, does not affect the program.
2	Frequency interval	integer, 0..5;	0 = no change; 1 = auto adjust; 2 = 0.01 - 10 Hz 3 = 0.1 - 100 4 = 1 - 1000 5 = 10 - 10.000
3	Points	integer, 0..4;	0 = no change; 1 = 64 2 = 128 3 = 256 4 = 512
4	Signal	integer, 0..7, 11..14;	0 = no change; digital signal generator: 1 = square; 2 = triangle; 3 = sine; 4 = harmonics; 5 = discrete sine; simulated: 6 = EEG; generated: 7 = 'white noise'; read from disk (via menu): 11 = ECG; 12 = EEG; 13 = EMG; 14 = Test;
5	Electrode material	integer, 0..18;	0 = no change; 1 = Pt black / 0.9% NaCl 2 = Pt black / 0.9% Saline 3 = Pt Ir black/ 0.9% saline 4 = Ag / 0.9% NaCl 5 = Pt / 1.46N H_2SO_4 6 = Ag / 0.1N $AgNO_3$ 7 = Pt / 1% H_2SO_4 8 = Stainless steel / 0.9% saline 9 = Stainless steel / in vivo (dog) 10 = Au / 1.46N H_2SO_4 11 = Pt / 0.025N HCL 12 = Au / 0.01N KBr 13 = Au / 1% H_2SO_4 14 = Pt / 1.1% H_2SO_4 15 = Pt / 0.25N H_2SO_4 16 = Pt / H_2SO_4 17 = Pt / 0.9% NaCl (needle) 18 = Ag/AgCl /0.9% NaCl

6	Electrode area	integer, 0..8;		
			0	= no change;
			1	= 1.0 E-4 cm^2
			2	= 5.0 E-4 cm^2
			3	= 1.0 E-3 cm^2
			4	= 5.0 E-3 cm^2
			5	= 1.0 E-2 cm^2
			6	= 5.0 E-2 cm^2
			7	= 1.0 E-1 cm^2
			8	= 2.0 E-1 cm^2
7	Skin Type	integer, 0..10;		
			0	= no change;
			1	= Extremely dry / Rb = 10000 Ohms.
			2	= Very dry / Rb = 8000 Ohms.
			3	= Quite dry / Rb = 6000 Ohms.
			4	= Dry / Rb = 4000 Ohms
			5	= Bit wet / Rb = 3000 Ohms.
			6	= Wet / Rb = 2000 Ohms
			7	= Sweaty skin / Rb = 1500 Ohms.
			8	= In vitro / 0.9% NaCl / Rb =1300 Ohms.
			9	= Using electrode-gel / Rb =750 Ohms.
			10	= Perforated skin / Rb = 0 Ohms
8	Amplifier	integer, 0..5;		
			0	= no change;
			1	= 100 kΩ // 18 pF;
			2	= 1 MΩ // 22 pF;
			3	= 10 MΩ // 40 pF;
			4	= 100 MΩ // 1 nF;
			5	= 200 MΩ // 500 pF.
9	Low-pass filter	integer, 0..6;		
			0	= no change;
			1	= filter off;
			2	= 15 Hz
			3	= 30
			4	= 41
			5	= 70
			6	= 300
10	Graphs	integer, 0..8;		
			0	= no change;
			1	= signal time;
			2	= signal spectrum;
			3	= electrode impedance;
			4	= amplifier impedance;
			5	= transfer function;
			6	= signal & system time;
			7	= signal & system spectrum;
			8	= CMRR.

References

Bekey, G.A.: "Theory vs. reality in biological modeling and data analysis". In: Vansteenkiste, G.C. and P.C.Young (Eds.): *Modeling and Data Analysis in Biotechnology and Medical Engineering.*
North-Holland Publishing Company/IFIP, 1983.

Cobbold, R.S.C.: *Transducers for biomedical measurements; Principles and applications.* Wiley-Interscience, New York, 1974.

During, J., A. den Hertog and F. Schot: *Metal Electrodes.*
Institute of Medical Physics, TNO report no. 2-2-58/2,
Utrecht, 1962. (In Dutch).

Ferris, C.D.: *Introduction to Bioelectrodes.*
Plenum Press, New York, 1974.

Geddes, L.A.: *Electrodes and the measurement of bioelectric events.*
Wiley-Interscience, New York, 1972.

Kemp, B.: *Model-Based Monitoring of Human Sleep Stages.*
PhD thesis, Twente University of Technology, Enschede, 1987.

Miller, H.A. and D.C. Harrison: *Biomedical Electrode Technology.*
Academic Press, 1974.

Ödman, S.: *On Medical Electrode Technology with special reference to long-term properties and movement-induced noise in surface electrodes.*
PhD thesis, Linköping, 1980.

Oosterom, A. van: *Cardiac potential distributions.*
PhD thesis, Amsterdam, 1987.

Papoulis, A.: *Circuits and Systems, a modern approach.*
Holt-Saunders Japan Ltd., Tokyo, 1981.

Plonsey, R.: *Bioelectric Phenomena.*
McGraw-Hill, New York, 1969.

Webster, J.G.: *Medical Instrumentation, Application and Design.*
Houghton Mifflin Comp., Boston, 1978.

Appendix

1 Specification of the signal files supplied

ECG-Files

Healthy male volunteer, in rest, sample frequency 481.6 Hz,
amplitude normalized around mean, standard ECG electrodes. Einthoven leads.
(Thanks are due to Dr. R.M. Heethaar, Department of Medical Physics, Utrecht
University, the Netherlands).

Filenames:
ECG1.SIG   Lead   I
ECG2.SIG   Lead   II
ECG3.SIG   Lead   III

EMG-Files

Healthy male volunteer, biceps brachii, exerting 50 N anteflection force, EMG
measured .5 s after force maximum was reached, sample frequency 1000 Hz,
amplitude normalized around mean, filtered 6 dB/oct at 1 Hz and 24 dB/oct at 1000
Hz.Electrodes: bipolar, gold-plated brass, $\Phi = 7.5$ mm, distance between electrodes
21.5 mm.
(Thanks are due to Mr. R. Happee, Department of Mechanical Engineering, Delft
University of Technology, the Netherlands).

Filenames:
EMG1.SIG, EMG2.SIG, EMG3.SIG

2 Contents of the default demo experiment file

Menu settings saved from ELECTRO.EXE
        1
        1
        7
        1
        1
        1
        1
        1
        1

10
The Heart as a Pump:
The Program "CARDIO"

Fredericus B.M. Min

Abstract

This chapter presents the results of CARDIO, a simulation program on blood pressure regulation under normal and abnormal conditions. The model underlying this program allows the simulation of pathological conditions, such as myocardial infarct, renal artery stenosis or renal insufficiency. On the other hand therapeutic interventions in abnormal conditions can also be simulated. The program allows the application of drugs like cardiac glycosides, diuretics or vasodilators.

10.1 Aim of Instruction

The computer simulation program CARDIO enables students to experiment with the basic principles of blood pressure regulation for a healthy person. The program simulates an experimental laboratory where hemodynamic research can be done, variables can be registered, and where interventions in the model can be made without real-life complications. The computer simulation program CARDIO (CS-program) is often the first instance in the curriculum in which the student can check his understanding of human circulation in the context of realistic problems. With this CS-program he can do a series of experiments, such as the simulation of heart failure, hypertension, renal failure, exceptional blood loss, selective venous constriction, and selective arterial constriction. These interventions can be selected by the student or chosen as a case by the teacher. Through the presentation as a case the student can be asked to make a diagnosis and to operate therapeutically, starting the hemodynamic picture on the graphical screen. The student has the ability to introduce medicines like heart glycoside (digitalis), vasodilators, diuretics, sympathicolytica, or noradrenaline.

10.2 Model of the Cardiovascular System

The model of the CS-program CARDIO is based on a model of Coleman (1980) of the cardiovascular system of a healthy human being. It consists of five compartments: the heart, the vascular system, the intra- and extracellular fluid volumes, the nerve reflex system, and the kidney. The model contains a number of relations between parameters in these compartments and consists of about 60 variables.
Coleman's model of blood pressure regulation and the cardiovascular system distinguishes three important regular quantities:

1. The extent of the vascular resistance is chiefly determined by metabolic and neural factors. In this way it is possible to keep the circulation of blood within the different tissues in the model as constant as possible (autoregulation) while adapting to system needs. The preservation of the flow within a number of tissues is one of the most important physiological priorities of the intact organism.
2. The mean arterial pressure is acutely regulated by the nerve reflexes, notably the baroreceptor reflex (blood pressure regulation by the renine-angiotensine system is not included in CARDIO). In the long term blood pressure is regulated by the kidney function, which controls the extracellular fluid volume.
3. The cardiac output is primarily determined by the venous return. Moreover, cardiac output can be directly influenced by the function of the heart as a pump.

Figure 10.1 contains a graph of the most important relations in the model. The interventions that are possible in this model and their points of application are mentioned in it. The most important variables of the model and their normal values are:

- Mean Arterial Pressure (AP)   100 mmHg
- Cardiac Output (CO)   5000 ml/min
- Total Peripheral Resistance (TPR)   0.02 mmHg.min/ml
- Urine Output (UO)   1 ml/min
- Heart Rate (HR)   70/min
- Blood Volume (BV)   5000 ml
- Right Atrial Pressure (RAP)   0 mmHg
- Sympathic nerve activity (chemoreceptor) (CHEMO)   0
- Sympathic nerve activity (baroreceptor) (BARO)   100% (=1)
- Sympathic Autonomic Outflow (SYMPS)   100% (=1)
- Arterial O_2 Pressure (PO_2)   100 mmHg
- Blood Urea Nitrogen (BUN)   10 mg%
- Extra Cellular Fluid Volume (ECFV)   15000 ml
- Mean Circ. Filling Pressure (MCFP)   7 mmHg
- Glomerular Capillary Pressure (GP)   60 mmHg
- Pressure Gradient for Venus Return (DELP)   7 mmHg

- Renal Artery Resistance (RAR)	1.67 mmHg.min/ml
- Urea Formation (BUNI)	10 mg%
- Urea Excretion (BUNO)	10 mg%
- Edema (ED)	0
- Resistance of the venous return (RVR)	0.0014 mmHg.min/ml

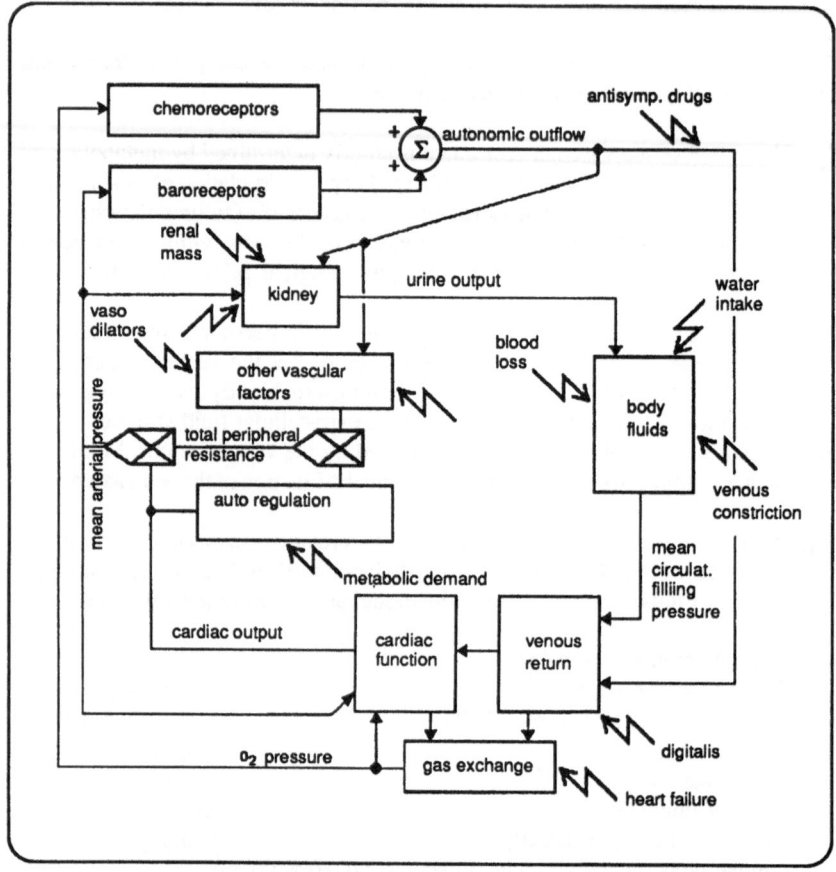

Figure 10.1.   Model of the blood pressure regulation available in the CS-program
CARDIO, identifying possibilities for intervention

10.3 Structure of the Model

The model underlying the CS-program CARDIO was further developed from Coleman's model and tested at the University of Limburg at Maastricht. The description of this model will be discussed relative to Pascal notation and to analogue notation, as well as to the manner in which these notations complement each other.

10.3.1 The Heart

In the model the basic Cardiac Output (COB) – i.e., the cardiac output without the influence of the autonomic nerve system – is a function of the right atrial pressure (RAP). The right atrial pressure (RAP) is calculated from the mean circulatory filling pressure (MCFP) and the pressure gradient for the venous return (DELP) via the relation:

RAP: = MCFP - DELP;

The mean circulatory filling pressure (MCFP) is a function of the blood volume (BV) and the potentiality of the vascular system, namely the veins, to fill themselves with blood.

MCFP: = 0.0047*(BV - BV0);

With the influence of venous constriction (VENCON) and the autonomic nerve system this becomes

MCFP: = 0.0047*(BV - BV0/(VENCON*(2*AO + 0.8)));

The cardiac output (CO) is, as is mentioned before, determined by the basic cardiac output (COB) and a heart function parameter (HF). This heart function (HF) is a function of the right atrial pressure (RAP), the mean arterial pressure (AP), the sympathic nerve activity (AO), possible cardiac glycoside (DIGI), and the basal heart strength (HSB). This is calculated as follows:

HF: = PO2M*APM*HSB*AO*(1+0.5*DIGI);

10.3.2 The Circulation

Many tissues in the body function optimally when the flow through them remains constant within certain limits. The regulating mechanism that controls this tries to realize it through changes in the resistance. This autoregulation mechanism varies in each tissue in respect to the "time constant" and the "gain".

In this model the inclusion of an overall description of the autoregulation phenomena has been attempted. The sum of all individual tissue flows (the cardiac output, CO) occurs in the following context: An abrupt change occurs in the cardiac output (CO), which triggers through an instantaneous total autoregulation an abrupt change in the basal total peripheral resistance (TPRB), the value of which in turn is determined by a non-linear function (F11). The factual adjustment of the resistance happens with a certain delay, the time constant of which is determined by the velocity of the autoregulation. In the model, factor TPRD is assumed for this. This results in the following formulas:

TPRB: = F11(CO*(TPR/TPRM)-5000*MD);

TPRD: = TPRD + (TPRD*TPRDK - TPRD*PRDK)*dt;

TPR: = 1/(50*AVF + 1/TPRM);

The resistance to venous return (RVR) is partly a constant function and partly a function of the total peripheral resistance (TPR). When it is influenced by an AV fistula, a good approach is given by:

RVR: = 0.025*TPR + 0.0009/(1 + 0.8*AVF);
The arterio-venous fistula (AVF) is normally 0 and the total peripheral resistance is
normally 0.02 mmHg.min/ml, so that RVR is normally 0.0055 mmHg.min/ml. The
mean arterial pressure is the product of the total peripheral resistance and the cardiac
output:

AP: = CO*TPR;

10.3.3 Intra- and Extracellular Fluid Volume

The extracellular fluid volume (ECFV) is directly influenced by the fluid volumes
that are taken in by the body minus those which leave it:

ECFV = ∫("intake"-"output").dt

In the model, ECFV(0) = 15000 ml and the intake and output is 1 ml/min. The water
intake of the body (WIN) is set to 1 ml/min. The urine output (UO) is 1 ml/min and
the loss of blood (BL) is normally 0 ml/min. The extracellular fluid volume (ECFV)
is calculated in milliliters:

ECFV: = ECFV + (WIN - UO - BL)*dt;

The extracellular fluid spreads over the plasma and the interstitium. In the model the
blood volume (BV) varies according to the function F1 while edema formation (ED)
varies according to the function F10, dependent on the extracellular fluid volume.

10.3.4 The Nerve Reflex System

The baroreceptor influence after adaptation (BARO) is equal to:

BARO: = BAROP-ADP;

while the adaptation (ADP) is normally 0. The time constant of this reflex is
BAROK. The total baroreceptor reflex part is therefore:

ADP: = ADP + (BAROK*(BARO - 1))*dt;

The autonomic nervous activity (AO), the value of which is normally adjusted to 1,
is dependent on the chemo- and baroreceptor activity, a possible adrenergic
blockade (BLOCK), a possible pheochromocytoma (PHEO) – which develops an
excessive sympathic activity – and a possible alpha-adrenergic influence of
norepinephrine (NOREPI). The heart frequency is directly a linear function of AO:

SYMPS: = (BARO + CHEMO)*(1 - BLOCK)*(1 + PHEO);
AO: = 0.5*SYMPS + 0.5*NOREPI;
HR: = 70*AO;

10.3.5 The Kidney

The kidney function is dependent on the composition of the blood, aldosterone, neuronal factors, angiotensine, vasopressin, and the mean arterial pressure. In this model the basic urine output (UOB), normally 1 (ml/min), is a function of the mean arterial pressure (AP), the basal renal arterial resistance (RARB), the autonomic output (AP), and a vasodilation factor (DILAT).

The basic urine output is the urine output before other influences act upon it, such as kidney mass and diuretic influences (DI). The urine output then becomes:

UO: = UOB*RM*(1 + 2*DI);

The nitrogen concentration in the blood depends on the renal mass (RM), the protein-intake (BUNI), and the basic urine output:

BUNI: = 10*(0.6 + 0.4*PRDIET);

QBUN: = QBUN + (BUNI - BUNO)*dt;

Part of the Pascal source code is given in the appendix. The complete analogue scheme is given in Figure 10.6.

10.4 Results

In the CS-program CARDIO, the normal values for the mean arterial pressure (AP), the cardiac output (CO), the total peripheral resistance (TPR), the urine output (UO) and the heart frequency (HR) with the given aggressive values and without any intervention, are: AP =100 mmHg; CO=5000 ml/min; TPR = 0.02 mmHg.min/ml; UO =1 ml/min and HR =70 min.

These five important hemodynamic variables can be supplemented by the extracellular fluid volume (ECFV), the blood volume (BV), or the pressure of the right atrium (RAP). These variables are usually difficult to measure. In the exercises described here, such variables play an important role. The CS-program CARDIO gives not only visual feedback on the student interventions but also textual feedback.

The following messages can appear on the screen when a particular variable reaches above or below a certain limit:

40 < PO2 < 80 "Doctor, I can't get any air."

35 < PO2 < 40 "Doctor, I don't really know what's happening to me anymore. I lose my bearings."

30 < PO2 < 35 Patient becomes blue.

PO2 < 30 Patient looses consciousness

AP   < 85 "Doctor, I feel so dizzy."

ED   > 1 "Doctor, my feet are swollen, my shoes pinch."

In one version of the CS-program CARDIO these texts were given in audio fashion using the speech chip of the MacIntosh computer. In the visual output, the head of the patient leans to the left when he dies.

10.4.1 Heart Failure

In figure 10.2 the hemodynamic effect of a light heart failure is simulated. In the time frame shown in the figure the contraction strength of the heart muscle is reduced to 70%. The cardiac output decreases sharply while the baroreceptor reflex mechanism tries to compensate the fall in the blood pressure occurring in this model. This activity can be seen by the rise of the heart rate.

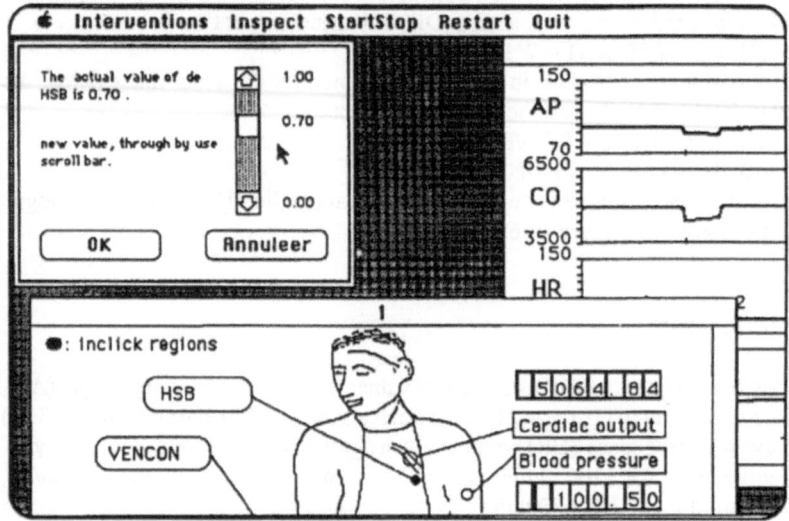

Figure 10.2.   Case of a light heart failure. The basal heart strength (HSB) reduces suddenly to
70%. After some hours digitalis is given. Adjustment of the scale:
time-axis from 0 to 5 days (1 stripe is 1 day).
1st graph: mean arterial pressure (AP) from 70 to 150 mmHg;
2nd graph: cardiac output (CO) from 3.5 to 6.5 l/min.

If the simulation is continued, there are, after thorough inspection of the model variables, no significant deviations to be found after two weeks in a number of measured and registered variables. However, other variables do change. Careful analysis of the simulation data with the inspection options shows that there is renal retention of fluid. After 14 days the ECFV has risen to 17000 ml and the blood volume (BV) to 5400 ml. Thus the inability of the heart to pump an adequate amount of blood per time unit is compensated by an overfill of the vascular system by which the cardiac output keeps its normal value. In reality this also happens in case of a light heart failure. We then speak of a "compensated heart failure". By this we mean that the heart function – in spite of the diminished contraction strength of the heart muscle – is restored. When the student intervenes in the model shortly after the compensation power of the heart muscle has diminished by 30%, a digitalis therapy can be started which can prevent the overfill of the vascular system and, by its primary effect, intensify the contraction strength of the heart muscle. After a short while all variables appear to have returned to their normal values before the sudden heart failure. Following this, with a careful dosage of digitalis (DIGI is plus/minus 0.6 units), the condition of no significant increased ECFV and BV and a normal right atrial pressure (RAP = 0 mmHg) can be achieved.

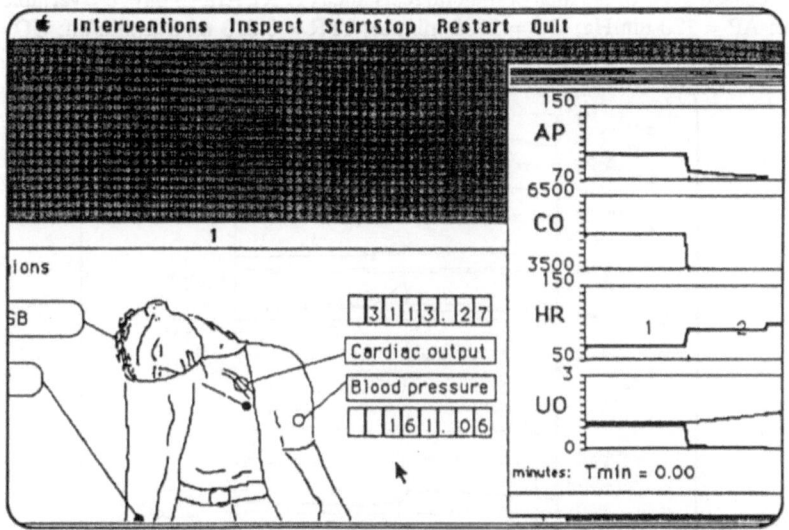

Figure 10.3. Case of a severe heart failure. The basal heart strength (HSB) suddenly drops to 30%. The patient dies. In this case no compensation is possible, but when a sufficient amount of digitalis is given the patient will not die. Scale adjustment as in figure 10.2 and in 4th graph: urine output (UO) from 1 to 3 ml/min; 3rd graph: heart rate (HR) from 50 to 150 l/min.

In Figure 10.3 the hemodynamic effect of a severe heart failure is simulated. At the given point of time the contraction strength of the heart muscle (HSB) has diminished to 30%. The cardiac output drops much faster than in the first case. The reflexive increase of the peripheral resistance and the fall in the urine production is also sharper. If there is no therapeutic intervention, a patient with this hemodynamic pattern will soon die of the type of suffocation which can occur with lung oedema. If the student intervenes in the model shortly after the heart insufficiency occurred and starts a digitalis therapy (DIGI > 0), he may safeguard this patient against the chance of oedema. A diureticum can also reduce a possible increase of the extracellular fluid (DI > 0). Finally, in this simulation the student can also experiment with a vasodilator (VASCON), together with the digitalis therapy (DIGI).

10.4.2 Hypertension

Figure 10.4 shows, by a renal arterial constriction (RARB = 1.2), a stable high blood pressure initiated after 7 days. We have here a simulation of a renal artery stenosis. The excretory function of the kidney is reduced and fluid retention occurs. The mean arterial pressure (AP) rises by the increased cardiac output, and the autoregulation has caused a rise of the total peripheral resistance (TPR). After 7 days the variables are: AP = 120 mmHg; CO = 5100 ml/min; TPR = 0.023 mmHg.min/ml; UO = normal and HR = normal.

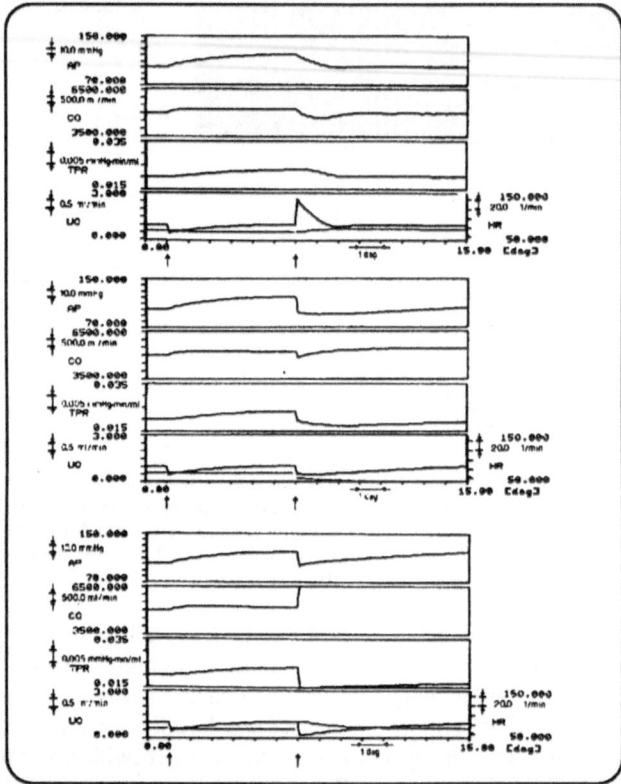

Figure 10.4. Case of hypertension. On day 1 RARB, renal arterial constriction, becomes 1.22. After six days the mean arterial pressure (AP) is 120 mmHg and the total peripheral resistance (TPR) has increased to 0.0235 mmHg.min/ml.

    a. "Diureticum therapy" (DI=0.9) gives a normal AP and CO and a temporary raised diureses (UO < 1) on day 15.

    b. "Sympathicalolyticum therapy" (BLOCK=0.65) gives a normal AP and a decreased total peripheral resistance (TPR).

    c. A vasodilator only (VASCON = 0.5) gives a reduction of the blood pressure for only 1 or 2 days. At the same time the CO is lastingly increased, higher than the fixed maximal value. Division of the scale as in Figure 10.2. (CARDIO on the RLCS-system, 1982)

The student can treat this form of hypertension, caused by a renal artery stenosis, with three antihypertensiva used singular in combination. The program has the possibility of:
1. a diureticum (DI)
2. a vasodilator (VASCON)
3. a sympathicolyticum (BLOCK)

The influence of the doses is shown in Figure 10.6 in the dose-response curves. Hypertension can also be simulated by acute kidney insufficiency (lower renal mass (RM) to 10%) or by a pheochromocytoma (PHEO). However, this case is limited to a renal artery stenosis. If the student chooses a diureticum as antihypertensivum, he has to choose for DI = 1, through which the patient becomes normotensive (100 mmHg) after about 2 days.

In case the choice is made for a vasodilator, which does not effect the kidney, VASCON = 0.15 is sufficient to keep the blood pressure normal for a few days (AP = 100 mmHg). The cardiac output is then raised. However, it does not lead to a permanent fall of blood pressure. With a sympathicolyticum as an antihypertensivum, BLOCK = 0.65 is sufficient to normalize the blood pressure after seven days. The cardiac output is then significantly raised to 5400 ml/min.

10.5 Discussion

The model of the CS-program CARDIO is one of the first models implemented in the RLCS-system of the University of Limburg in Maastricht.

With this CS-program, experience is acquired in certain curriculum blocks, namely, in the block "heart and lungs", in the block "fatigue", and in the block "pain on the chest". Causal relationships, within cases regarding to heart failure, hypertension, and therapies for various afflictions, are developed by several students working together. Presenting a task in this way, by putting it in the form of a patient case, appeared to be desirable. In the average time of one or two hours in which students use a CS-program in a block, they can only put one or two problems under discussion. The cases "light" and "severe heart failure" usually give occasion to check in books how the body can compensate a heart insufficiency by holding fluids. If the "digitalis therapy" is absent or insufficient, the student sees the risk, getting the message that the patient has "swollen feet" at a certain moment.

The case "hypertension" often allows the student to study the working mechanisms of antihypertensiva more closely. An important aspect of this case is that the student also learns about some relationships among working doses of antihypertensiva and that vasodilators cannot function chronically as a monotherapy for hypertension.

The CS-program CARDIO has been tested in various versions, particularly in two terminal version where there has been much experimentation. It has frequently been demonstrated before external and internal experts in order to stimulate discussions and opinions about the design of the system (Min et al., 1982).

The program has been used by several generations of students at the University of Limburg at Maastricht from May 1980 until now. In 1987 there were a total of 3000 student sessions with the CS-programs CARDIO, FLUIDS and the well-known Canadian CS-program MacDOPE. Other universities in Europe use other versions of CARDIO, but the extent of this usage in their regular curriculums is not known. The CS-program CARDIO has versions on PDP11/03 (MINC computer; The RLCS-system), VAX 11/750 (RLCS-system, VAX version and THESIS, VAX version), MacIntosh (MacTHESIS) and Atari 1040 ST with Alladin emulator.

A MacIntosh version of the program CARDIO is available on request from the author of this book.

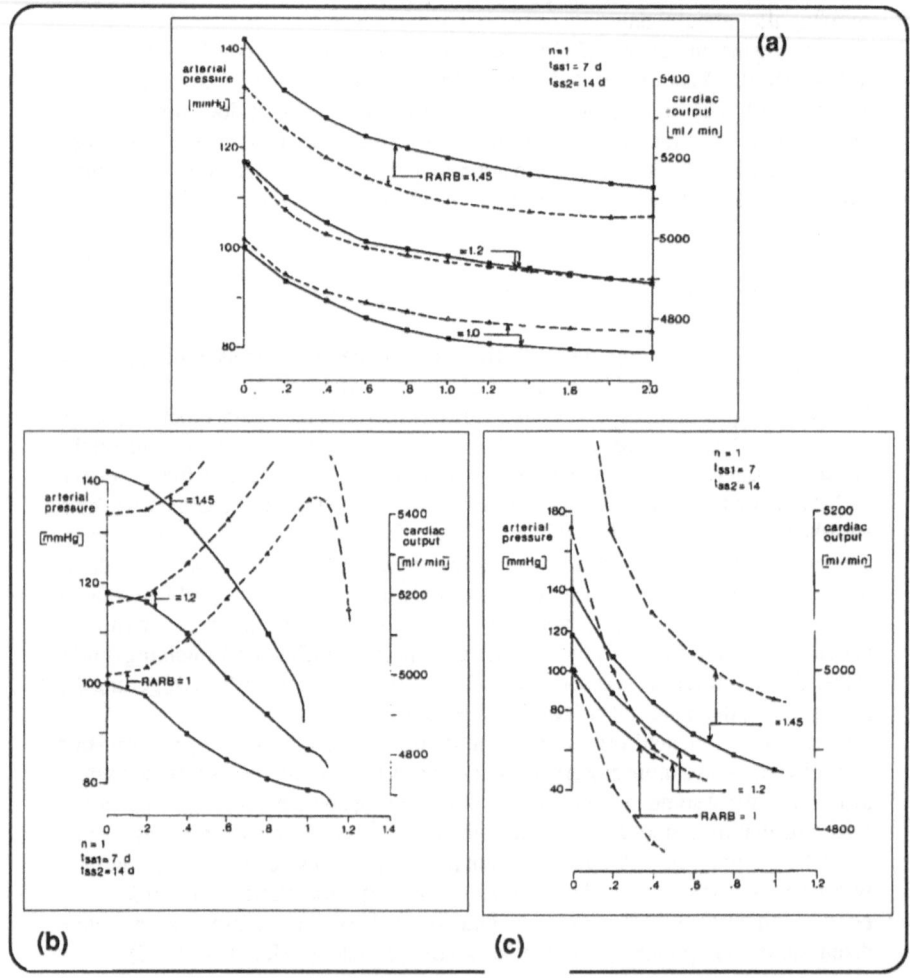

Figure 10.5. Dose-response curves for antihypertensiva: a) diureticum; b) "reserpine" and c) vasodilator

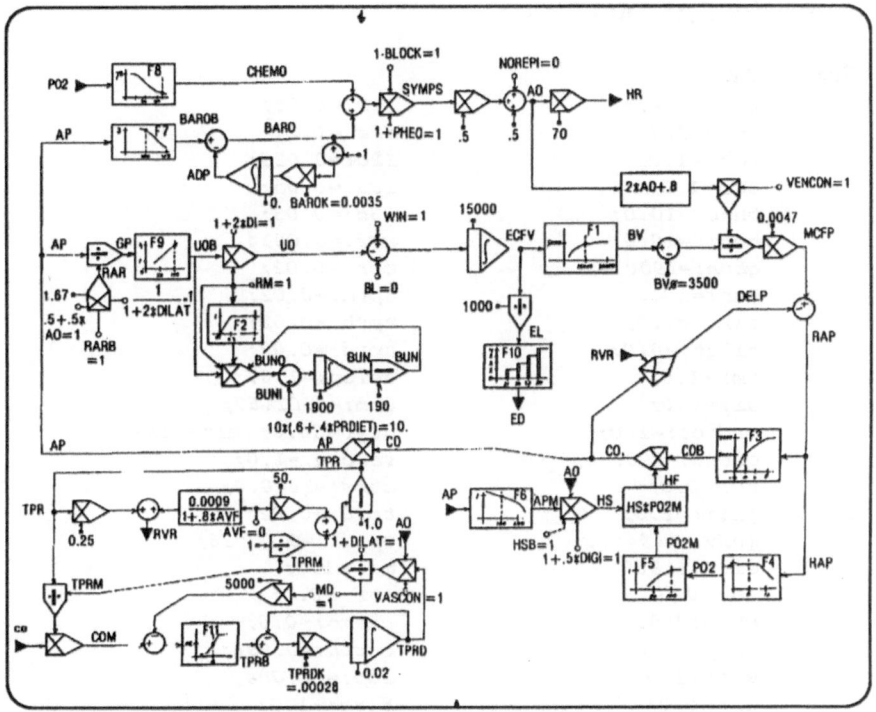

Figure 10.6 Complete analogue scheme of the model, according to Coleman
1980 (Min 1982)

Acknowledgement - The author gratefully acknowledges the contribution of Dr. T.G.
Coleman, University of Mississippi, USA to the CS-program CARDIO.

Appendix CARDIO

Starting values:

rar:=1.67;	i1c:=0.05;
gp:=59.9;	i2e:=0.2;
uob:=1.0;	i2c:=0.025;
uo:=1.0;	i3e:=0.05;
buno:=10.0;	i3c:=0.05;
buni:=10.0;	rvr:=0.0014
qbun:=1900.0;	tpr:=0.02;
bun:=10.0;	tprm:=0.02;
rarb:=1.0;	tprb:=0.02;
dilat:=0.0;	tprd:=0.02;
rm:=1.0;	delco:=1.67;
di:=0.0;	com:=5001.67;
prdiet:=1.0;	dtg:=30.0; {minutes}
qbun0:=1900.0;	vascon:=1.0;
t0:=0.0;	dtas:=1440.0;
gigi:=1.0;	tpr0:=0.02;
ecfv:=14969.0;	tprdk:=0.00028;
el:=14.969;	md:=1.0;
ed:=0.0;	avf:=0.0;
bv:=4989.7;	chemo:=0.0;
mcfp:=7.0;	barob:=0.996;
win:=1.0;	adp:=-0.004;
bl:=0.0;	baro:=1.0;
ecfv0:=15000.0;	symps:=1.0;
vencon:=1.0;	ao:=1.0;
bv0:=3500.0;	hr:=70.0;
hs:=1.0;	pheo:=0.0;
hf:=1.0;	block:=0.0;
po2m:=1.0;	norepi:=0.0;
cob:=5001.67;	barok:=0.00035;
co:=5001.67;	adp0:=0.0;
po2:=100.0;	err:=0;
rap:=0.0;	dum:=0.0;
delp:=7.0;	gdag:=5.0;
ap:=100.0;	t:=0.0; {minutes}
apm:=0.0;	tmi:=0.0;
digi:=0.0;	tma:= 5 * 1440.0;
hsb:=1.0;	t_in_cursor:=0.0; {minutes}
l1e:=0.05;	

*)   CARDIO is available on MS.DOS (implemented in THESIS), on VAX (in the RLCS-system) and
       on Macintosh (in MacTHESIS version 3.0).
**)   For more information about the model: write to F.B.M. Min, University of Twente, P.O. Box 217,
       7500 AE ENSCHEDE (The Netherlands).
***) The function FUNC contains 5 pairs of x,y-coordinates of a "not-linear function block" and
       interpolate a input value (x) to an output value y=f(x). In case of overflow of one of the values an
       error message comes about.

Model of CARDIO in THESIS (MS.DOS version) and MacTHESIS.

T. Coleman (University of Mississipi, USA)

Version F.B.M. Min
Thanks to P. van Schaick Zillesen, B. Reimering & H. van Kan

```
tuur := t/60;
tdag := tuur/24;
if (po2 < 80.0) and (req1=0.0)    THENreq1:=+1.0;
if (po2 < 80.0) and (req1=+1.0)   THENGiveMessage(1);
if (po2 < 80.0) and (req1=+1.0)   THENreq1:=-1.0;
if (po2 > 80.0) and (req1=-1.0)   THENreq1:=0.0;
if (po2 < 40.0) and (req2=0.0)    THENreq2:=+2.0;
if (po2 < 40.0) and (req2=+2.0)   THENGiveMessage(2);
if (po2 < 40.0) and (req2=+2.0)   THENreq2:=-2.0;
if (po2 > 40.0) and (req2=-2.0)   THENreq2:=0.0;
if (po2 < 35.0) and (req3=0.0)    THENreq3:=+3.0;
if (po2 < 35.0) and (req3=+3.0)   THENGiveMessage(3);
if (po2 < 35.0) and (req3=+3.0)   THENreq3:=-3.0;
if (po2 > 35.0) and (req3=-3.0)   THENreq3:=0.0;
if (po2 < 30.0) and (req4=0.0)    THENreq4:=+4.0;
if (po2 < 30.0) and (req4=+4.0)   THENGiveMessage(4);
if (po2 < 30.0) and (req4=+4.0)   THENreq4:=-4.0;
if (po2 > 30.0) and (req4=-4.0)   THENreq4:=0.0;
if (ap < 85.0) and (req5=0.0)     THENreq5:=+5.0;
if (ap < 85.0) and (req5=+5.0)    THENGiveMessage(5);
if (ap < 85.0) and (req5=+5.0)    THENreq5:=-5.0;
if (ap > 85.0) and (req5=-5.0)    THENreq5:=0.0;
if (ed >= 1.0) and (req6=0.0)     THENreq6:=+6.0;
if (ed >= 1.0) and (req6=+6.0)    THENGiveMessage(6);
if (ed >= 1.0) and (req6=+6.0)    THENreq6:=-6.0;
if (ed < 1.0) and (req6=-6.0)     THENreq6:=0.0;
if (HSB < 0.25) and (req7=0.0)    THENreq7:=+7.0;
if (HSB < 0.25) and (req7=+7.0)   THENGiveMessage(7);
if (HSB < 0.25) and (req7=+7.0)   THENreq7:=-7.0;
if (HSB > 0.25) and (req7=-7.0)   THENreq7:=0.0;
dydt := win-uo-bl;
ecfv:=ecfv+dydt*dt;
if ecfv <= 0 THEN err := -1;
if ecfv > 30000 THEN err := 1;
model_goed:=TRUE;
error_number:=0;
bv=FUNC(ecfv,0.0,0.0,0.0,0.0,15000.0,5000.0,
            20000.0,6000.0,30000.0,7500.0,err);
buni:=10*(0.6+0.4*prdiet);
if rm < 0 THEN err := -2;
if rm > 10 THEN err := 2;
bunc=ob*rm*bun*FUNC(rm,0.0,0.0,0.0,0.0,0.0,0.0,
                     0.3,1.0,10.0,1.00,err);
```

```
dydt :=buni-buno;
qbun :=qbun+dydt*dt;
bun :=qbun/190.0;
dydt:=tprb*tprdk-tprd*tprdk;
tprd :=tprd+dydt*dt;
dydt:=barok*(baro-1.0);
adp :=adp+dydt*dt;
n1 := 1;
loop1end :=FALSE;
REPEAT {start loop1}
   BEGIN
     tprm=ao*vascon*tprd/(1.0+dilat)*(md);
     tpr:=1.0/(50.0*avf+1.0/tprm);
     rvr=0.0009/(1.0+0.8*avf)+0.025*tpr;
     mcfp=0.0047*(bv-bv0/(vencon*(0.2*ao+0.8)));
     n2 := 1;
     loop2end :=false;
     REPEAT {start loop2}
        rap :=mcfp-delp;
        if rap < -4 THEN err := -3;
        if rap > 12 THEN err := 3;
        cob=FUNC(rap,-4.0,0.0,-4.0,0.0,0.0,5000.0,
                        6.0,10000.0,12.0,12000.0,err);
        po2=FUNC(rap,0.0,100.0,0.0,100.0,0.0,100.0,
                        5.0,100.0,20.0,0.0,err);
        n3 := 1;
        loop3end :=FALSE;
             REPEAT {start loop3}
          if po2 < 0.0 THEN err := -4;
          if po2 > 200.0 THEN err := 4;
          po2m=FUNC(po2,0.0,0.0,50.0,0.5,65.0,0.85,
                        90.0,1.0,200.0,1.0,err);
          hf := hs*po2m;
          co := hf*cob;
          ap := co*tpr;
          if ap < 0.0 THEN err := -5;
          if ap > 250.0 THEN err := 5;
          apm :=
          FUNC(ap,0.0,1.3,0.0,1.3,0.0,1.3,200.0,0.7,250.0,0.0
          ,err);
          hsc:=apm*hsb*ao*(1.0+0.5*digi);
          if Abs((hsc-hs)/hs)>13e THEN
              hs:=(1.0-13c)*hs+13c*hsc
          else
              loop3end :=true;
          n3 := n3 + 1;
          if (n3 > 100) and (not loop3end) THEN err := 10;
        UNTIL (err <> 0) or (loop3end); {-------loop3}
```

```
    DELPC := co*rvr;
    if ABS(delpc-delp) > 12e THEN
            delp:=(1.0-12c)*delp+12c*delpc
    else loop2end := true;
    n2 := n2 + 1;
    if (n2 > 200) and (not loop2end) THEN err := 10;
UNTIL (err <> 0) or (loop2end);  {--------loop2}
if ap < 50.0 THEN err := -6;
if ap > 150.0 THEN err := 6;
barob :=
FUNC(ap,50.0,3.0,50.0,3.0,50.0,3.0,100.0,1.0,150.0,0.0,err
);
baro := barob-adp;
if po2 < 0.0 THEN err := -7;
if po2 > 200.0 THEN err := 7;
chemo=FUNC(po2,0.0,0.75,30.0,0.75,60.0,0.75,
                        90.0,0.0,200.0,0.0,err);
symps =(baro+chemo)*(1.0-block)*(1.0+pheo);
aoc:=0.5*symps+0.5+norepi;
if ABS((aoc-ao)/ao) > 13e THEN
    ao:=(1.0-11c)*ao+11c*aoc
else loop1end := true;
n1 := n1 + 1;
if (n1 > 100) and (not loop1end) THEN err := 10;
    END;
UNTIL (err <> 0) or (loop1end);  {--------loop1}
rar=1.67*(0.5+0.5*ao)*rarb/(1.0+2.0*dilat);
gp := ap/rar;
if gp < 30.0 THEN err := -8;
if gp > 90.0 THEN err := 8;
uob :=
FUNC(gp,30.0,0.0,45.0,0.2,60.0,1.0,80.0,4.0,90.0,5.0,err);
uo:=uob*rm*(1.0+2.0*di);
el := ecfv/1000.0;
ed := 0;
if (el > 21.0) and (el <= 24.0) THEN ed := 1.0;
if (el > 24.0) and (el <= 27.0) THEN ed := 2.0;
if (el > 27.0) and (el <= 30.0) THEN ed := 3.0;
if el > 30.0 THEN ed := 4;
com:=co*(tpr/tprm);
delco:=com-5000.0*md;
if delco < -5000.0 THEN err := -9;
if delco > 5000.0 THEN  err := 9;
tprb=FUNC(delco,-5000.0,0.0,-5000.0,0.0,
            -500.0,0.01,500.0,0.03,5000.0,0.04,err);
hr := 70.0*ao;
```

References

Coleman, T.G.: *Simulation of biological systems.* Dept. of Physiology and Biophysics, Univ. of Mississipi, Jackson, internal paper, 1977.

Min, F.B.M. and H.A.J. Struyker Boudier : "Computer simulation in problem oriented medical learning at the University of Limburg". *Comp. and Educ.*, Vol. 6, 1982, pp. 153-158.

Min, F.B.M.: *Computersimulatie en wiskundige modellen in het medisch onderwijs: Het RLCS-systeem.* PhD Thesis (in Dutch), Rijksuniversiteit Limburg, Maastricht, 1982.

Min, F.B.M. and H.A.J. Struyker Boudier : "The RLCS system for computer simulation in medical education".
Simulation & Games, vol. 16, no. 4, 1985, pp. 429-440.

Min, F.B.M., M. Renkema, B. Reimerink and P. van Schaick Zillesen: "MacTHESIS, a design system for educational computer simulation programs". *Proceedings of the EURIT 86 Conference*, Eds.: Tj. Plomp, J. Moonen, Pergamon Press, 1987.

Verhagen, P.: "Computersimulation and mathematical models in the medical curriculum by F.B.M. Min".*Educ.Comp.Techn. J.*,Vol.31, no.1,1983, pp. 56-57.

11
A Model of the Baroreflex-Controlled Circulation with Emphasis on the Baromodulation Hypothesis

Dietmar P.F. Möller and Karel H. Wesseling

11.1 Introduction

Continuous 24-hour registrations of blood pressure show diurnal blood pressure variations. Experimentation in normal awake subjects shows pronounced blood pressure changes during mental stress, painful stimuli or mild static exercise.

From physiological literature we know that blood pressure is controlled by a multitude of systems, among these the baroreflex is fast and has high gain. Studies in circulatory models with a baroreflex at similar gains show that blood pressure is stable, regardless of changes by factors of two from normal in heart rate and peripheral resistance or 500 ml changes in blood volume.

Still even these big changes in the circulatory parameters do not explain the normal blood pressure variability. This conflict we called the "baroreflex paradox".

From control theory we know that to change a controlled variable one should use set point changes of the servo-loop. Set point changes occur in the baroreflex adaptation, but it is a passive process. The process is also too slow, having a time constant of around 10 hours. The only other control factor is the loop gain. Modulation of the baroreflex loop gain (baromodulation) by factors of 2 from the control value indeed causes blood pressure to shift by the required amounts. Baromodulation by a factor of 4 has been observed in normal subjects. Baromodulation is therefore postulated as the cause of short-term blood pressure variability.

11.2 Theory of Circulation Models

For understanding the complexities of the circulation we will begin by studying a simple model of the closed circulation without a baroreflex or any other explicit control system. The model circulation consists of a single-chambered, non-pulsatile heart that is only sensitive to preload in a linearized Frank-Starling type manner. The resultant heart function curve states that cardiac output (F_H) is linearly proportional to the venous pressure (P_V) at its input. The heart pumps its output into an aortic compliance (C_a) or "Windkessel" from which the blood flows through a single peripheral resistance (R_a) to return to the venous side with also a "Windkessel" compliance (C_V). The complete model is shown in Fig. 11.1. Note the presence of two flows in the system: F_H is the cardiac output, F is the total peripheral perfusion flow returning to the venous system. When the circulation is stationary, cardiac output and venous return are equal: ($F_H = F$).

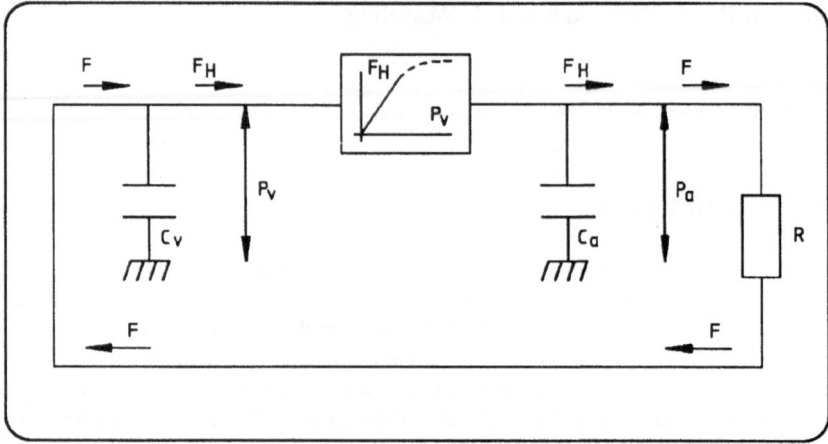

Figure 11.1. Block diagram of a simple model of the circulatory system

The model shown in Fig. 11.1 is based on the following highly simplified concept: A volume of blood, Va, is stored in the arterial compliance, C_a, under a rise in pressure. For a young adult, the aortic or Windkessel compliance is about 1, i.e. every ml of blood pumped into it causes the pressure to rise 1 mmHg. Hence we find:

$$C_a = V_a/P_a \text{ (ml/mmHg)} \tag{11.1}$$

in which $C_a = 1$.

Another boundary of the circulation is that the peripheral outflow is proportional to arterial pressure. Therefore, a pressure rise in the arterial system of 1 mmHg causes an increase in peripheral flow of 1 ml/s or 60 ml/min. The proportionality factor is the so-called peripheral resistance R_a:

$$R_a = P_a/F \text{ (mmHg.s/ml)} \tag{11.2}$$

This is a simplification, since the venous pressure at the other end of the peripheral resistance is ignored. More precisely, therefore:

$$R_a = (P_a - P_v)/F \tag{11.3}$$

The venous compliance is defined in the same way as the arterial compliance, except its value is much greater:

$$C_v = V_v / P_v \tag{11.4}$$

in which C_v is assumed to be 30 ml/mmHg. In other words, much more volume is needed to let the venous pressure rise a certain amount than is needed in the arterial system. However, if a venous pressure of 2 mmHg and an arterial pressure of 100 mmHg are assumed, about 50 to 100 ml each are stored in either compliance:

$$V_v = C_v P_v = 30 \cdot 2 = 60 \text{ ml};$$

$$V_a = C_a P_a = 1 \cdot 100 = 100 \text{ ml}.$$

The total compliance-stored volume in the model is:

$$V_t = V_v + V_a = 60 + 100 = 160 \text{ ml}.$$

The remainder of the blood volume, about 5.000 ml, is stored in the so-called unstretched volumes of the various parts of the circulation.

When the circulation is in steady state, cardiac output equals venous return, which means that $F_H = F$. Assuming that during a short period a 1% difference exists between both flows, F (venous return) is 1 ml/s greater than F_H (cardiac output). This results in an additional amount of blood affecting the venous compliance and the venous pressure.

This volume must have come from somewhere, and the only possible source in this model is the arterial compliance. This also causes a pressure change in the arterial system. As a result of these different pressure changes, the balance between cardiac output and venous return is a critical one. Even slight imbalances cause fluid shifts that seriously affect arterial pressure. To bring this effect under control the heart will pump more blood as venous pressure, and therefore the filling of the ventricle, increases. This is one way of phrasing the Frank-Starling effect. From various litery sources we estimate that a normal heart will increase its output by 50 ml/s from 100 to 150 ml/s under a venous pressure increase of only 1 mmHg. Thus, only a slight increase in venous pressure will cause a substantial increase in cardiac output; maintaining a careful balance at its input side between F and F_H yields:

$$F_H = GP_v \tag{11.5}$$

Hence cardiac output increases in proportion to the venous input pressure. The proportionality factor G has the dimension of a conductance, since it describes a transfer from input pressure to output flow that we call the contractility of the heart.

Finally, if there is a difference between inflow and outflow, the volume change in the compliances must be formulated. To do that clearly we must integrate this difference over time and add the accumulated difference to whatever volume was already stored:

$$V_v = V_v\ (0) + \int_0^t (F - F_H) d\,t \qquad (11.6)$$

As was seen, if venous return, F, is greater than cardiac output, F_H, their difference is positive and over time increases the total volume stored in the venous system, V_v, from its value at time zero, $V_v(0)$. Conversely, for the arterial system we have:

$$V_a = V_a\ (0) - \int_0^t (F - F_H) d\,t \qquad (11.7)$$

What flows into the veins flows out of the arteries, hence the minus sign before the integral.

11.3 Aspects to Be Considered

11.3.1 Limitations of the Model

The main purpose of this is to demonstrate the basic functioning of blood pressure stabilization under different disturbances. For that reason the model is kept as simple as possible, but nevertheless accurate and realistic within its limitations. These limitations are summarized in the following paragraphs:
- There is no fluid exchange taken into account between the circulation and the interstitial compartment.
- The right heart and the lung circulation are neglected, and therefore so is the cardio-pulmonary reflex.
- The venous muscle pump effect is also absent in this simple model, and one of the four effectors of the baroreflex was neglected.

11.3.2 Recent Developments and Outlook for the Future

The hemodynamic behavior in different kinds of circulatory orthological and pathological states can be studied with the models developed. Moreover the model allows more quantitative studies concerning control mechanisms to achieve insight into the mechanisms of the system itself, e.g. stabilization of model arterial blood pressure effectivity within the baroreflex loop. Therefore the model closes the gap between control engineers and biomedical scientists, allowing them to speak in the same terms and apply the same methods to a wide range of control systems research.

One of these applications is the baromodulation hypothesis. Baromodulation deals with the fact that blood pressure is not constant, but shows fluctuations at low frequencies with periods from 10 to 100 seconds and also circadian variability [Möller,1982]. Therefore the extended model of the circulatory system in Fig. 11.3 shows baromodulation, incorporated as variable gain component in the afferent connection between the baroreceptor and vasomotor center. This is not to say that if such a variable gain component existed it could not be located in the baroreceptors or elsewhere in the loop, for example even in the sinus node of the heart. For our present purpose the precise location is irrelevant.

In the model we change the gain from a nominal value of 1 over an order of magnitude in either direction from 0.1 to 10 to explain the effect of the baromodulation hypothesis [Wesseling,1982,1983].

11.3.3 Sensitivity Analysis

Sensitivity is a general aspect that arises when mathematically describing a physiological system. A given mathematical description of the circulary system is based on a set of known parameters, which completely describes the dynamic behaviour of the system. But how will the observed behaviour be affected by changes in the values of the system parameters around their nominal values?

This will be an important question in case studies of pathophysiological situations because the parameters are out of order, and simulation should show the causality between normality and pathology.To solve this problem, a computation must be made for each set of parameters versus the respective variable or parameter, and this we call "sensitivity analysis".

The sensitivity analysis of a circulatory system model consists, in principle, in the comparative study of its time behaviour for different sets of parameter values by a computer on which the mathematical model is implemented.

Section 11.4.3 describes a sensitivity analysis based on a physiologically closed model of the circulatory system.

11.4 Model Description

11.4.1 Mathematical Model

From equations (11.1) to (11.7) we get the following four equations describing the model shown in Fig. 11.1 (one for each component):

Heart: $F_H = G \cdot P_v$ $\hspace{4cm}$ (11.8)

Arteries: $P_a = V_a / C_a = \dfrac{1}{C_a} \left\{ V_a(0) - \displaystyle\int_0^t (F - F_H) dt \right\}$ $\hspace{1cm}$ (11.9)

Periphery: $F = \dfrac{1}{R_a}(P_a - P_v) = \dfrac{1}{R_a} \left\{ P_a - \dfrac{F}{R_a \cdot G} \right\} = P_a / R_a$ $\hspace{0.3cm}$ (11.10)

Veins: $P_v = V_v / C_v = \dfrac{1}{C_v} \left\{ V_v(0) + \displaystyle\int_0^t (F - F_H) dt \right\}$ $\hspace{1cm}$ (11.11)

The total blood volume can be calculated as follows

$$V = V_a + V_v = C_a \cdot P_a + C_v \cdot P_v \hspace{3cm} (11.12)$$

11.4.2 Simulation models

Equations such as these shown above can be solved by a computer that allows elementary operations such as addition, subtraction, multiplication, division and integration. The simulation model can be developed based on a block diagram algebraic representation.

Figure 11.2 shows the block diagram algebraic representation of the mathematical model of the circulation in Fig. 11.1.

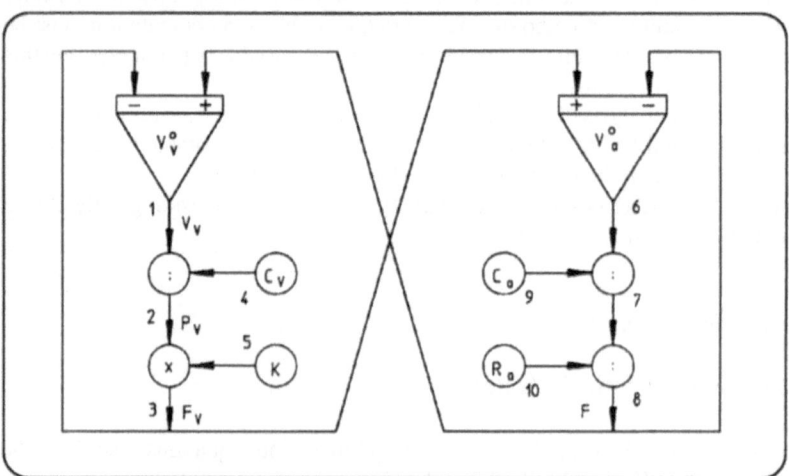

Figure 11.2. Block diagram algebraic representation of Fig. 11.1

The simulation models you will become acquainted with show you the influences on hemodynamic behavior with and without the effect of the baroreceptor reflex. Therefore we must introduce the baroreceptor loops.

Figure 11.3 shows the realization of sensing and stabilization of the arterial systemic pressure (PAS) by the baroreceptor via the afferent pathway to centers in the medulla and via the efferent pathway to the positioning elements, peripheral resistance (Ra), the heart and the venous tonus.

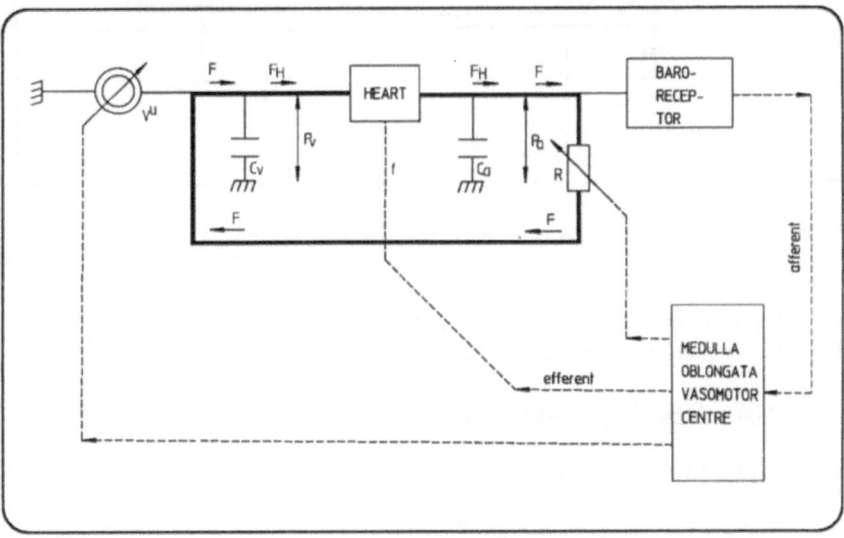

Figure 11.3.   Block diagram of the extended model of the circulatory model, including baroreceptor feedback loops

11.4.3 Sensitivity Analysis

Based on the theoretical description of the parameter sensitivity analysis methods in Chap. 1, section 1.3, we will now apply the results of this section to the parameter sensitivity analysis of the physiologically closed model of the circulatory system (Fig. 11.4). C_{ap}, C_{vp} and R_p describe the pulmonary circulation, while C_{as}, C_{vs} and R_a are the respective components that describe the systemic circulation.

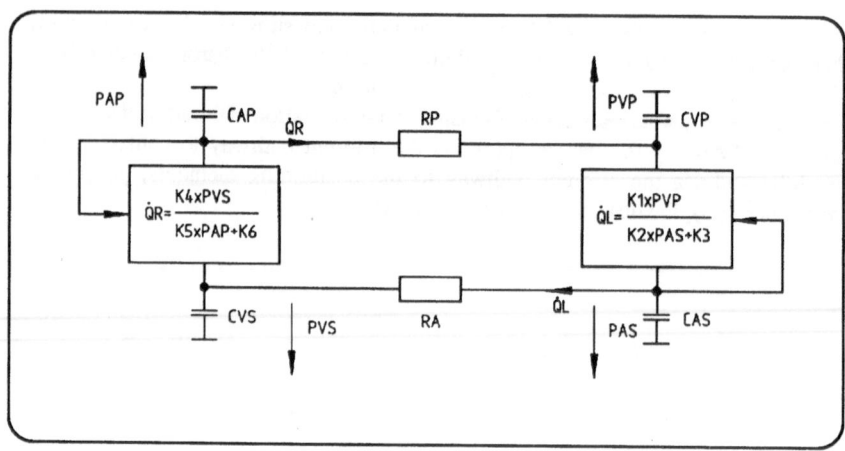

Figure 11.4 Physiologically closed model of the circulatory system

To make use of computer-aided solution, the model shown in Fig. 11.4 must be described in mathematical notation as follows:

$$\underline{\dot{X}}(t) = \underline{A}(\underline{\Theta}) \cdot X(t) + \underline{B}(\underline{\Theta}) \cdot \underline{U}(t) \qquad (11.13a)$$

where the state vector $\underline{X}(t)$ is given by:

$$\underline{X}(t) = (PAS, PVS, PAP, PVP)^T \qquad (11.13b)$$

and the parameter $\underline{\Theta}$ by:

$$\underline{\Theta}: (C_{as}, C_{vs}, C_{ap}, C_{vp}, R_a, R_p, K1, K2, K3, K4, K5, K6)$$

where K1 through K6 are parameters describing the Frank-Starling mechanism. Hence we get the following differential equations:

$$
\begin{aligned}
X1 : \dot{PAS} &= \frac{1}{CAS \cdot RA}(PVS - PAS) + \frac{K1 \cdot PVP}{CAS\,(K2.PAS+K3)} \\
X2 : \dot{PVS} &= \frac{1}{CVS \cdot RA}(PAS - PVS) + \frac{K4 \cdot PVS}{CVS\,(K5.PAP+K6)} \\
X3 : \dot{PAP} &= \frac{1}{CAP \cdot RP}(PVP - PAP) + \frac{K4 \cdot PVS}{CVS\,(K5.PAP+K6)} \\
X4 : \dot{PVP} &= \frac{1}{CVP \cdot RP}(PAP - PVP) + \frac{K1 \cdot PVP}{CAS\,(K2.PAS+K3)}
\end{aligned}
\qquad (11.14)
$$

The partial derivatives for the individual components of the respective pressures are sets of four equations each (16 in total) shown below in the example of PAS:

$$\frac{\partial X1}{\partial PAS} = -\frac{1}{CAS.RA} \cdot \frac{CAS . K2 . K1 . PVP}{[CAS (K2 . PAS + K3)]^2}$$

$$\frac{\partial X2}{\partial PAS} = -\frac{1}{CVS.RA}$$

$$\frac{\partial X3}{\partial PAS} = 0$$

$$\frac{\partial X4}{\partial PAS} = \frac{CVP . K2 . K1 . PVP}{[CVP (K2 . PAS + K3]^2}$$

(11.15)

The partial derivatives for the individual components of the parameter vector are also sets of four equations each, shown here in the example of CAS:

$$\frac{\partial X1}{\partial CAS} = \frac{1}{CAS^2}\left(\frac{PAS}{RA} - \frac{PVS}{RA} - \frac{K1 . PVP}{K2 . PAS + K3} \right)$$

$$\frac{\partial X2}{\partial CAS} = 0$$

$$\frac{\partial X3}{\partial CAS} = 0$$

$$\frac{\partial X4}{\partial CAS} = 0$$

(11.16)

In matrix notation the sensitivity vector becomes:

$$
\begin{bmatrix} SPASi \\ SPVSi \\ SPAPi \\ SPVPi \end{bmatrix} =
\begin{bmatrix}
\frac{\partial X1}{\partial PAS} & \frac{\partial X1}{\partial PVS} & \frac{\partial X1}{\partial PAP} & \frac{\partial X1}{\partial PVP} \\
\frac{\partial X2}{\partial PAS} & \frac{\partial X2}{\partial PVS} & \frac{\partial X2}{\partial PAP} & \frac{\partial X1}{\partial PVP} \\
\frac{\partial X3}{\partial PAS} & \frac{\partial X3}{\partial PVS} & \frac{\partial X3}{\partial PAP} & \frac{\partial X3}{\partial PVP} \\
\frac{\partial X4}{\partial PAS} & \frac{\partial X4}{\partial PVS} & \frac{\partial X4}{\partial PAP} & \frac{\partial X4}{\partial PVP}
\end{bmatrix}
\begin{bmatrix} SPAS \\ SPVS \\ SPAP \\ SPVP \end{bmatrix}
*
\begin{bmatrix} \frac{\partial X1}{\partial \Theta_i} \\ \frac{\partial X2}{\partial \Theta_i} \\ \frac{\partial X3}{\partial \Theta_i} \\ \frac{\partial XY}{\partial \Theta_i} \end{bmatrix}
$$

(11.17)

As an example for CVS we calculate the sensitivity vector:

$$S \frac{PAS}{CVS}$$

The differential equation is:

$$S \frac{PAS}{CVS} = \frac{X1}{CVS} \cdot \left(\frac{1}{CAS.RA} + \frac{K2.K1.PVP}{CAS(K2.PAS + K3)^2} \right) S \frac{PAS}{CVS} + \frac{1}{CASRA} S \frac{PVS}{CVS}$$

$$+ \frac{K1}{CAS(K2\,PAS + K3)} S \frac{PVP}{CVS} \tag{11.8}$$

From this equation it will be seen that calculating the sensitivities requires a specific computer program, as explained in section 1.1.3. As an example, results for the parameters that describe the contractility of the heart include a total of 24 dependencies!

$$S \frac{PAS}{K2} > S \frac{PVP}{K2} > S \frac{PAP}{K5} > S \frac{PAP}{K2} > S \frac{PVS}{K2} > S \frac{PAP}{K5} \quad ...$$

Contractility is the cardiac force for pumping blood through the vascular system.

In section 11.5.2 we suggest you do an indirect sensitivity analysis, based on the simulation of the model shown in Fig. 11.4.

11.4.4 Parameters and Units

B:	baroreceptor function curve
CA, CAS:	arterial compliance in ml/mmHg
CAP:	arteriopulmonary compliance in ml/mmH
CV, CVS:	venous compliance in ml/mmHg
CVP:	venous-pulmonary compliance in ml/mmHg
F, QL:	arterial flow in ml/min
FH, QR:	pulmonary flow in ml/min
F, HF:	heart frequency in 1/min
K1:	the characteristic quantity of left heart compliance in ml/mmHg
K2:	the characteristic quantity of left heart contractility in 1/mmHg
K3:	the time characteristic quantity of the heart action without dimension
K4:	the characteristic quantity of right heart compliance in ml/mmHg
K5:	the characteristic quantity of right heart contractility in 1/mmHg
K6:	the time characteristic quantity of the heart action without dimension
PAP:	arteriopulmonary pressure in mmHg
PA, PAS:	arterial pressure in mmHg
PVP:	venous-pulmonary pressure in mmHg
PV, PVS:	venous pressure in mmHg
RA:	peripheral resistance in mmHg ml/s
RP:	pulmonary resistance mmHg ml/s
VA:	arterial blood volume in ml
VS:	stroke volume in ml
VU:	venous tone in ml
VV:	venous blood volume in ml

11.4.5 Program Operation

The operation of the models used is fairly self-explanatory based on the general structures, shown in Figures 11.1, 11.3 and 11.4. We only make the following remarks in order to assist you in understanding the operation.

- The respective case study must be chosen in the pull down menu structure. There are 9 cases available, BARO 1 to BARO 9. The first simulation experiment, BARO 1, shows the transient period of arterial pressure P, arterial volume VA and venous volume VV, filling up the circulation by infusing a venous volume. When starting the program, the simulation runs with a parameter set within normal range. This can be changed by the user.
- When changing from the first case to another model, you will become acquainted with step response of arterial pressure, heart frequency (heart rate), peripheral resistance and arterial flow with and without baroreflex.
- You may change more than one model parameter at a time. The program re-calculates the new steady states, and the new values appear. Moreover, you may change the graphic mode and/or the total time range.
- If you prefer the SI-Standards for the pressures in kPa instead of mmHg, you may implement the following relationship:

SI, CON, 0.1330
PSI, MUL, SI, P

11.5   Experiments

In general it is not possible to get closed analytical solutions for the sets of non-linear equations that describe non-linear mathematical models of the circulatory system. One way to obtain the solution, and thus to understand the dynamic properties of the mathematical models, is to use a simulator.
Implementation of the models in BIOPSI is based on the parameter sets of the respective model. These can be printed out using the command

PSI.L, SM

11.5.1   Exercises

To give you an idea of the haemodynamic behavior of the circulation under different situations we are undergoing a step-by-step approach to studying these effects.

BARO 2 shows the step responses of arterial pressure without baroreflex; BARO 3 shows them with baroreflex. As a specific exercise in BARO 3, the influence of the baromodulation gain factor MOD can be studied. This is shown in Fig. 11.5 the response with MOD = 0.2.

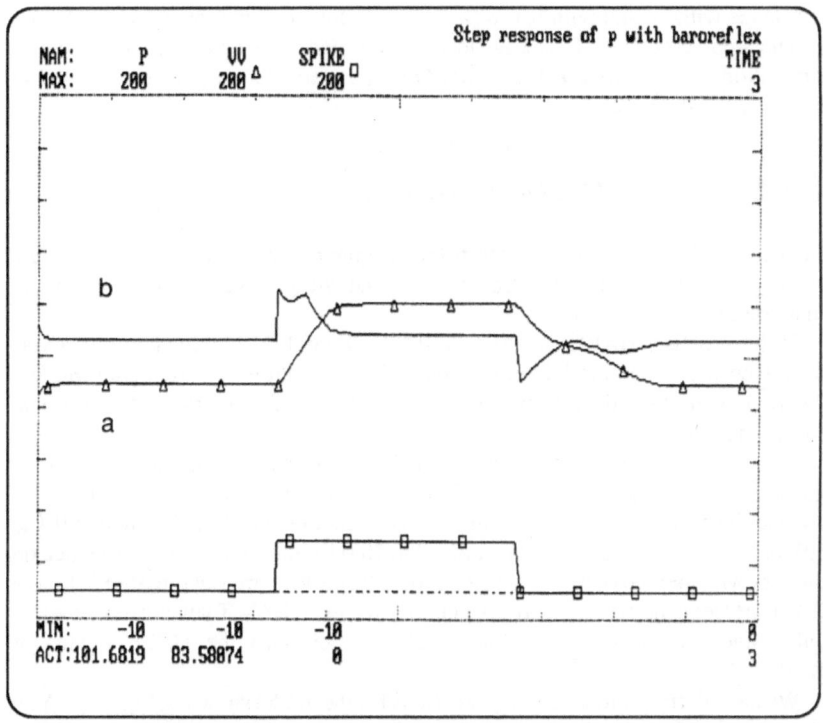

Figure 11.5. Step response of arterial blood pressure with baroreflex and baromodulation gain MOD = 0.2; spike = step function of system excitation

Comparing Fig. 11.5b to Fig. 11.5a, it can be seen that the arterial pressure P is stabilized at a higher value. This demonstrates the fast effect of the baroreflex under a system excitation as a unit step, called spike, applied at time 1 at the pressure and lasting for 1 s. Pressure variations from the nominal pressure value depend on baromodulation.

Setting MOD = 0, the pressure will stabilize at the nominal level under excitation.

The amplitude of the spike, called size, is 20 mmHg. The transient over- and undershoot of the arterial pressure has its explanation in the different time constants of the circulatory system elements.

The increase of the venous volume VV in Fig. 11.5b depends on the flow-ratio under baromodulation with respect to the pressure-volume relationship of the venous compliance: VV increases, if either P or $V = \int_0^t Vdt$ increase. The level of VV will change, when setting MOD = 0.

BARO 4 shows the response of heart rate without baroreflex; BARO 5 shows the same response with baroreflex. The modulation gain factor is MOD = 1.

BARO 6 and BARO 7 show the response of step responses of the peripheral resistance without and with baroreflex. The modulation gain factor is MOD = 2.

The step responses of arterial flow without and with baroreflex is shown in simulation models BARO 8 and BARO 9. The modulation gain factor used in BARO 9 is MOD = 2.

11.5.2 What You Can Do on Your Own

Now that we are acquainted with the different equations that make up the models, we can begin more research-oriented exercises that will give you more insight in the function of the real system.

In the exercises BARO 3, BARO 5, BARO 7 and BARO 9 the baroreflex was overlayed with a baromodulation effect, realized as baromodulation gain MOD. At first you should investigate the baroreflex effects without additional baromodulation.

To get a feel for baromodulation, you should build up a simulation model for general baroreflex gain modulation. The effects of baromodulation are inhibition and facilitation. These should be simulated continuously within the range of 0.1 to 1.0 for inhibition, 1.0 as nominal, and 1.0 to 10.0 for facilitation. The influence on the following hemodynamic variables should be proved: arterial pressure (P), flow (F), heart frequency (FRQ) and peripheral resistance (RA). If you are not sure you will solve this problem accurately, please look into the HELP menu for BAROMOD.

We have derived the sensitivity functions for the model shown in Fig. 11.5. You should undertake a sensitivity analysis to check the results given in section 11.4.3. Use the model BARO 5 and vary the respective parameters -CAS, CVS, K1, etc.- in a physiologically relevant range.

11.6 Conclusions

Although the models presented here are very basic and describe only transient behavior and steady states, they are very instructive. Moreover, they have already proven to be accurate and realistic in explaining the real system.

These models are helpful tools for education and research activities in the entire field of circulatory system analysis.

Appendix

Example of model listing.

BARO3: Step response of p with baroreflex

Block	Type	Input1	Input2	Input3	Par1	Par2	Par3
PULSDO	BNG	TIME			.0000	-1.000	2.000
PULSUP	BNG	TIME			.0000	1.000	1.000
CV	CON				10.00		
HR	CON				1.250		
INF	CON				10.00		
MOD	CON				.5000		
OFF	CON				.5000		
RP	CON				1.000		
SIZE	CON				20.00		
WIDTH	CON				.0000		
F	DIV	P	R				
PV	DIV	DIF	CV				
BAR	FNG	P			.0000	280.0	-12.00
PF	FNG	VA			.0000	2250.	-51.00
VS	FNG	PV			.0000	40.00	-11.00
VA	INT	F	FH		424.1	-1.000	1.000
VV	INT	FH	F		80.00	-1.000	1.000
BEF	MUL	MOD	BOF				
FH	MUL	VS	FRQ				
SPIKE	MUL	NORSPIKE	SIZE				
BOF	SUM	BAR	OFF		1.000	-1.000	
BRG	SUM	BEF			1.500		
DIF	SUM	VV	VUC		1.000	-1.000	
FRQ	SUM	BEF	HR		-1.000	1.000	
NORSPIKE	SUM	PULSUP	PULSDO		1.000	1.000	
P	SUM	PF	SPIKE		1.000	1.000	
R	SUM	BRG	RP		-1.000	1.000	
VUC	SUM	BEF			1500.		

References

Möller, D.P.F., J.J. Settels and K.H. Wesseling: "Incorporation of the baromodulation hypothesis into a model of the circulatory adaptation to exercise". In: *TNO Progress Report No. 8*, Eds.: W.T. van Beekum, B. van Eijnsbergen and A. Kamp, 1982, pp. 128 - 133.

Wesseling, K.H.: "A baroreflex paradox solution". In: *TNO Progress Report* No. 8, Eds.: W.T. van Beekum, B. van Eijnsbergen and A. Kamp, 1982, pp. 152 - 164.

Wesseling, K.H. and D.P.F. Möller: "Baroreflex simulation study: Possibilities of current simulation languages on small computers". *Funkt. Biol. Med.* Vol. 2, 1983, pp. 85-93.

12
Pumping and Wall Mechanics of the Left Ventricle

Ben J. Jansen, Jos E.C.M. Aarts, and Matheus G.J. Arts

12.1 Introduction

In this chapter we examine a model of pumping of the heart and its relation to the wall mechanics of the left ventricle.

In the present model, the left ventricular pump is considered to be a muscular wall enclosing a cavity. As in the realistic situation, the pump function is determined by the parameters:

1. heart rate
2. left atrial pressure (preload)
3. peripheral resistance (afterload) and
4. contractility

Generally, these parameters are controlled by sympathetic and vagal stimuli, which are part of the blood pressure controlling system. In the present set-up the latter controlling loop is opened by considering the above mentioned parameters as input signals of left ventricular function. Blood pressure and cardiac output are output signals. The educational purpose of the model is to find out how these input signals must be changed in order to adapt to changes in hemodynamic load of the left ventricle. Thus, the student is supposed to do what is normally done by the physiological controlling system for blood pressure. Simulations and adaptations deal with the control situation, lowered peripheral resistance due to factors such as running and hemorrhagic shock, and lowered contractility due to myocardial infarct. The simulation shows hemodynamic variables dynamically as a function of time. Pressure-volume loops obtained during separate cardiac cycles may also be studied.

The physiology of the heart as a mechanical pump is extensively treated in reference [Milnor, 1974].

12.2 Set-up of the Model

The left ventricle is a muscular wall. Muscle fiber stress in combination with left ventricular wall geometry determines left ventricular pressure. Muscle fiber shortening is determined by fiber stress and the mechanical properties of the myocardial material. Changes in muscle fiber length are associated with changes in left ventricular volume. The myocardial material has passive and active mechanical properties. The passive component is simulated by a time-invariant, non-linear, elastic anisotropic material. The active component is of major importance during systole. Active muscle fiber stress increases with muscle fiber length, and decreases with velocity of muscle fiber shortening.
Left atrial pressure determines left ventricular filling prior to contraction. Peripheral resistance determines aortic pressure, which loads left ventricular ejection.
The model is developed by one of the authors (T. Arts) and extensively described in reference [Arts, 1979].

12.3 Description of the Model

12.3.1 Simplifying Assumptions of the Model

In the simulation the only factor describing geometry is the ratio of cavity volume to wall volume. If this ratio is low, the wall is relatively thick as compared to cavity diameter. As a result, at a certain fixed left ventricular pressure level wall stress is relatively low. Many models presented in literature describe cardiac geometry in more detail.

Myocardial fiber stress and sarcomere length (repetitive length between muscular striations) are each represented by a single value, ignoring possible inhomogeneities of mechanical loading of the tissue in the wall. Modern insights tend to recognize a homogeneous distribution of mechanical load over the various structures of the left ventricular wall. In case of local ischemia due to coronary artery obstruction, this condition of homogeneity is no longer satisfied. In that case, the parameter values associated with mechanical load of the myocardial tissue should be considered as an average, and the description might be less accurate.

Developed fiber stress normally increases with sarcomere length and contractility affects this relation. A change in contractility is simulated by a change in the dependence of activation factor as a function of sarcomere length. Macroscopically, this dependency is associated with a change in the relation between left ventricular pressure and volume during an isovolumetric contraction.

Increase of contractility is associated with an increase of intracellular calcium concentration during systole. The relation between calcium concentration and developed stress shows a saturation effect that is reached earlier at greater

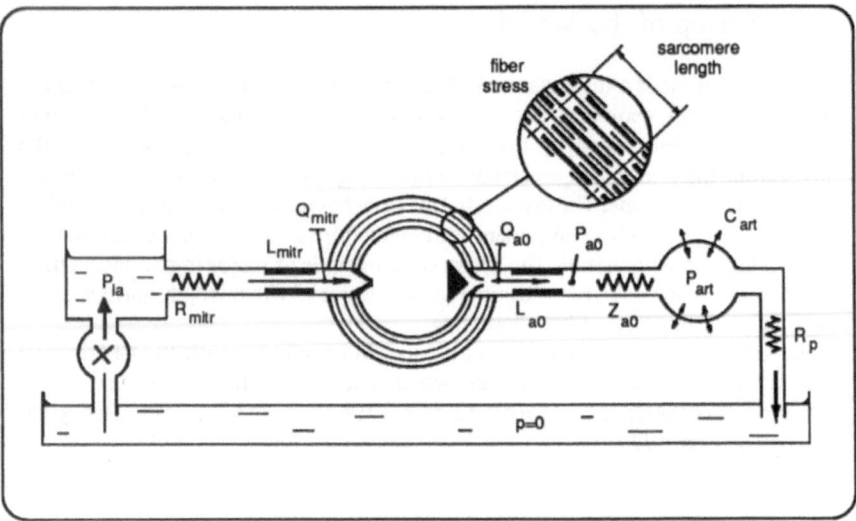

Figure 12.1. Mechanical model of the left ventricle and electrical analogon

sarcomere length where stress is already high. In the simulation this saturation effect is also incorporated. When contractility increases, fiber stress also increases, resulting in a lowering of the sarcomere length at which fiber stress saturation occurs.

As shown in Figure 12.1, the mitral inflow system is simulated by a diode in series with an inertia and a resistance. The aortic outflow system is simulated by a diode in series with an inertia, a characteristic impedance of the aorta, and an arterial compliance proximal to a peripheral resistance.

12.3.2   Structure of the Model

The model (Figure 12.1) is set up as a system of four differential equations. The basic parameters associated with this system are left ventricular volume VLV, mitral flow QMITR, aortic flow QAO, and arterial pressure PART. Knowing the values of these parameters and time, all relevant parameters in the model may be calculated, including the derivatives of the basic parameters.

The derivative of left ventricular volume equals the difference between mitral inflow and aortic outflow. Sarcomere length SL is calculated from left ventricular volume by

$$SL/SL0 = (1 + 3 \text{ VLV/VWALL})^{1/3} \qquad (12.1)$$

where SL0 is a reference sarcomere length and VWALL is wall volume. Using this equation, the time derivative of sarcomere length can be calculated from the time derivative of left ventricular volume. When applying an empirical model of the mechanical properties of myocardial material with many parameters, fiber stress SF is calculated as a function of sarcomere length, its time derivative and time. At this point, heart rate and the contractility parameter are introduced. Successively, left ventricular pressure PLV is calculated by

$$PLV/SF = (1 + 3 \text{ VLV/VWALL})^{-1} \qquad (12.2)$$

Aortic pressure PAO is calculated from arterial pressure PART, aortic flow QAO and aortic characteristic impedance ZAO by

$$PAO = PART - QAO * ZAO \qquad (12.3)$$

If the valve is open, for the time derivatives of aortic flow it holds

$$d/dt(QAO) = (PLV - PAO)/LAO \qquad (12.4)$$

with LAO being aortic inertia. Similarly for the time derivative of mitral flow it holds

$$d/dt(QMITR) = (PLA - PLV + QMITR*RMITR)/LMITR \qquad (12.5)$$

with PLA being left atrial pressure. If a valve is closed, flow and the derivative of flow are set to zero. If the pressure drop over a closed valve becomes positive, the valve opens.
For the time derivative of arterial pressure it holds

$$d/dt(PART) = (QAO - PART * RP)/CAO \qquad (12.6)$$

with RP and CAO being peripheral resistance and arterial capacitance respectively. The actual value of PART has no physiological meaning, but is needed for a model description of the aortic input impedance. Thus all time derivatives of the basic parameters are known and an integration step is performed.
Most parameters are set to a fixed value, several of which are listed in Table 12.1. Table 12.2 shows the default values and the possible range of variation of heart rate, left atrial pressure, peripheral resistance and contractility. The latter two parameters are normalized to the control values.

Table 12.1. Values of fixed parameters in the simulation model

PARAMETER	SYMBOL	VALUE (default)	UNIT
aortic inertia	LAO	$8 \ 10^{+4}$	$Pa \ m^{-3}s^2$
aortic characteristic impedance	ZAO	$5 \ 10^{+6}$	$Pa \ m^{-3}s$
peripheral resistance	RP	$11 \ 10^{+7}$	$Pa \ m^{-3}s$
arterial capacitance	CART	$3 \ 10^{-8}$	$Pa^{-1}m^3$
mitral inertia	LMITR	$2 \ 10^{+4}$	$Pa \ m^{-3}s^2$
mitral resistance	RMITR	$1 \ 10^{+5}$	$Pa \ m^{-3}s$

Table12.2. Default values and ranges of simulation variables

VARIABLES	SYMBOL	DEFAULT	RANGE	UNIT
duration heart cycle	-	800	>300	ms
left atrial pressure	PLA	0.4	-0.2 to 3	kPa*
peripheral resistance	RP	100%	>25%	-**
contractility	-	100%	<300%	-**

*   1 kPa = 7.5 mmHg
** normalized to default value

12.3.3 Working with the Simulation

The simulation starts with the steady-state default situation. At the beginning of each cardiac cycle (end-diastole), the program may be interrupted and the following four variables may be changed interactively:

- duration of cardiac cycle
- left atrial pressure
- peripheral resistance
- contractility

The default values and possible range are shown in Table12.2. Under some, generally extreme circumstances, the simulation may become unstable. In that case the simulation should be reset.
The graphics screen shows as a function of time:

- left ventricular pressure
- aortic pressure
- left ventricular volume
- aortic volume flow

Furthermore, mean aortic pressure and aortic volume flow are shown, respectively. In another mode of graphic representation, one of these variables may be plotted as a function of another in a phase diagram. For instance, left ventricular pressure may be plotted as a function of left ventricular volume, thus forming pressure-volume loops.

12.4 Experiments

12.4.1 The Default Demonstration

When starting the simulation, initial conditions are set to default, and the steady state control situation is calculated and shown on the graphical display. Mean aortic pressure and volume flow appear to be 12 kPa (= 90 mmHg) and 100 ml/s, respectively.

12.4.2 Exercises

A few situations are shown below, in which blood pressure control demands adaptations of cardiac function. Generally, in the physiological situation, feedback is performed with minimal impact on the physiological system. Cardiac output is a demand of the system, and can be modulated only as the very last resort. Aortic pressure should be normal in order to guarantee undisturbed perfusion of the various organs. Left atrial pressure should be normal in order to avoid oedema in the lungs and the rest of the body. Heart rate and contractility should be minimal in order to minimize the metabolic need of the heart. When dealing with the special cases below, these physiological requirements should be kept in mind.

a. *Running*
    When a person is running, systemic peripheral pressure drops dramatically. In the simulation, peripheral resistance is set to 50% of default. Try to find out how to adapt heart rate and contractility in order to keep mean aortic pressure at the original level. If this seems impossible, then left atrial pressure may be changed. Repeat the same procedure with a peripheral resistance setting to 25 %.
b. *Hemorrhagic shock*
    In case of hemorrhagic shock by bleeding, circulating blood volume decreases. As a result, left atrial pressure drops. In the simulation, set left atrial pressure to 0 kPa. Now try to find out how to adapt heart rate and peripheral resistance in order to keep mean aortic pressure at the original level. (Changing the peripheral resistance of the cardiovascular system simulates peripheral vasoconstriction.) Repeat the same procedure with a left atrial pressure setting to - 0.2 kPa.

c.   *Myocardial ischemia*

In case of myocardial ischemia, effective contractility decreases. In the simulation, contractility is decreased to 70%. Try to find out how to adapt heart rate in order to keep mean aortic pressure at the original level. If this does not work, then left atrial pressure and peripheral resistance may be changed. Repeat the same procedure with a contractility setting of 50%.

12.5 What you can do on your own

Besides the exercises shown above, the model may be used to simulate many other hemodynamic situations, such as the effect of a coronary occlusion. Also, phase plots may be made, instead of studying the time course of the various signals. For instance, the above mentioned cases may be repeated while left ventricular pressure is plotted as a function of left ventricular volume.

References

Milnor, W.R.: "The Heart as a Pump". In: Mountcastle, V.B. (Ed.) Medical Physiology, Vol. 2, 14th ed., St. Louis: The C.V. Mosby Company, 1980, pp. 986-1006.

Arts, T., Reneman, R.S. and Veenstra, P.C.: "A Model of the Mechanics of the Left Ventricle", Ann. Biomed. Eng., Vol. 7, 1979, pp. 299-318.

13
Heart Rate Regulation During Physical Load

Jiří Potůček

13.1 Aim of Instruction

A model was developed of heart rate during physical load comprising rapid feedforward and slow feedback components. Using this model, the organism's basic control abilities for heart rate during defined physical load (ergometric load, etc.) can be learned. The nonlinear version of the model is also presented. The parameters obtained from its linear part are independent of the input load.

One of the most important problems in this project is parameter estimation of the model. This procedure is described in detail and the process of optimization of the parameter vector is illustrated in a practical example.

Different experiments with this model are presented:

a. The basic behavior of normal subjects (see 13.5.1).
b. Top athletes (cyclists) exhibit a decrease in K_{ref} and T_i parameters and an increase in W_r (see 13.5.2).
c. The results from a group consisting of six patients with ischeamie heart disease, with one to three yearly histomyocardial infarctions, are presented in 13.5.3.
d. Administration of drugs (methypranol, atropine) causes significant changes in some parameters (see 13.5.4).

13.2 Theory

Physical load can greatly increase the working muscle's demands for oxygen transfer and removal of metabolic waste products. Blood circulation must adapt itself to meet these requirements. Circulation adaptation is an integral part of a wider range of adaptation mechanisms, including the metabolic sphere and respiration.

The following two main mechanisms are involved in circulatory adaptation:

a)   increase in cardiac output (minute volume)
b)   redistribution of cardiac output (i.e. increased blood flow in muscles and decreased flow in other tissues, for example through the splanchnic bed). Increase of oxygen extraction is an auxiliary mechanism (increase of the a-v oxygen difference).

Cardiac output is the product of stroke volume and heart rate. Trained individuals show a relatively higher stroke volume and lower heart rate than untrained ones, on every level of oxygen consumption. The heart rate is therefore only an incomplete marker of cardiac output. However, its measurement is very simple and easy causing no distress. The main advantage of investigation of heart rate during exercise is its linear rise in steady state within a certain range and its close positive correlation between the intensity of exercise and oxygen consumption. Within this range it is a system with a linear static characteristic. Investigation of the heart rate in the transient state (at the start of exercise or during modification of its intensity) shows two components. One is fast and has an almost constant value, while the other is slower and more variable.

We consider that the underlying mechanisms of the fast component are the inhibition of the vagus at the start of exercise, and the restored action of the parasympathetic inhibition after the exercise. The fast component disappears after surgical or pharmaceutical vagotomy. The slow component has a complex neurohumoral character, and its basis is the increased or decreased orthosympathetic activity. In addition to the cardiac orthosympathicus, factors involved in the slow component include circulating catecholamines, irradiation from the thermoregulatory and respiratory centres to the cardioexitation centre, and biochemical influences (pO_2, pCO_2, pH), among others.

The simulation model of heart rate regulation during exercise was developed on the basis of the described physiological phenomena. This exercise allows estimation of the model's parameters.

13.3   Aspects to Be Considered

13.3.1   Limitations of the Model

From results obtained from the same subjects exposed to increasing test loads, it follows that for the load 0.5 - 2.5 W/kg B.W. the results are almost identical. This is not true for the T_i parameter describing the feedback sympathetic component. This is hard to determine at low loads (0.5 W/kg), where the influence of the slow component is minimal. Figures 13.1 and 13.2 show some selected measurements. Fig. 13.1 depicts the situation connected with determinability of T_i at low loads.

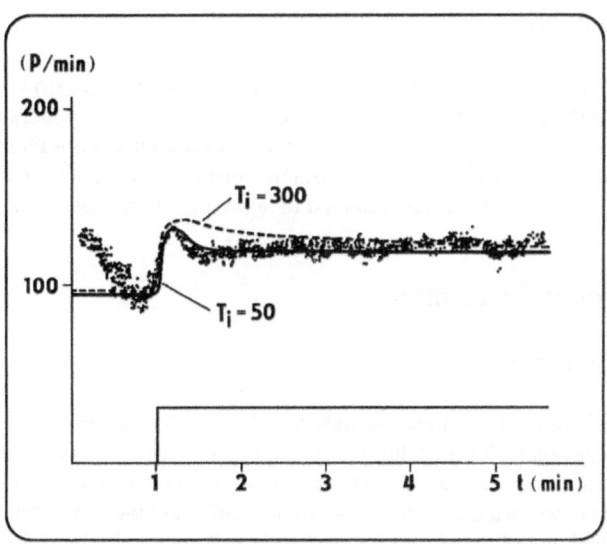

Figure 13.1. Graphical result connected with determinability of T_i at low loads

Fig. 13.2 represents the situation at a load of 3.5 W/kg, which causes a total exhaustion of the proband after a couple of minutes. This figure demonstrates the onset of the "heart-rate debt", i.e. the difference between the actual and the required value of heart rate.

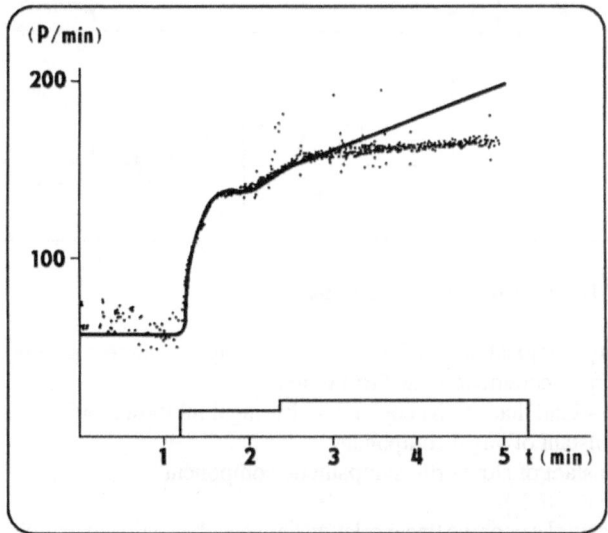

Figure 13.2. Graphical representation of the situation at a load of 3.5 W/kg B.W.

13.3.2 Outlook

The model proved useful in several experimental situations. However, future experiments should also comprise female control groups and homogenous samples of different age categories. The analysis of bio-oscillations is not included in the model structure. Evaluation of the constants must be done under strictly defined conditions. The best results are obtained in tests that are repeated several times.

13.4   Model Description

13.4.1 Structure

The block diagram of the model is illustrated below. The model postulates a rapid component of vagal inhibition during exercise (state variable x_1) and a slower acting feedback complex, orthosympathetic component (state variable x_2). The rapid component works on the switched-on-switched-off principle, i. e. independently of the degree and type of load ($u_1=1$ for $u_2>0$ and $u1=0$ for $u_2=0$). The slow component is a feedback loop with an integral controller with a time constant T_i. A set-point of the heart rate is derived from the actual value of the load and equals $Kref \cdot u_2$. The input-output relation is determined by four parameters that are uniquely determinable from the response of heart rate to load u_2 (see literature).

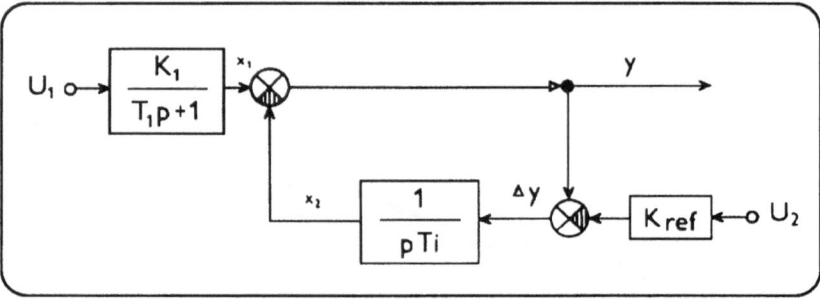

Figure 13.3. The block diagram of the model

From the physiological point of view the following parameters are important:
K_{ref} - heart rate increment related to 1W load
$W_r=K_1/K_{ref}$ - load that can be coped with by vagal inhibition only
T_1 - time constant of vagal component
T_i - time constant of slow orthosympathetic component

The hypothesis of a positive linear relation between the degree of exercise and heart rate holds only to a limited extent. If the test load exceeds a certain limit (u_{max} - load at which the maximum heart rate is reached), some difference between actual and requested heart rate arises. Let this difference be denoted "heart-rate debt", which correlates well with the "oxygen debt" in peripheral tissues.

Certain anomalies in the relation between heart rate and the degree of load can be observed also at very low load (u_{min}), where the feedback component does not apply or (even) operates "antagonistically", i.e. decreases the initial heart rate to a steady-state level. It is therefore necessary to use the input load between (u_{min}, u_{max}) for parameter estimation. These limitations to the model structure are represented schematically in the diagram below.

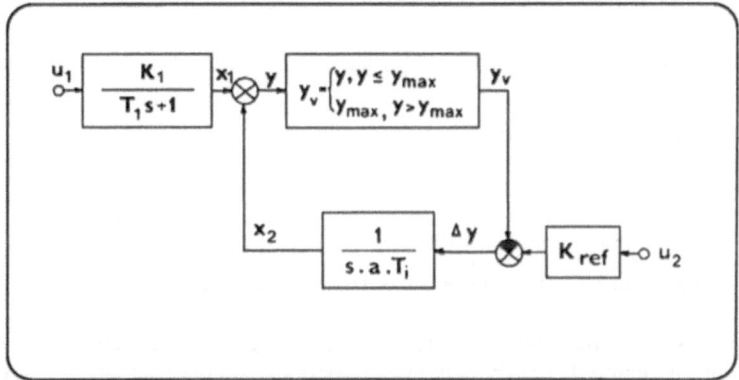

Figure 13.4. The block diagram of the model including limitations

13.4.2 Signals

A two step load test (duration about 6 min.) was performed on a bicycle ergometer.

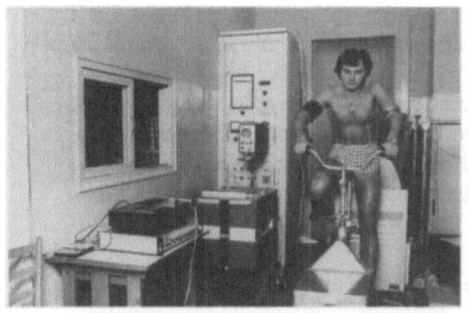 

Figure 13.5. Laboratory conditions

Simultaneously, at least one lead of the ECG curve (the Gibson lead because of movement artifacts) and information about the load were on-line processed. The graphical output signals from one volunteer (a veteran long-distance runner) can be seen in the diagram below:

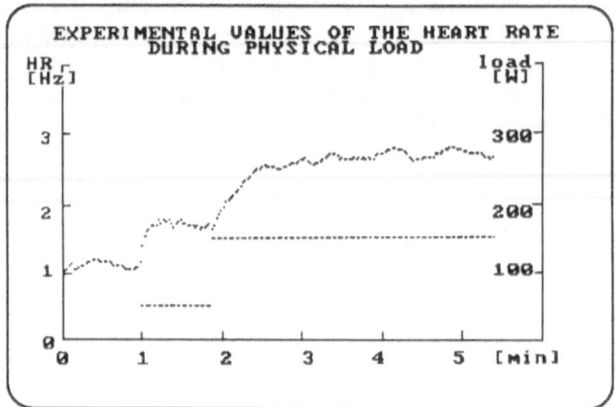

Figure 13.6.   Output heart rate and input load for a long distance runner

It follows from the above observations that parameters of the feedforward component (mainly T_1) can be determined from values of the heart rate measured in the first minute only, while changes in the feedback component parameters have practically no influence on the heart rate during the first minute. This implies that optimization of parameters can be broken down into two steps:

Step 1: Optimization of parameters K_1 and T_1 from values of the heart rate in the first minute for suboptimal values of parameters K_{ref} and T_i.

Step 2: Optimization of parameters K_{ref} and T_i determined in Step 1.

A cycle between Steps 1 and 2 is repeated for as long as the parameters change. Let it be noted that breaking down optimization into two steps not only saves computing time but is necessary in this case.
The parameters of the feedforward component (mainly T_1) are reduced to only a few residuals at the beginning in the simultaneous optimization of all four parameters. This is why their influence on the value of the criterion J is negligible in comparison with parameters K_{ref} and T_i of the feedback component.

This can be even more accurate if the load is chosen as follows: It can be easily derived that Laplace transforms of the trajectory sensitivity functions $\eta_2 = dy/dK_{ref}$ and $\eta_4 = dy/dT_i$ are:

$$\eta_2 = u_2/T_i.s = 1)$$

$$\eta_4 = K_1 \cdot u_1 \cdot s/(T_1 \cdot s \cdot 1).(T_i \cdot s+1) - K_{ref} \cdot u_2 \cdot s/(T_i \cdot s+1)^2$$

The load u_2 should be chosen so that η_2 and η_4 are as small as possible within the first minute of load application (for more detail see literature). For practical purposes, an approximation of the above type of load is usually used:

$$u_2 = 0.5 - 0.75 \text{ W/kg of body weight } (0 < t < 60 \text{ sec})$$
$$u_2 = 1.5 - 2.0 \text{ W/kg of body weight } (t > 60 \text{ sec})$$

13.4.3   The Parameter Estimation Problem

The task is to find such values of parameters K_1, K_{ref}, T_1 and T_i for which the sum of squares of differences between theoretical (y) and experimentally measured (y_{exp}) values of heart rate at measured time instants is minimal. This problem leads to a minimization of the criterion (n-number of experimentally determined values):

$$J(K_1, K_{ref}, T_1, T_i) = \sum_{i}^{n} [y(t_i) - y_{exp}(t_i)]^2, \quad i = 1, \dots, n$$

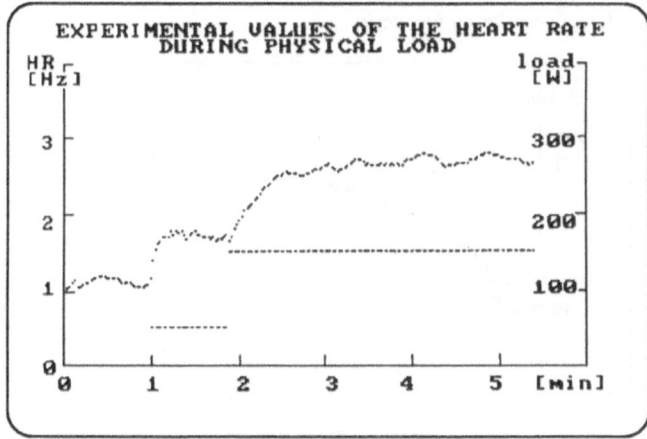

Figure 13.7.   Starting point for optimization

A preliminary analysis of several heart rate responses has shown that time constant T_i of the feedback component is about five timeshigher than time constant T_1 of the feedforward component. As the input u1 equals 1 during the whole period for which the load u_2 is being applied, a response of the feedforward component reaches its steady state almost in the first minute. From parameter sensitivity functions (see diagram below) we can deduce that:

a)   time constant T_1 of the feedforward component and parameter K_1 affect the heart rate in the first minute only

b)   parameters K_{ref} and T_i have practically no influence on the heart rate in the first minute (see sensitivity functions)

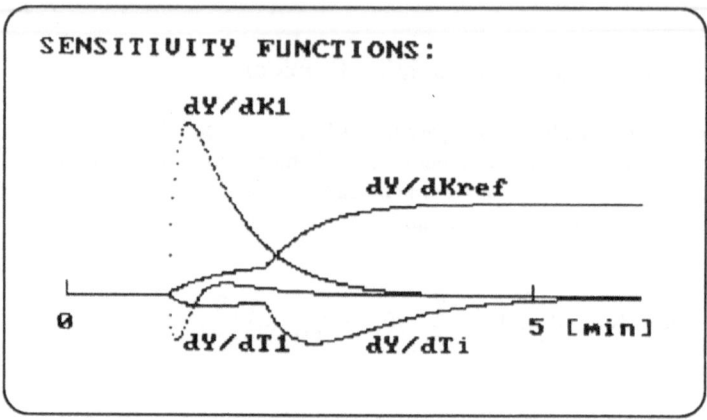

Figure 13.8.   Output sensitivity functions

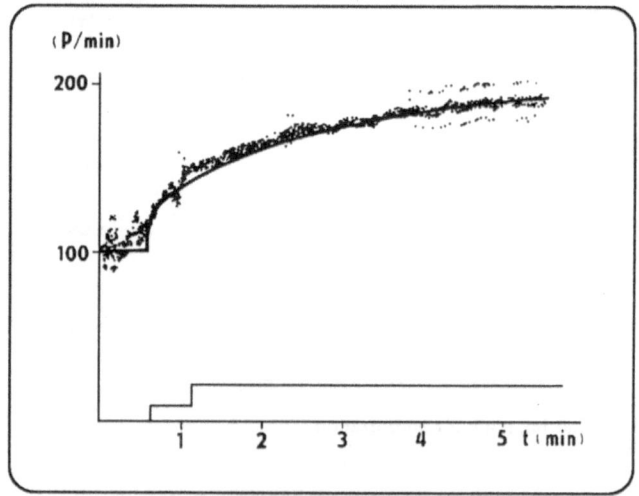

Figure 13.9. Graphical result from optimization

13.4.4 Possibilities and Limitations

The trial carried out on 15 men aged 18-20 was designed to identify the basic reference values of the selected model parameter describing heart rate regulation during exercise. Conceivably, several factors implicated in its regulation during exercise affect these constants, factors that were described, by Astrand and others many years ago. The values of those parameters can therefore be expected to depend not only on the degree of training and the actual fitness level, but also on age, sex, environment and so on. Therefore the significance of any model parameter increases within each individual or group if it is evaluated twice or several times at a defined period and under exactly defined conditions.

The basic parameter of fitness is the K_{ref} parameter, i.e. the heart rate increment related to 1W load. The lower the K_{ref}, the higher the fitness of an individual or group. This observation also confirms a highly significant hyperbolic correlation of Kref to the Harvard Index the result of the standard proven step-test that characterizes fitness by the decrease of the absolute values of heart rate during recovery after a maximal or submaximal load. The second parameter is W_r, the load that can be coped with only by vagus inhibition. The higher the fitness, the higher is W_r. This conclusion also confirms the significant hyperbolic correlation between K_{ref} and W_r.

The two parameters are determined by the resting heart rate, which shows the degree of sympaticotonia and/or vagatonia under basic conditions before exercise. The third parameter of interest is the time constant of the feedback component, T_i. The higher the fitness, the slower is T_i. T_i and K_{ref} behaved in a similar fashion (see positive correlation). Because W_r also includes K_1 (the latter not evaluated separately), no attempt was made to determine time constant T_1 owing to its poor determinability. The hypothesis of a positive linear relation between the degree of load and heart rate holds true only to a limited extent. In most people, the upper limit of heart rate is closely related to the so-called oxygen ceiling or anaerobic threshold, i.e., to the maximum available aerobic metabolic turnover. In terms of energy, a load increase beyond some level is covered anaerobically. Oxygen consumption and transport via blood are closely linked with heart rate. Oxygen debt, with a magnitude limited and reflected by "heart-rate debt" (difference between the required and actual heart rate), arises when oxygen consumption exceeds the upper limit.

The better the physical fitness, the higher the load covered aerobically and the higher the maximal oxygen consumption and maximal heart rate. Such persons also show a higher oxygen consumption and heart-rate debt compared with untrained individuals. To evaluate the performance capacity of the proband, it is therefore necessary to measure not only the maximum heart rate but also the magnitude of the heart-rate debt, which is connected with individual anaerobic capacity.

Certain anomalies in the heart rate/load relation may also be found at very low test load exertion. In such a situation, the main mechanism that increases the heart rate is vagal inhibition (rapid component).

The orthosympathetic component can also work antagonistically, i.e. it can decrease heart rate originally regulated by the rapid component to the required lower level. It follows that, for exact determination of the model constants, that is an exercise load should be used neither too low nor too high.

In untrained persons, the optimal exercise level ranges between 0.75 and 1.5 W/kg; in trained athletes the exercise level should lie under the anaerobic threshold.

Very poor physical fitness is to be expected in postinfarctic patients with anginal pain on effort. Moreover, all patients were from a higher age category, and some of them were obese. This need not be true for younger patients. A very low load test can be applied in some, but the test must sometimes be discontinued owing to anginal pain in others. All this can make model parameters even more difficult to determine.

Yet all the patients have one feature in common: a very low participation of the rapid component (low W_r). This holds also for patients showing no significant difference in the Kref parameter against controls.

In postinfarction patients, a low W_r is a sign of sympaticotonia. Our experimental group received no drugs for three days before the test exercise. As a rule, almost all these patients take beta-blockers. Therefore vegetative disbalance need not be very obvious. These patients are likely to perform poorly on load tests. What is missing is the rapid component, which cannot be compensated by the blocked slow component.

13.5   Experiments

13.5.1   Default Demo

The basic experiments were carried out on 15 healthy volunteers aged 18-20 with normal body weight (90-110% of the ideal b.w. according a standard load on a bicycle ergometer), i.e. 60 s with 0.5 W/kg b.w. and then 4-6 min. with 1.5 W/kg b.w. The graphical output of the simulation model with the mean values of the parameters is presented in the following figure.

13.5.2 Experiment on Top Athletes

This experiment was performed on junior top cyclists in the age range 18-20, Broca's index 81.3 of the ideal body weight. The loadtest (60s with 0.5 W/kg b.w. and 4-6 min. with 2-4 W/kg of body weight) was performed in the morning at the beginning and end of a three-week period of intensive training.

Compared with controls in cyclists, in whom ergometric load is optimal for exertion, all the three main characteristics of high physical fitness were expressed in the initial period of training: a low increase of heart rate per minute (low K_{ref}), a high value of load covered by vagal inhibition only (high W_r), and a very short time constant of the other sympathetic component (low T_i). An intensive three-week training with loads

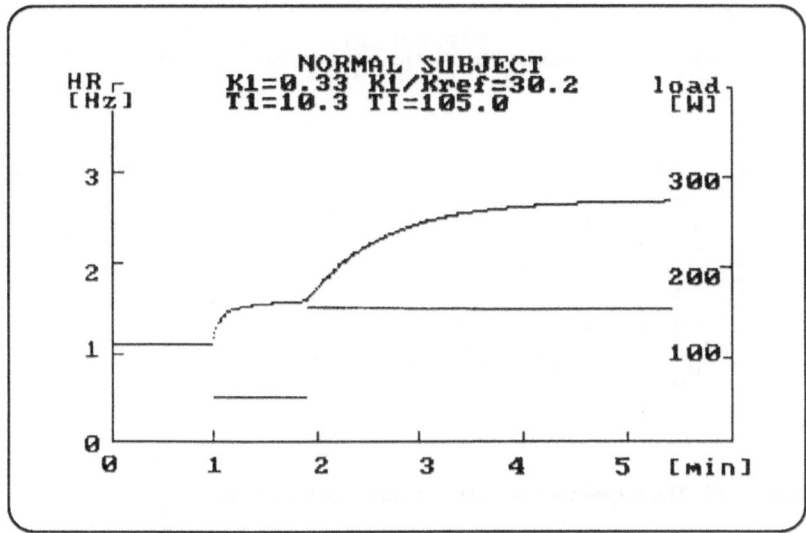

Figure 13.10.   Graphical output (normal subject)

ranging from 120-150 km elicited a more significant expression of vagal inhibition. At the end of the training period, the rapid component averaged 150 W (compare with fig. 13.11 !). Very similar results occurred in another experiment on six marathon runners in whom two weeks of intensive training produced only one significant change, an increasing of W_r value from $61.71 + 22.4$ to $79.21 + 14.1$ W. This observation suggests that a high W_r value is characteristic of a top cyclic endurance type of load. The W_r value was found to be very useful for predicting the proband's instantaneous condition.

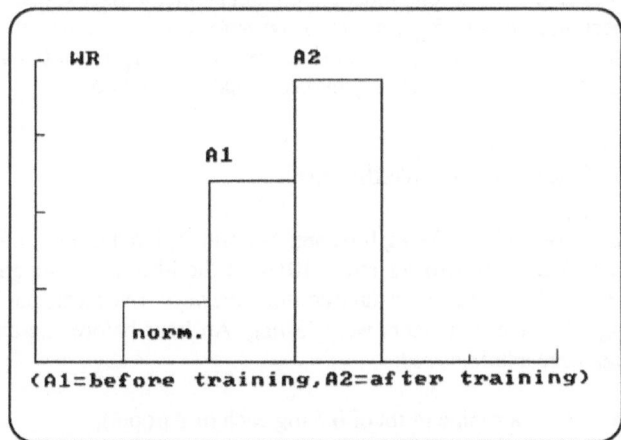

Figure 13.11. The changes of W_r on normal subject and top athlete before and after training

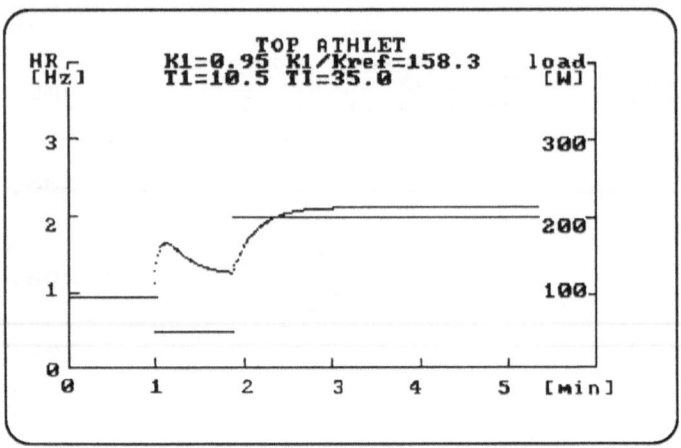

Figure 13.12. The graphical output from the model for top athlete

13.5.3   Experiment on Patients with Ischaemic Heart Disease

The group consisted of six patients with ischaemic heart disease, with one to three yearly histomyocardial infarctions, age range 43-55 years, Broca's index 118.5 of ideal body weight. The load was 60sec with 0.5 W/kg b.w. and 240-300s with 1W/kg b.w.

In two cases the test was interrupted (at the fourth and fifth minute) because of severe angina pain. Moderate angina pectoris was present in five of six patients. The data recorded in six IHD patients showed that all of them had a significantly lower contribution of the fast component, manifesting itself through a very low W_r value. This holds also if the second most important parameter that expresses heart rate increase per unit of loading K_{ref}, is not too low or not different from controls. On the average, the K_1 value was significantly lower whereas T_i did not change as compared to controls. T_1 was practicaly undeterminable (but $T_1 < 6$).

13.5.4 Experiments Using Medication

Three load tests (60s with 0.5 W/kg b.w. and 4-5 min. 1.5 W/kg b.w.) on seven volunteers aged 18-21 with Broca's Index 101% of the ideal body weight were performed on a bicycle ergometer on three consecutive days. Tests were performed in the morning when the volunteers were fasting. An hour before exercise, the following drugs were administered:

Experiment A: 2 mg atropine (4 tbl of 0.5 mg each of Atropin);
Experiment M: 40 mg methypranol (4 tbl of 10 mg each of Trimepranol Spofa);
Experiment C: (controls) 4 tbl of placebo.

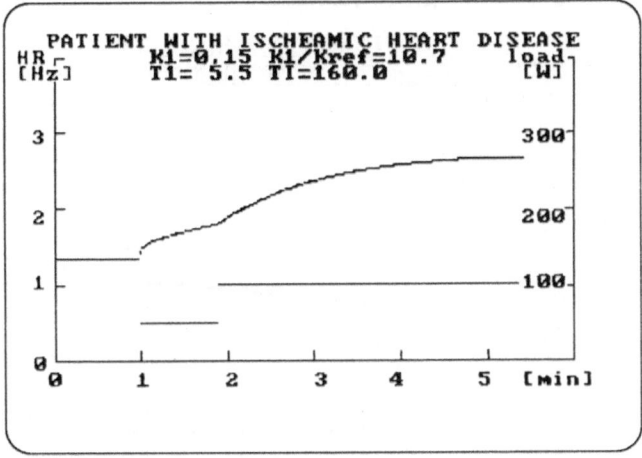

Fig 13.13. Patient with ischeamic heart disease

The sequence of experiments A, M and C was randomly interchanged.

The pharmacological experiments were carried out with parasympathetic blockers, namely atropine and methypranol, a Czechoslovakian beta-blocker. It is supposed that these drugs significantly influence the balance between the para- and orthosympathetic components in heart rate regulation during exercise. Even a very low dose of atropine elicited a significant rise in resting heart rate. Under these circumstances, the rapid regulation component at the beginning of loading, which is based on vagal inhibition, must be decreased or not exist at all.

In this experiment, the W_r parameter was about 50% lower than in the controls. The other parameters were not changed. Conversely, methypranol induced a minor, but significant decrease of basic heart rate. An incomplete pharmacological blockade of the orthosympathetic nervous system resulted in a significant change of only the K_{ref} parameter. This result corresponds with previous results, because sensitive adjustment of heart rate to the required level is a complex sympathetic function.

It should be stressed that, in this event, the lower K_{ref} value compared to controls is not indicative of higher physical fitness.

On the other hand, the working capacity after the application of a beta-blocker was distinctly lower. Therefore each parameter should be evaluated with a view to physiological conditions and the potential influence of drugs.

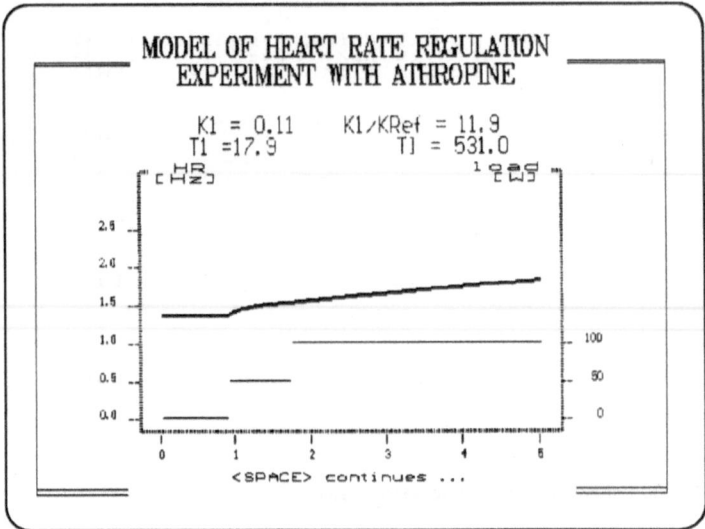

Figure 13.14. (a) Graphical output of the experiment using medication

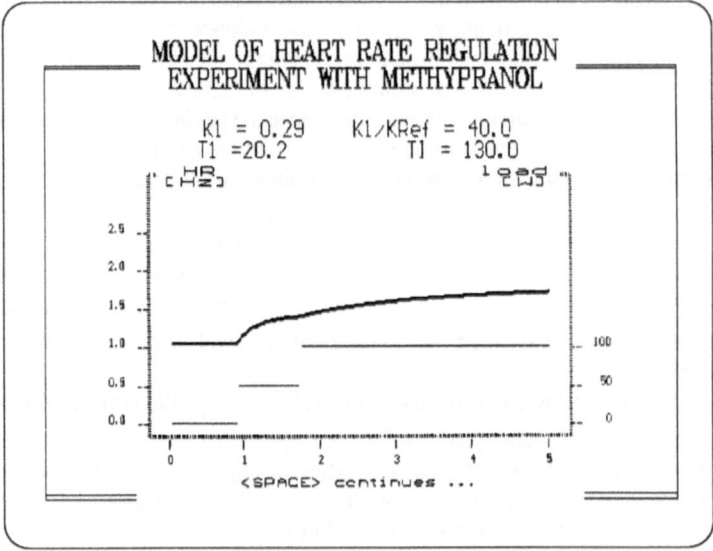

Figure 13.14. (b) Graphical output of the experiment using medication

References

Al-Dahan, M.I., Leaning M.S., Carson E.R., Hill D.W. and Finkelstein L.: "The validation of complex unidentifiable models of the cardiovascular system". In: *Proc. 7th IFAC/IFORS Symposium on Identification and System Parameter Estimation*, Eds.: S.A. Billings and P. Young, Pergamon Press, Oxford, 1985.

Astrand P.O.: *Experimental studies of Physical working capacity in relation to sex and age*. Muwsgaard, Kopenhagen, 1952.

Brodan V., Hajek M. and Kuhn E.: "An analog model of pulse rate during physical load and recovery". *Physiolog, Bohemoslov.*, 20, 1971, pp. 189-198.

Best C.H. and Taylor N.B.: *The physiological basis of medical practice*. Bailiere and Cox, London, 1955.

Bekey G.A.: "Parameter estimation in biological systems - a survey". In: *Proc. 3th IFAC Symposium on Indentification and System Parameter Estimation*, North Holland, 1973, pp. 1123-1130

Carson E.R., Cobelli C. and Finkelstein L.: *The mathematical modeling of metabolic and endocrine systems*. John Wiley and Sons, New York, 1983.

Carson E.R. and Cramp D.G.: *Computers and control in clinical medicine*. Croom Helm London, 1985.

Cobelli C., Carson E.R., Finkelstein L. and Leaning M.S.: "The   validation of simple and complex models in physiology and   medicine". *American Journal of Physiology*, 246, 1984, pp. 259-266.

Cramp D.G. and Carson E.R. Eds.: ·*Measurement in medicine, Vol. 1: The circulatory system*. London, Croom Helm, 1986.

Finkelstein L. and Carson E.R.: *Mathematical modelling of dynamic biological systems*. John Wiley and Sons, New York, 1979.

Hajek M., Potucek J. and Brodan V.: "Mathematical model of heart rate regulation during exercise". *Automatica*, Vol. 16, 1980, pp. 191-195.

Hyndman B.W.: *A digital simulation of the human cardio-vascular system and its use in the study of sinus arrhythmia*. Ph.D.Thesis, Imperial College, London, 1979.

Johnson T.J., Brouha L. and Galagher J.R.: "Dynamic physical fitness in adolescence: use of step - test in evaluation. *Yale, J., Biol. Med.* July, 1980, pp. 781-785.

Leaning M.S., Pullen H.E., Carson E.R. and Finkelstein L.: "Modelling a complex biological system: The human cardio-vascular system. The methodology and model description". *Trans. Inst. Meas. System Control*, Vol. 5, 1983, pp. 71-86.

Leaning M.S., Pullen H.E., Carson E.R., Al-Dahan M., Rajkumar M. and Finkelstein L.: "Modelling a complex biological system: The human cardio-vascular system. Model validation, Reduction and development". *Trans. Inst. Meas. System Control*. Vol. 5, 1983, pp. 87-98.

Neill W.A.: "Regulation of cardiac output". In: *Clinical cardiovascular physiology*, Levins H.J. (Ed.), Grune and Stratton, New York, San Francisco, London, 1976.

Potucek J., Brodan V. and Hajek M.: "Reliability range of model parameters and its application to biological models". *Applied mathematical modelling*, June, 1979.

Potucek J., Brodan V. and Hajek M.: "Hybrid computer analysis of the heart rate during physical load". *Proc Simulation of Systems Conf 1979*, North-Holland Publishing Company, 1980.

Seliger V. and Wagner J.: "Evaluation of heart rate during exercise on a bicycle ergometer". *Physiol. Bohemoslov*. 18, S., 1969, pp. 41-18.

14
Catheter-Manometer Systems

Frank J. Pasveer and Rogier P. van Wijk van Brievingh

14.1 Aim of Instruction

This section will give an introduction to wave reflection phenomena occurring in situations where wavefronts, travelling through a medium, encounter reflective boundaries. Part of the wave is reflected, whereas, dependent on the properties of the two media, another part of the wave will travel across the boundary and continue on its way.

Electrical line transmission theory will be applied to demonstrate how blood pressure waves travel through a fluid-filled tube, at the distal end of which a manometer records the arriving pressure waves. It will be shown how irregularities in the transmission path distort wave shape. The effect is analyzed with a square pulse of relative short duration or with a stepwise change in pressure as an input signal.

The simulation allows the observation of wave behaviour along the transmission line as if this were a transparent tube.

Also, reflection phenomena at the proximal end, at the distal end, and at an irregularity somewhere along the line (such as an air bubble) will be simulated.

The instruction aim of this section is to demonstrate the physical properties of a catheter-manometer system for blood pressure measurement, especially the distortion caused by the hydraulic transmission of pressure waves and the effects of discontinuities.

14.2 Theory and Reference to Standard Textbooks

Pressure is one of the basic physiological parameters that can be measured in the human organism. The values obtained as well as the waveforms recorded give indications of various static and dynamic conditions in the perfused tissue and the state of the body's control mechanisms.

Blood pressure is of special interest both in the diagnosis and the therapeutic treatment of the critically ill.

Direct access to the arterial pressure waveform is gained by the cannulation of an artery and insertion of a fluid-filled catheter.

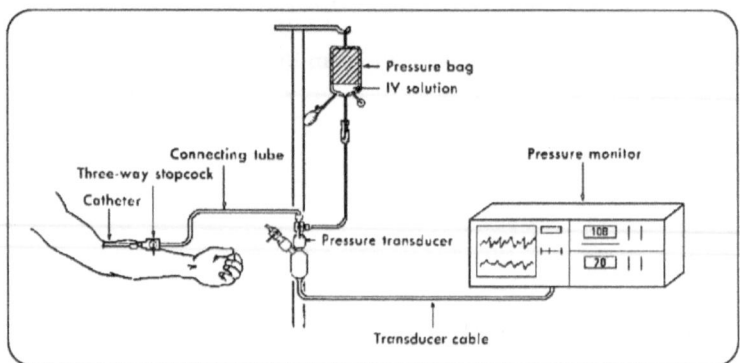

Figure 14.1.   Application of a catheter-manometer system

Because catheter-manometer systems are cheap, flexible and easy to handle in medical diagnostics, they are widely used. Compared with the catheter-tip internal pressure transducer, however, their dynamic response is inferior, especially when quality assurance measures are slack and air bubbles are present within the system. In that case, the measured blood pressure signals may even be inadequate for clinical interpretation.

The pressure at the distal end of the catheter is hydraulically coupled to an external pressure transducer, which converts it to an electrical signal. This transformation requires a movement of compliant membrane in the transducer. Most of the measuring errors in the intravascular method arise from the hydraulic coupling [Jensen, 1982]. For faithful reproduction of the blood pressure waveform, the most relevant property of the fluid-filled manometer system is its compliance (volume displacement per pressure change). It is composed of the inherent compliance of the transducer membrane itself and the added compliance of the dome, tubing, stopcocks, and possibly minute amounts of air trapped in the system. To ensure that the pressure wave is faithfully recorded, the system should have the same sensitivity at all frequencies within the relevant band and introduce a phase shift which is directly proportional to frequency [Latimer, 1968, 1980].

14.2.1   The Electrical Analogon

In the description of wave front propagation through the catheter, the theory of electrical transmission lines can be applied because of the similarity of the equations governing the process. Thus, an electrical analogon is used for describing the hydraulic transmission line with voltage representing pressure and current representing volume flow. In a first approximation, the hydraulic transmission line can be described by the parameters:

inertance,   L,   related to the kinetic energy of the fluid column, per unit of length

$$L = \rho/A \qquad \text{(plug flow assumed)} \qquad (14.1a)$$

compliance, C,   related to the potential energy stored in the elastic catheter wall, per unit of length

$$C = \Delta V/\Delta P \qquad (14.1b)$$

resistance, R,   related to the energy dissipation due to fluid viscosity, per unit of length

$$R = 8\pi\eta/A \qquad \text{(laminar flow assumed)} \qquad (14.1c)$$

with: ρ = fluid density
             η = fluid dynamic viscosity
             V = volume
             P = pressure
             A = lumen area

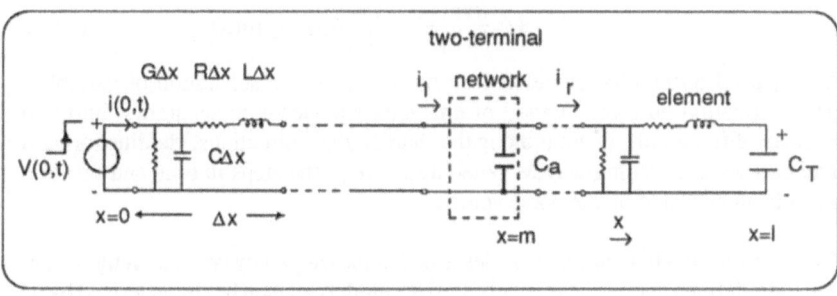

Figure 14.2. Electrical transmission line

The pressure transducer can be modeled by its diaphragm compliance, being equal to the reciprocal of Young's modulus of its material. Air bubbles are accounted for as compliances also; the compressibility of air is dominant to that of the other components. The quantities mentioned above get "recognizable" values if an analogy with electrical quantities is made: if one microliter of air is equivalent to one microfarad of electrical capacitance, R and L will be given in ohms and henry's respectively. The transverse conductance G can be neglected.

14.2.2 Transmission Line Theory

The analog scheme can be described by a hyperbolic partial differential equation in
$v = v(x,t)$ for the propagation of the voltage along the line (the so-called "Telegraph
Equation"):

$$\frac{\partial^2 v}{\partial t^2} = \frac{1}{LC} \cdot \frac{\partial^2 v}{\partial x^2} - \frac{R}{L} \cdot \frac{\partial v}{\partial t} \quad (G = 0) \tag{14.2}$$

with $c = 1/\sqrt{LC}$ the wave velocity.

In order to facilitate the numeric solution, the PDV is divided into two coupled
partial differential equations in the voltage v and the current i.

$$C \frac{\partial v}{\partial t} + \frac{\partial i}{\partial t} = 0$$

$$\tag{14.3a}$$

$$\grave{O}\grave{O}\frac{\partial i}{\partial t} + \frac{\partial i}{\partial x} = -Ri$$

The boundary conditions for the transducer and the air bubble are:

$$C_T \frac{\partial v(l,t)}{\partial t} = i(l,t) \tag{14.3b}$$

$$C_a \frac{\partial v(m,t)}{\partial t} = i_1 (m,t) - i_r (m,t) \tag{14.3c}$$

Such partial differential equations can be solved by the separation of variables.
Transformation with the method of characteristics leads to separation into two
ordinary differential equations along the characteristic directions, yielding forward
and backward travelling waves. When integrating, the steps in time and position
must be interrelated by the wave velocity.

The transmission line may also be described in the *frequency domain*. With the parameters defined above, the transmission line model can then be characterized by its
characteristic impedance, Z, and its *transmission factor*, γ ,which are both
complex functions of frequency:

$$Z = \sqrt{(R + j\omega L) / j\omega C} \tag{14.4a}$$

$$\gamma = \sqrt{(R + j\omega L) \cdot j\omega C} \tag{14.4b}$$

The terminations and discontinuities are accounted for by *reflection factors*,
functions of frequency, dependent on the local impedance Z_T and the characteristic
impedance Z:

$$r = \frac{Z_T - Z}{Z_T + Z} \qquad (14.4c)$$

If the transmission line is terminated with $Z = Z_T$, $r = 0$; the case of characteristic termination, without reflections. Termination with a capacitor is non-characteristic, as its impedance is a pure reactance and Z is almost real. Thus, the primary effect of the transducer is a phase change of the reflected wave.

The proximal end of the catheter is "connected" to a pressure source of low internal impedance, as the flow in the catheter is negligible and will not effect the patient's blood pressure [Latimer, 1980]. Therefore, the reflection coefficient at the input termination equals -1, meaning phase reversal.

By the Laplace method, the partial differential equation is transformed into a form suitable for direct algebraic solution. In the frequency domain, the solutions are expressed in forms in which the sum of two separate terms occur:

$$V(x,s) = A \cdot e^{-\gamma x} + B \cdot e^{+\gamma x}$$

$$\qquad (14.5)$$

$$I(x,s) = \frac{1}{Z}(A \cdot e^{-\gamma x} - B \cdot e^{+\gamma x})$$

The terms with e^{-x} describe wave fronts traveling in the forward direction, those with e^{+x} refer to the backward direction. These equations allow the determination of the *transfer function*, which describes the signal transfer in terms of amplitude and phase as functions of frequency.

The Laplace method, however, is not suitable for simulation of wave phenomena in the time domain. In order to obtain a solution in time functions, the differential equations for voltage and current are transformed as shown below:

$$\frac{\partial}{\partial t}\begin{pmatrix} v \\ i \end{pmatrix} = \begin{pmatrix} 0 & -\dfrac{1}{C} \\ -\dfrac{1}{L} & 0 \end{pmatrix} \frac{\partial}{\partial x}\begin{pmatrix} v \\ i \end{pmatrix} + \begin{pmatrix} 0 & 0 \\ 0 & -\dfrac{R}{L} \end{pmatrix}\begin{pmatrix} v \\ i \end{pmatrix} \qquad (14.6)$$

In these equations, the $\begin{pmatrix} v \\ i \end{pmatrix}$ vector can be transformed into a $\begin{pmatrix} \Phi \\ \Psi \end{pmatrix}$ vector on characteristic directions, by means of the substitution:

$$\begin{pmatrix} v \\ i \end{pmatrix} = T\begin{pmatrix} \Phi \\ \Psi \end{pmatrix} \qquad (14.7)$$

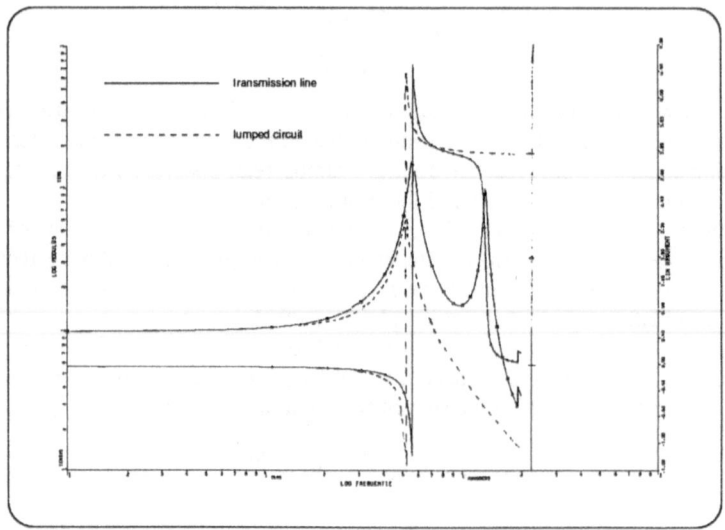

Figure 14.3. This figure also shows the modulus and phase of a corresponding second-order
         type lumped-circuit model. The inadequacy of the latter is evident due to the
         absence of the second resonance peak and the very deviating behaviour of the
         phase characteristic.

Through application of rules from linear algebra this leads to the following matrix
equation:

$$\frac{\partial}{\partial t}\begin{pmatrix}\Phi\\\Psi\end{pmatrix} = \begin{pmatrix}\lambda_1 & 0\\ 0 & \lambda_2\end{pmatrix}\frac{\partial}{\partial x}\begin{pmatrix}\Phi\\\Psi\end{pmatrix} + D\begin{pmatrix}\Phi\\\Psi\end{pmatrix} \tag{14.8}$$

Here, D is a matrix expressing the coupling between the Φ and Ψ terms.
When we introduce the characteristic directions $dx/dt = \lambda_1$ and $dx/dt = \lambda_2$, the matrix
equation, written out in its components, leads to the following set of ordinary
differential equations in terms of Φ and Ψ as functions of time:

$$\frac{d\Phi}{dt} = -d_1\Phi + d_2\Psi$$

$$\tag{14.9}$$

$$\frac{d\Phi}{dt} = -d_1\Psi + d_2\Phi$$

Having solved these equations for a particular instant of time t and of position x
along the transmission line, the pressure and flow are then found by:

$$v = \Phi + \Psi \tag{14.10a}$$

$$i = \sqrt{\frac{C}{L}}\,(\Phi - \Psi) \tag{14.10b}$$

Thus, the propagation of pressure and flow is decomposed into two distinct waves: the first front traveling forward and the second one traveling backward. At the start of the wave front propagation, only one forward traveling wave is present. Once the front encounters a boundary, i.e. the manometer, part of the energy is reflected and travels backward. The sum of both waves constitutes the output pressure at all moments thereafter.

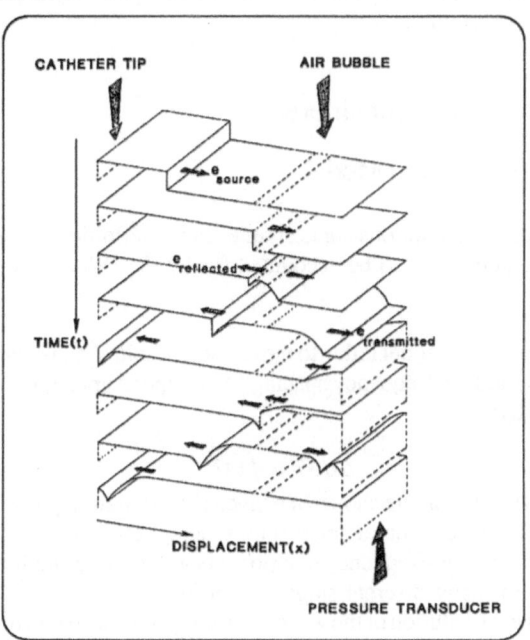

Figure 14.4.   Wave propagation

14.2.3 Compensation

Compensation of distorted blood pressure signals can be achieved by analog means [van Wijk van Brievingh, 1980]; an electrical analogon may be used for parameter estimation [Taylor, 1986].

A method for on-line compensation for the distortion caused by the hydraulic transmission line has been given by [Latimer, 1968]. A second, identical catheter is connected, in parallel with the measuring catheter, to the transducer, external to the patient. As the tip of this compensation catheter is closed, reflected waves arrive at the transducer with opposite phase relative to incoming waves, thus mostly cancelling the distortions.

This elegant principle is not practicable in clinical practice, chiefly because of adjustment time: hydraulic resistances in the form of needle valves need to be matched to catheter properties.

But the principle can be useful in correcting distorted pressure signals with the help of simulation. If it is possible to estimate the parameters of the catheter with which the signal has been obtained, e.g. from response on a "whip" caused by mechanical excitation external to the patient, then correction of the signal may be performed by a simulated Latimer system.

14.3   Aspects to Be Considered

14.3.1 Limitations of the Model

According to the assumptions underlying their definitions, the frequency dependence of L and R has to be accounted for by the Womersley corrections [Vierhout,1966].

Another limitation in the model is due to the method of solution just mentioned. The time spacing (Δ t) and the frequency spacing (Δ f) depend upon one another. Their relation is given by:

$$\Delta t/\Delta x = \sqrt{LC} \qquad (14.11)$$

This formulation states that the choice of a given time step Δt requires a spatial step of Δx. In general it means that a small time spacing, chosen to obtain solutions containing sufficient high-frequency components will require small spacing steps also. This leads to many discrete steps, which in turn require time-consuming computations for each solution of the waveform along the line. It is at this point that a compromise has to be found between computation time and accuracy in the wave representation. We have chosen relatively fast solutions for purposes of demonstration and thus have made concessions in respect to accuracy.

Another limitation of this model occurs when real measurement data are used. In this case the ratio between time spacing (Δt) in the simulation and the sampling interval of the measured data becomes very important. The sampling interval of the measurement data is equal to 2 milliseconds, whereas the time steps in the catheter-manometer model are 200 microseconds. So, ten linear interpolations are required between two measurement points. The repetition frequency of the heartbeat is in the order of 1 Hz. In the case of left-ventricular pressure measurements, this means an interval of about 500 milliseconds between two pressure pulses during which 5000 computation steps in the model are executed to transport the negligible diastolic part of the signal. Thus it may be concluded that this model is very time ineffective when calculating on real measurement data.

14.3.2   Outlook

With faster computers becoming available at lower prices, such time limitations may soon be overcome.
The simulation method described may be applied to the compensation of pressure signals measured in the clinic.

14.4   Model Description

14.4.1   Foundations and Assumptions

The functions Φ and Ψ represent waves in the forward and backward directions respectively, and are implemented in the calculation scheme as depicted in below.

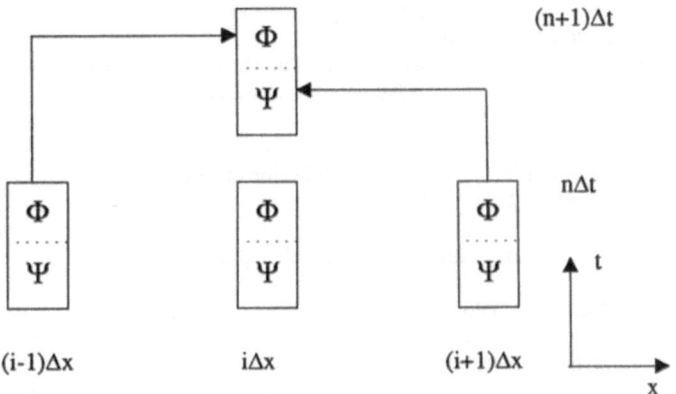

The time steps Δt are counted by n and are depicted vertically; the space steps Δx are counted by i and shown horizontally.
At time $n\Delta t$ the values of Φ and Ψ in each cell are known. To calculate cell $(i\Delta x,(n+1)\Delta t)$, the values of Φ and Ψ in cells $((i-1)\Delta x,n\Delta t)$ and $((i+1)\Delta x,n\Delta t)$ are used. So, Φ at $(n+1)\Delta t$ and $i\Delta x$ is estimated from Φ and Ψ at the $i-1^{th}$ cell.
Ψ is calculated backward in the same fashion from the $i+1^{th}$ cell.

14.4.2 Structure

Two arrays have to be reserved in computer memory for the new values of Φ and Ψ, to be derived from their previous values. After finishing a complete computation along the line, all recent Φ and Ψ values are kept. At this stage the program now adds the Φ and Ψ numbers in order to draw the waveform along the line. Since this calculation loop is repeated at fixed intervals of time, the reader is presented with a traveling wave.

The following Pascal source code illustrates the implementation:

```
procedure transml (var isize:integer;var vexcit:real);

{   parameters:                                                      }
{   isize    number of compartments of transmission line            }
{   vecxit   excitation voltage, fed at input of line               }

var i,j: integer;            {   i and j local variables            }

begin

{   F and Y                 global arrays, contain F and Y values   }
{   along the line, initially set at -1                             }
{   huF and                 auxiliary arrays in which the new F and }
{   huY                     Y distribution is calculated at the next}
{   delta t step.                                                   }

    huF[1]:= vexcit-Y[1];    {   input reflection of zero           }
    huY[1]:= d1*Y[2]+d2*F[2];

{   d1 and d2 are coefficients to determine huF and huY            }

    for j:=2 to (isize-1) do
    begin
       huF[j]:=d1*F[j-1]+d2*Y[j-1];
                            {     estimated from backward           }
       huY[j]:=d1*Y[j+1]+d2*F[j+1];
                            {   similarly from forward              }
    end;
                    {     distal part of line, here for            }
                    {     capacitive termination                   }
vo:=F[isize]+Yisize];
huF[isize]:=d1*F[isize-1]+d2*Y[isize-1];
huY[isize]:=ch1m*vo+ch2m*huF[isize];
end;
                    {     end procedure transml                    }
```

14.4.3 Parameters and Signals

The parameters R, L and C of the catheter per unit of length are considered constants in the differential equations to be solved. They are calculated by a separate initialization program in which the user can enter catheter specifications in terms of length, lumen diameter and temperature.

The user can choose between step response and impulse response. The latter is excited by a pulse of a fixed duration which is an integer multiple of the basic integration time step.
It is possible to demonstrate various reflection conditions at the distal end of the line. The user can experiment with an open, a short-circuited and a characteristically terminated line. In the first two cases different backward traveling waves are observed. In the last case no wave will travel backward.
In order to analyze the effects of a capacitive termination, the transducer compliance can be chosen as a fraction of the line compliance. The same goes for the compliance of the air bubble, the position of which can be selected in steps of 1/16 of the length.

14.4.4 Possibilities and Limitations

The transmission line model used is restricted by its computation time, accuracy of the solutions and in the use of real-time measurement data. The most important limitation is the computation speed of the P C. Due to this restriction the partial differential equations, making up the transmission line model, must be solved with such an accuracy that the wave front propagation will be demonstrated clearly at all times. In cases where more accurate solutions are needed, smaller steps in the space parameter is needed, which will lead to longer computation time. Thus, we have made a compromise between the user's patience in observing time-consuming phenomena and the accuracy of the solutions presented.

The Womersley corrections are formulated in the frequency domain. The frequency dependency of R and L of the transmission line cannot readily be transformed into the time-domain algorithm.
In the case of the Latimer correction, the coupling should take place through characteristic impedances. These cannot be simulated by an ordinary differential equation, which precludes inclusion in the simulation. As an approximation, the characteristic resistance of the line, $R_c = \sqrt{\dfrac{1}{C}}$ is used instead.

$$Z = \sqrt{(R + j\omega L)/j\omega C} = \sqrt{\frac{L}{C}\left(1 + \frac{R}{j\omega C}\right)}$$

$$\approx \sqrt{\frac{L}{C}} \text{ for } \omega >> \frac{R}{C}$$

(14.12)

14.5   Experiments

14.5.1   The Hydraulic Transmission Line

Simulation experiments with the transmission-line model can be performed with presentation of either the manometer response as a function of time or of the wave profiles along the line.
The following options can be chosen:

> Input signal: unit impulse or unit step
>
> Termination: 1. Characteristic impedance
> 2. Open-end
> 3. Short-circuited
> 4. Capacitive (pressure transducer)

The compliance of the transducer can be chosen as a factor of the line capacitance (0.1 < factor < 10). Of particular interest is the situation in which the transmission line is terminated by a capacitor as an analogon of the manometer. As is to be expected, the capacitor introduces a "charging" effect at the end of the line, leading to exponentially increasing and decreasing time behavior.
Choices can be made from the input screen and graphs may be printed.

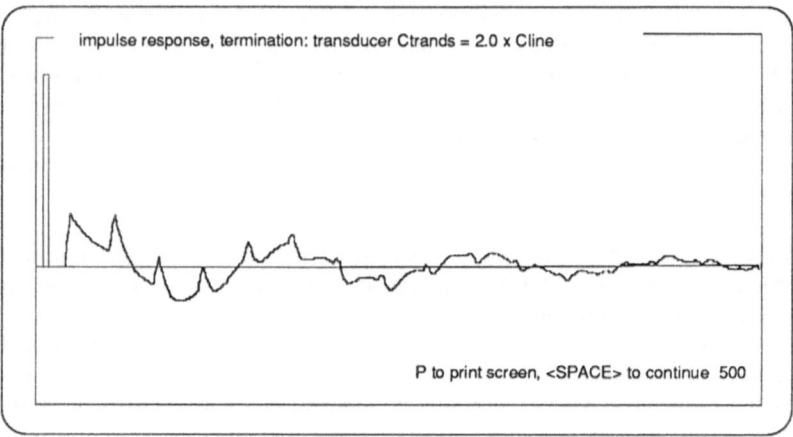

impulse response, termination: transducer Ctrands = 2.0 x Cline

P to print screen, <SPACE> to continue 500

Figure 14.5a

14.5.2 Air Bubble

The effect of an air bubble somewhere in the catheter can also be studied with a unit impulse or a unit step as input signals.
The air-bubble position can be chosen in steps of 1/16 of the line length from the input screen (2/16 < step < 14/16).
The size of the air bubble is expressed as a factor times the line capacitance (0.1 < factor < 10).

Due to multiple reflections of the pressure waves at the ends of the line as well as at the air bubble (in both directions !), the response shows an asymmetric behavior. Choices can be made from the input screen and graphs may be printed.

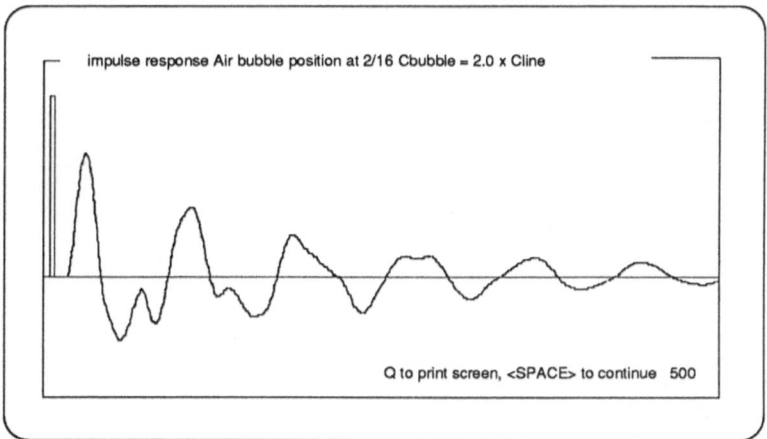

impulse response Air bubble position at 2/16 Cbubble = 2.0 x Cline

Q to print screen, <SPACE> to continue   500

Figure 14.5b

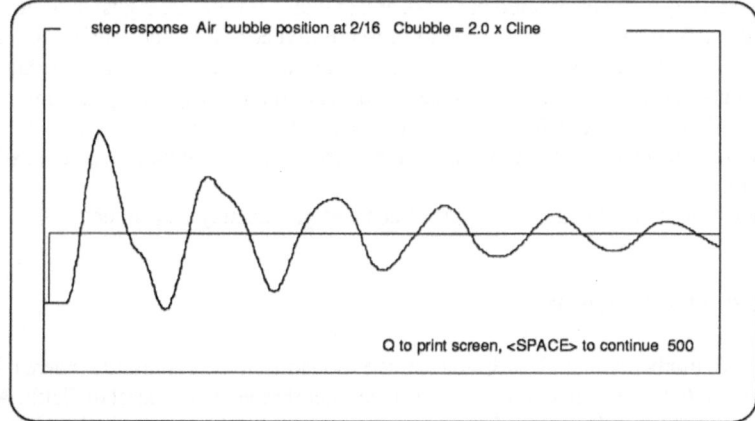

step response Air bubble position at 2/16   Cbubble = 2.0 x Cline

Q to print screen, <SPACE> to continue   500

Figure 14.5c

14.5.3 Compensation with Latimer Twin Catheter

The Latimer correction principle uses an identical second catheter with a closed end, characteristically coupled to the transducer in parallel with the measuring catheter. Due to the short-circuited end, the compensation line reflects waves in counterphase. Thus, at the transducer two reflected (and distorted) pressure waves arrive so that distortions are canceled in the steady state. The response on unit impulse and unit step input signals can be studied with the transducer capacitance as an independent parameter, to be chosen as a factor times the line capacitance (0.1 < factor < 10). No air bubbles are assumed to be present here.

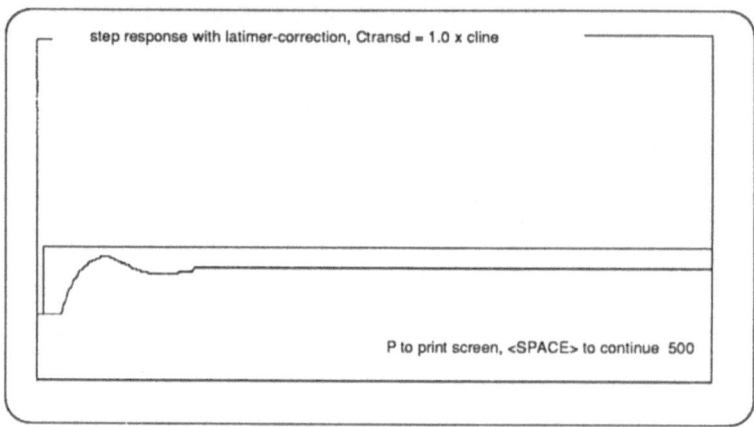

Figure 14.5d

Note that in the step response curve the system almost appears to be critically damped (only a few oscillation periods), and that the steady-state signal magnitude is about 2/3 of the input level, due to pressure division in the coupling network (characteristic resistors). As it is not possible to model the characteristic impedance of a transmission line with a lumped circuit (i.e. the corresponding differential equation), a "characteristic resistance" has been used in the model.
Of course, the effect of the transducer impedance (also non-characteristic) is still present.
Choices can be made from the input screen and graphs may be printed.

14.5.4   Conclusions

Detailed mathematical description of the transmission of pressure waves in compliant tubes terminated with a transducer membrane is in general difficult. An electrical analogon in terms of a transmission line allows instructive simulation of traveling waves as a function of time and place. The effect of air bubbles can be shown with variable bubble position. The principle of distortion compensation by means of a compensation catheter is elucidated. The curves presented give insight into the underlying processes that can not otherwise be easily obtained.

References

Brok, S.W. and E.E.E. Frietman: "Mathematical Modeling of a Catheter Manometer System". *Proc 16th IAESTEAD International Conference on Identification, Modeling and Control*, Paris, 1987.

Jensen, Ø../: *Measuring Blood Pressure and Avoiding the Technical pitfalls.* The National Hospital of Norway, 1982.

Latimer, K.E.: "The transmission of sound waves in liquid-filled catheter tubes used for intravascular blood-pressure recording". *Med. & Biol. Engng.* Vol.6, pp. 29-42. Pergamon Press, 1968.

Latimer, K.E. and E. van Vollenhoven: "Blood pressure monitoring catheter manometer systems: response to pressure variations, design, calibration, and testing analysis". In: D.N.Ghista (Ed.): *Biomechanics of Medical Devices*, Marcel Dekker, New York, 1980.

Taylor, B.C., D.M.Ellis, and J.M.Drew: "Quantification and Simulation of Fluid-filled Catheter/Transducer Systems". *Medical Instrumentation* 20(3), pp 123-129, (1986).

Vierhout, R.R.: *The response of Catheter Manometer Systems Used for Direct Blood Pressure Measurements.* PhD Thesis Catholic University of Nijmegen, The Netherlands, 1966.

Wijk van Brievingh, R.P. van, J.C.Bernouw, R.M.Heethaar, M.Schmelz,T.van der Werf, and A.N.E.Zimmerman: "Automatic Electronic Correction of Catheter Manometer Dynamic Characteristics in situ: Procedures and in vivo trials". In: D.N.Ghista, E. van Vollenhoven, W.J.Yang, H.Reul and W.Bleifeld (Eds): *Cardiovascular Mechanics; Procedures and Devices*, Witzstrock, New York, 1980.

Wijk van Brievingh, R.P. van and F. Pasveer: "Simulation of Catheter-Manometer Dynamic Response". In: D.P.F. Möller (Ed.): *System Analysis of Biomedical Processes*, Proc. 3rd Ebernburger Working Conference, series: Advances in Systems Analysis, Vol. 5, Vieweg, 1989.

15
Regulation of Respiration

Erik W. Kruyt, Aad Berkenbosch, and Richard J.M.G. de Zwart

15.1   Introduction

The model of the chemical regulation of respiration as presented in this chapter is an example of the functioning of a biological control system. It illustrates the basic mechanisms of steady-state respiratory control and shows the influence of a limited set of parameters on the ventilation and blood gas tensions.

The model is implemented as a student controlled simulation program. The student can put forth a hypothesis and change model parameters, and the computer program supplies feedback by calculating and presenting the resulting situation.

15.2   Theory

The respiratory system is a negative feedback system that we can regard as a "metabolic regulator".

In such a regulator we can recognize two distinctive processes – the controlled system or plant and the controlling system or controller (Figure 15.1) – which are arranged in a feedback pattern. The purpose of the regulator is to keep the plant outputs close to some desired values.

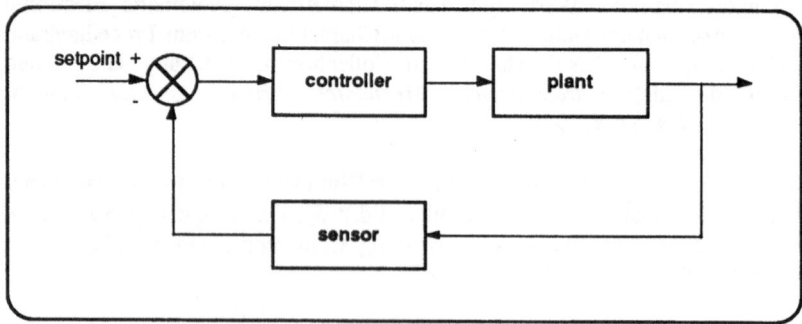

Figure 15.1. The feedback system

Although we have no way of knowing the actual purpose or goal of the biological control system, we can describe it as if it is goal directed to ensure CO_2, O_2 and H^+ homeostasis in arterial blood and brain tissue.

In biological systems, as opposed to man-made control systems, it is in general impossible to identify an independent setpoint or a comparator. We therefore include the setpoint and the comparator in the controller.

The input to the plant in the control system is the expiratory ventilation[1] ($\dot{V}_E$), and the outputs are the arterial CO_2 tension (P_{aCO_2}), oxygen tension (P_{aO_2}), and hydrogen ion concentration ($[H^+]_a$). Thus the plant must consist of the pulmonary gas exchanger and the blood buffer system.

The controller includes all the processes from the central[2] and peripheral[3] sensory mechanisms through the ventilatory apparatus (Figure 15.2).

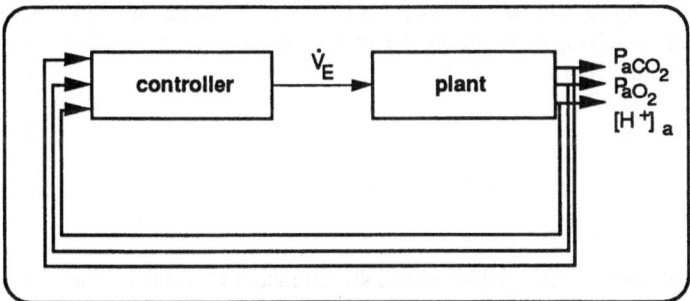

Figure 15.2. The respiratory control system

Now we try to find the *plant* equations that describe the dependency of the gas tensions on the ventilation during steady–state conditions.

1. The ventilation ($\dot{V}_E$), which we can measure at the mouth and is the result of the ventilatory apparatus, is composed of the dead space ventilation ($\dot{V}_D$), which does not contribute to the gas exchange in the lungs, and the alveolar ventilation ($\dot{V}_A$):

$$\dot{V}_A = \dot{V}_E - \dot{V}_D \tag{15.1}$$

2. The plant equations can now be derived from the mass-balance of the CO_2 and O_2 gasses. The amount of CO_2 produced by the body tissue ($\dot{V}_{CO_2}$) must in the steady state be equal to the amount of CO_2 removed through the expired air. The amount of O_2 consumed ($\dot{V}_{O_2}$) must be equal to the amount of O_2 supplied through the inspired air:

---

[1] The expiratory ventilation is the quantity of gas a person expires in a certain time. It is also referred to as "minute volume" and is expressed in liters per minute.
[2] The central chemoreceptors and respiratory centre are located in the medulla oblongata.
[3] The peripheral chemoreceptors are located in the aortic and carotid bodies.

$$\dot{V}_{CO_2} = \dot{V}_A \, (F_{ACO_2} - F_{ICO_2}) \qquad (15.2)$$

$$\dot{V}_{O_2} = \dot{V}_A \, (F_{IO_2} - F_{AO_2}) \qquad (15.3)$$

Where F stands for 'fraction', the index A denotes 'alveolar' and the index I denotes 'inspiratory'.

The fractions can be converted into tensions when the barometric pressure (P_B) is known:

$$P_{aCO_2} = F_{ACO_2} \cdot P_B \qquad (15.4)$$

$$P_{AO_2} = F_{AO_2} \cdot P_B \qquad (15.5)$$

3. In healthy subjects the diffusion of CO_2 and O_2 gasses through the alveolar membrane is rapid. Therefore the alveolar gas tensions are a good approximation of the arterial gas tensions:

$$P_{aCO_2} = P_{ACO_2} \qquad (15.6)$$

$$P_{aO_2} = P_{AO_2} \qquad (15.7)$$

4. The amount of CO_2 produced and the amount of O_2 consumed are almost the same. The ratio, $\dot{V}_{CO_2}/\dot{V}_{O_2}$, is called the respiratory quotient, RQ.
5. The dependence of $[H^+]_a$ on $\dot{V}_A$ is indirectly known by using (15.2), (15.4), (15.6) and the equations describing the blood buffer system:

$$[H^+]_a = f(P_{aO_2}, P_{aO_2}, Hb, [HCO_3^+]) \qquad (15.8)$$

where Hb is the hemoglobin content and $[HCO_3^-]$ is the bicarbonate concentration in the blood.

$[H^+]_a$ and P_{aCO_2} are closely coupled through the blood buffer system. Therefore we cannot discriminate between the effects of changes in P_{aCO_2} and those of changes of $[H^+]_a$ caused by changes in the P_{aCO_2}. So we only need the $[H^+]_a$ as a parameter to reflect changes in the concentrations of non-volatile acids.

We see that the plant equations found are based on physical and chemical processes which hold both in vitro and in vivo.

The *controller* has the arterial gas tensions as input and the expiratory ventilation as output. There are many processes in between these inputs and the output, and they cannot all be described functionally. We therefore apply a purely empirical way of describing the controller and use the simplest equations, which are compatible with the observations.

The ventilatory response to changes in P_{aCO_2} at constant P_{aO_2} appears to be linear, at least in the ventilation range considered:

$$\dot{V}_E = S \cdot (P_{aCO_2} - B) \tag{15.9}$$

in which S is the P_{CO_2} sensitivity and B a constant.
According to Lloyd and Cunningham [Lloyd, 1963], S depends on the P_{aCO_2} in the following way:

$$S = D + AD/(P_{aO_2} - C) \tag{15.10}$$

In our group, however, [Berkenbosch, 1986] found an exponential dependency of the sensitivity to the P_{aO_2} tension. We write

$$\dot{V}_E = S \cdot P_{aCO_2} - K \tag{15.11}$$

in which S can be split into a sensitivity of the peripheral and central chemoreceptors and both the sensitivity and K depend in an exponential way on the P_{aO_2} tension:

$$\dot{V}_E = P_{aCO_2} \cdot \{S_p + G_p \cdot \exp(-D \cdot P_{aO_2}) + S_c\} - K_0 - G_c \cdot \exp(-D \cdot P_{aO_2}) \tag{15.12}$$

where S_p and G_p are the sensitivities of the peripheral chemoreceptors for CO_2 and O_2 respectively, S_c and G_c are the sensitivities of the central chemoreceptors for CO_2 and O_2, and D and K_0 are constants.

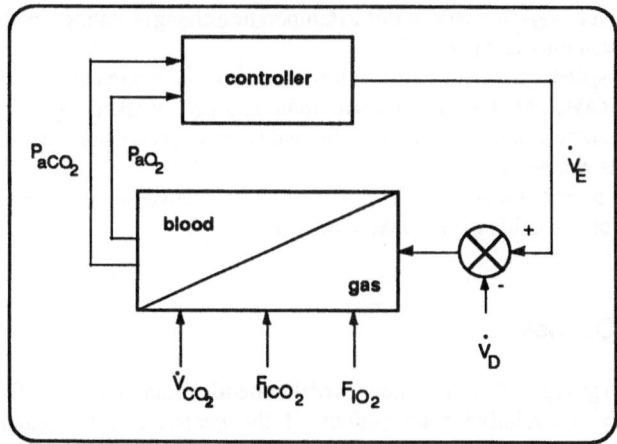

Figure 15.3. The model

Now that we know the plant and controller equations, we can close the loop of the feedback control system (Figure 15.3). In equilibrium the system will have to fulfill both sets of equations. Thus the resulting ventilation and arterial gas tensions can be found by solving the set of plant and controller equations. Both the CO_2 and the O_2 relations must be satisfied.

15.3   Aspects to be Considered

15.3.1   Limitations of the Model

The main purpose of our model is to demonstrate the basic functioning of the control of breathing during the steady state. For that reason the model is kept as simple as possible but nevertheless, within its limitations, accurate and realistic. In the next paragraphs these limitations are summarized.

-   The way the system comes to the steady state after changing a stimulus has not been modeled. Moreover the breathing cycle has been left out. Incorporating the breathing cycle would have made the model much more complex with many more parameters needed to identify the system, and this would detract from the essence of the model.
-   As in the biological system the model does not have setpoints. We cannot instruct the model to adjust the plant output to certain P_{aCO_2} and P_{aO_2} values. We can only change certain parameters of the plant and controller.
-   The model is not suited to study the effects of exercise. The CO_2 production and cardiac output have special feed-forward influences on the controller. However, V_{CO_2} variations due to temperature changes, eating and drug effects are taken into account.
-   The respiratory system adapts to low barometric pressure within a time span of a few days. This high altitude acclimatization is not taken into account.
-   We assume equality of the alveolar and arterial gas tensions, so no diffusion defects should be present.
-   In the present model the $[H^+]$ effects due to changes in the concentration of non-volatile acids are not taken into account.

15.3.2   Outlook

The Leiden group studies the dynamics of the chemical control of breathing and tries to quantify the relative contributions of the central and peripheral sensors. Moreover our aim is to discriminate between the separate $[H^+]$ and P_{aCO_2} effects. As part of this research we develop models incorporating the dynamics of the system.

15.4   Model Description

15.4.1   Physical Foundations and Assumptions

The physical foundations and equations used for plant and controller are as presented in (15.2) and (15.12).

To simulate the effects of drugs (e.g. narcotics) the model offers the opportunity to depress the sensitivity of the respiratory controller. In the model the S_p, G_p, S_c and G_c are scaled down with the depression factor specified. The K_0 is adapted in such a way that the CO_2 controller line rotates at a P_{aCO_2} of G_c / G_p. This is in correspondence with our observations.

15.4.2   Structure

The block diagram of the model is as presented in Figure 15.3. To determine the steady-state situation the system has to fulfill both the plant and the controller equations for both CO_2 and O_2. Because we cannot write the resulting values as explicit expressions of the parameters, we must find the solution by an iterative approach. We implemented the Newton-Raphson method for nonlinear systems of equations as described in [Press, 1986].
When the patient is artificially ventilated we can compute the P_{aCO_2} and the P_{aO_2} by simply solving the equations (15.2) and (15.3) respectively.

The model is implemented in Turbo Pascal. One of the reasons for not using the BIOPSI system is that we want to model and display more than one different x-axis at a time (the P_{aCO_2} and the P_{aO_2}).

15.4.3   Parameters and Values

P_B: the barometric pressure in kPa
$\%O_2$:   the percentage of O_2 in the inspiratory gas
$\%CO_2$:   the percentage of CO_2 in the inspiratory gas
$\dot{V}_{CO_2}$:   the amount of CO_2 produced by the body in l/min
$\dot{V}_D$:   the dead space ventilation in l/min
depression: the depression of the ventilatory center in %
RQ: the respiratory quotient, $\dot{V}_{CO_2} / \dot{V}_{O_2}$

As a result of these input parameters the model computes the following values:

P_{aO_2}:   the arterial partial O_2 tension in kPa
P_{aCO_2}:   the arterial partial CO_2 tension in kPa
$\dot{V}_E$:   the ventilation in l/min

You can mechanically ventilate the patient using a ventilator. In this case the ventilation becomes an input parameter instead of a resulting value.

In the model we specify the $\%O_2$ and $\%CO_2$ respectively, while in the formulas the fractions F_{IO_2} and F_{ICO_2} are used.

$$F_{IO_2} = \%O_2 / 100 \qquad\qquad (15.13)$$

$$F_{ICO_2} = \%CO_2 / 100 \qquad\qquad (15.14)$$

The $\dot{V}_{CO_2}$ is presented in l/min STP, which means it is the volume as if measured at a standard temperature of 273 K and standard pressure of 101 kPa. The actual value used in the model is:

$$\dot{V}_{co_2} = \dot{V}_{co_2}, STP \times \frac{310}{273} \times \frac{101}{P_{bar}} \qquad\qquad (15.14)$$

When you change RQ you change $\dot{V}_{O_2}$ implicitly while $\dot{V}_{CO_2}$ remains unchanged. This is because the $\dot{V}_{CO_2}$ is considered a primary parameter.

Along with the parameters presented above, all controller parameters can be listed and modified. See equation (15.12) for a description of these parameters. Keep in mind that when you want to change the central CO_2 sensitivity you must not only change S_c but also adjust K_0 in such a way that the intercept with the P_aCO_2 axis remains constant. In short, you must keep $S_c.K_0$ constant.

15.4.4 Operation of the Program

The operation of the model is mainly self-explanatory. We only make the following remarks to assist you in understanding the operation.

-   The program tries to open a file DRIVER in which some data are stored concerning the graphics configuration used. When the file DRIVER is missing the program prompts you for the graphics configuration of your system and creates the file DRIVER. When starting the program for the first time, you therefore have the opportunity to tune the program to your configuration. If you want to change the configuration just delete the file DRIVER and start the program.
-   When changing parameters the model does not automatically calculate the resulting values. This way you have the opportunity to change more than one parameter at a time. After you are satisfied with the parameters chosen, you instruct the program to recalculate the new steady state and the new values appear. Only then can you recall the new graphs.
-   When you are in the controller menu you can view and change the controller parameters. The parameters shown are not influenced by the specified depression. The model applies the depression factor each time it determines a

new equilibrium. When leaving the controller menu, the resulting situation is automatically recalculated.

- When starting to mechanically ventilate the patient and the patient is still breathing spontaneously, you should impose a ventilation of at least the spontaneous ventilation. If you impose too low a ventilation the message "I am fighting against the ventilator" is presented. You must depress the patient's respiratory centre (using drugs) before starting to ventilate him with a ventilation lower than the spontaneous ventilation.

15.4.5 Possibilities and Limitations

With this model we can study both the influence of one of the parameters and the simultaneous effect of changing more than one of the parameters on the resulting variables.

In this way we can simulate clinical trials such as supplying O_2, the effects of drugs (e.g. narcotics) or kidney dialysis. During kidney dialysis carbon dioxide is removed from the blood thus falsely lowering the V_{CO_2}. The pathophysiological process of drowning can also be simulated and we can see the effects of different actions that could be taken.

Suggested theories of respiratory system functioning can be simulated to see if they hold true with the present model.

When using the model we must be aware of its limitations. Aside from the limitations discussed in section 15.3.1, the current implementation of the model imposes some additional limitations:

- The model computes the resulting values of the arterial gas tensions and ventilation using the equations (15.1) through (15.7) and (15.12). Although the model is guarded to keep parameters from attaining physical impossible values, the resulting gas tensions and ventilation may become very unrealistic or even incompatible with the human body. The computer program will present a "clinical warning" when this occurs, and you can change the model parameters to return to a realistic situation.
- The dead-space ventilation, $\dot{V}_D$, is incorporated in a rather rudimentary way as a single parameter. In reality, though, two components contribute to the dead-space ventilation. One is caused by a fixed dead-space volume and therefore depends only on the breathing frequency; the second can be described as a certain fraction of the total ventilation. Because the breathing cycle is not modeled, we have not incorporated a more advanced model of the dead-space ventilation either.
- As pointed out in section 15.2, we consider the expiratory ventilation as output of the respiratory controller. In the biological system the respiratory controller provides the neuronal drive for the muscles of the mechanical respiratory system. So we assume a mere "unit transformation" between the neuronal drive and the expiratory ventilation.

This neural-mechanical link is disturbed in patients with pulmonary disease. When we want to apply the model to describe the respiratory system of a patient with pulmonary disease, we have to change the sensitivity of the controller to reflect the effects of the decreased lung function.

15.5   Experiments

15.5.1   Exercises

To give you a feeling for the behavior of the model, we apply a step-by-step approach to studying the influence of some of the separate parameters on the model.

First of all we'll look at the influence of the dead space ventilation ($\dot{V}_D$) on the P_{aCO_2} and the P_{aO_2}. In the first exercise we see the so-called metabolic hyperbola for CO_2 and O_2 respectively which will shift upwards to a higher $\dot{V}_E$ when you increase the dead-space ventilation and visa versa. This illustrates equations (15.1), (15.2) and (15.3).

Exercise 2 shows you the influence of the percentage of CO_2 and O_2 in the inspiratory gas on the arterial CO_2 and O_2 tensions and ventilation. This illustrates equations (15.2), (15.3), (15.4) and (15.5).

The controller equation is illustrated in exercises 3 and 4. In exercise 3 the CO_2 regulation is illustrated. The controller equation for CO_2 is linear (15.11). The sensitivity of the ventilation to CO_2 and K depends on the P_{aO_2} and the sensitivity of the chemoreceptors (15.12). The sensitivity of the chemoreceptors can be decreased by depressing the respiratory controller e.g. by administering drugs.

The O_2 controller is illustrated in exercise 4 (15.12). The sensitivity of the ventilation to O_2 depends on the P_{aO_2} and the depression of the respiratory controller.

15.5.2   What You Can Do on Your Own

Now that you are acquainted with the separate equations making up the model, you can start the integral model and get accustomed to the user interface of the program. In the following we propose a couple of interesting exercises that will give you more insight into the functioning of the respiratory system.

15.5.2.1   Ventilation and Barometric Pressure

Imagine you are sitting at ease on top of the Mont Blanc. The barometer indicates a pressure of 50 bar, which means there is only half as much oxygen available as at sea level.

- What do you expect the ventilation will be?
- Try to reason out the actions of the control system.
- Enter this barometric pressure of 50 bar in the model and recalculate the steady state. Compare the result with your expectations.

As can be seen, the ventilation has hardly changed. Due to the decreased amount of oxygen, you initially will breath more but the increased wash out of carbon dioxide suppresses the ventilation again. The result is that both P_{aCO_2} and P_{aO_2} are decreased while the ventilation is almost kept the same!

Now suppose you had been able to change the P_{aO_2} only. In that case you have to keep the P_{aCO_2} constant by applying more CO_2 in the inspiratory gas.

- Start the program once more, write down the P_{aCO_2} change the barometric pressure and recalculate the new equilibrium.
- Now change the inspiratory $\%CO_2$ in such a way that the P_{aCO_2} returns to its initial value.

You will see that you must administer quite a lot of CO_2 in the inspired air, but the ventilation will rise to a value you probably expected initially!

15.5.2.2 Ventilation and Dead Space

The dead space consists of the dead space in your lungs, mouth/nose cave and trachea and amounts to roughly 0.2 liter. This implies that at a respiratory frequency of 10 breaths per minute the dead space ventilation will be 2 liters per minute. The effect of increasing this dead space by only half a liter will result in an increase of $\dot{V}_D$ to 5 l/min!

Now assume you have a lung embolism. While the lung volume remains the same, there is no gas exchange in some compartments: there is no perfusion with blood although gas is flowing in and out of these compartments. They therefore contribute to the dead space. Assume that the dead space ventilation ($\dot{V}_D$) is increased by 2 l/min.

- What do you expect to happen with the ventilation?
- Try to reason out the actions the control system will take.
- Compare your answer to the results of the computer program by increasing $\dot{V}_D$ to 4 l/min.

The control system appears to completely compensate for the effects of the applied disturbance. By enlarging the dead space, the effective ventilation will diminish at first. The effect is an increase in P_{aCO_2} which stimulates the ventilation with such an amount that there is almost complete compensation for dead space ventilation.

15.5.2.3 Ventilation and Depression

The sensitivity of the control system can be highly depressed by drugs. This means that a higher P_{aCO_2} does not result in a higher ventilation. The effect is a lack of oxygen in the blood. People who die of an overdose of heroine frequently die from cessation of breathing!

- See what occurs by depressing the ventilatory center to 90% and 97%.
- Try to figure out what you must do if you see a victim who is blue from lack of oxygen caused by an overdose of sleeping drugs.
- Simulate the effect of your actions using the computer program.

If you decided to apply oxygen to the victim, you see that the situation has worsened. That is because the main stimulus to which the victim reacts is the low P_{aO_2}. When you apply oxygen, this stimulus diminishes and ventilation stops. What you should do is ventilate the patient. As long as the respiratory center is out of order, the patient has to be ventilated artificially.

15.6 Conclusions

Although the model presented is very basic and describes only the steady state, it is very instructive. Moreover it has already proven to be accurate and realistic in explaining clinical situations that were previously misunderstood.
In many educational, research, or even clinical situations it can be useful to study the basic mechanisms of the respiratory control mechanism in the steady state.

References

Grodins F.S. and Yamashiro S.M.: Respiratory Function of the Lung and its Control. Macmillan Publishing Co., Inc, New York 1978.

Lloyd B.B. and Cunningham D.J.C.: "A quantitative approach to the regulation of human respiration". In: The Regulation of Human Respiration, Eds.: Cunningham, D.J.C. and Lloyd B.B., Blackwell Scientific Publications Ltd, Oxford, 1963.

Cunningham D.J.C., Robbins P.A. and Wolf C.B.: "Integration of respiratory responses to changes in alveolar partial pressures of CO2 and O2 and in arterial pH". In: Handbook of Physiology, sect. 3, The Respiratory System, Vol II. Eds.: Cherniak N.S. and Widdicombe J.G., American Physiological Society, Bethesda, Maryland, 1984, pp. 475-528.

Berkenbosch A. and DeGoede J.: "Actions and interactions of CO2 and O2 on central and peripheral chemoceptive structures". In:Neurobiology of the Control of Breathing, Eds.: Euler C. von, and H. Lager-crantz, Raven Press, New York, 1986, pp 9-17.

Press W.H., Flannery B.P., Teukolsky S.A. and Vetterling W.T.: Numerical Recipes, The art of scientific computing.
Cambridge University Press, Cambridge, 1986.

Appendix

Parameters with default values and normal ranges

parameter	default value (at rest)	normal range	unit
P_B	101		kPa
$\%O_2$	21		%
$\%CO_2$	0		%
V_{CO2}	0.25		l/min
V_D	2.00		l/min
depression	0		%
RQ	0.8	0.7 - 1.0	
P_aO_2	14.8	10.6 - 13.3	kPa
P_aCO_2	5.15	4.5 - 6.0	kPa
V_E	7.6	3.5 - 10.0	l/min

16
A Model for Capnograms from the Bain Circuit

Jan E.W. Beneken and Dietmar P.F. Möller

16.1 Introduction

Continuous measurement of carbon dioxide concentration [CO_2] in the respiratory gases is called capnometry. Recording of continuous carbon dioxide concentration changes is called capnography or capnogram. Hence capnography is the study of the CO_2 waveform produced by monitoring respiratory gases.

The CO_2 waveform, the capnogram, can be divided into five phases:

Phase I -   clearance of anatomical dead space in the very beginning of expiration;
Phase II -   dead space air mixed with alveolar air during expiration;
Phase III -   alveolar plateau with end-tidal CO_2 at the end of the expiration phase;
Phase IV -   clearance of dead space air in the very beginning of inspiration;
Phase V -   inspiratory gas devoid of CO_2 equal to phase I, the expiratory phase starts.

The normal capnogram is shown in Figure 16.1.
    The ultimate goal of respiration is to maintain proper concentrations of oxygen, carbon dioxide, and hydrogen ions in the body fluids. A capnogram is therefore a valuable indicator of disturbed ventilatory mechanics and of various disproportions in ventilation and perfusion. It also indicates changes in overall metabolic rates, such as metabolism abnormalities, and can reveal sudden changes in regional tissue or organ perfusion based on circulatory disorders. Hence the capnogram contains important information for the anesthesiologist during his routine clinical work, like troublefree connection of the breathing circuit with the patient's airway. Therefore CO_2 measurement under anesthesia is a standard monitor and is not only used for generating the disconnection alarm. Hence the capnogram is a standard item for an anesthesia working place.

The program was implemented as the MSc thesis work of R.Takami, Department for Anaesthesiology, University of Florida Medical School, Gainesville, Fla., U.S.A

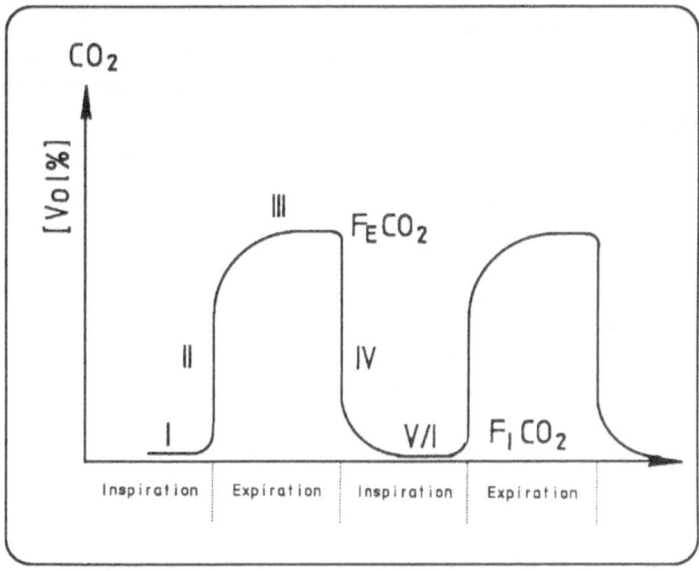

Figure 16.1. Normal capnogram. Ordinate: [CO$_2$] in Vol. %, abscissa: t in s. For explanation see text.

The partial pressure of gases in humidified mixtures can be calculated by the law of Henry and Dalton:

$$pCO_2 = FCO_2 \cdot (P_{amb} - PH_2O) \tag{16.1}$$

which equals it to the fraction of CO_2 (FCO_2) times the difference between the dry ambient pressure (P_{amb}) and the water saturation pressure (PH_2O). The end-tidal concentration of CO_2 gives a fairly good estimate of arterial partial pressure, based on the Henry-Dalton law.

During routine clinical work, the anesthesiologist is mainly interested in continuous measurement of carbon dioxide concentration in respiratory gases. The two measurement techniques in use are based on the CO_2 measure in the expiratory gases, using a sidestream or a main gas flow analyzer. The sidestream continuous breath measurement instruments extracts a small amount of gas flow near the mouth of the patient, which allows to fast response time end-tidal patient breath analysis, whereas the mainstream principle the gas flow is let through a measuring cuvette, near the mouth of the patient. In both instrument types CO_2 absorb energy in the infrared region of the spectrum, and an infrared detector senses the fluctuating infrared level and generates an electrical signal which corresponds to the CO_2 concentration, indicated.

When one breathes air containing carbon dioxide, the alveolar partial pressure paCO$_2$ and the partial tissue pressure tpCO$_2$ rise above normal. As a consequence, alveolar ventilation increases. During weaning, a patient gets enough CO_2 in order to stimulate the humoral regulation of respiration. Figure 16.2 sites the patient-anesthesia complex, where CO_2 reveals clinically significant problems and interactions are shown.

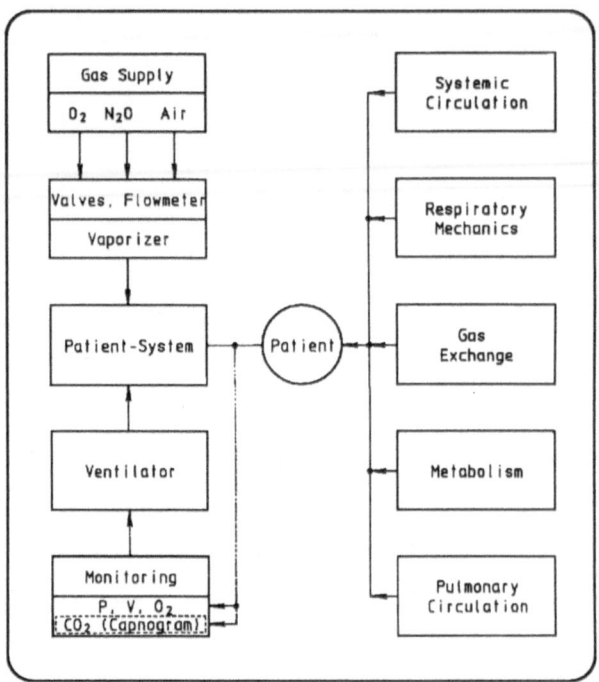

Figure 16.2.   The capnogram in the patient-anesthesia circuit complex

From Figure 16.2, we can see the most relevant uses of the capnogram:

-    to determine adequate ventilation by adjusting artificial ventilation and/or to assess the respective parameters of spontaneous breathing
-    to monitor the metabolic state; e.g. CO_2 associated with malignant hyperthermia- to monitor circulation; e.g. cardiac or hypovolemic shock, pulmonary embolism
-    to monitor the ventilatory circuit; e.g. patient disconnect, detection of kinked tube, etc.

16.2   Mechanical Ventilation and the Bain Circuit

When a patient is under anesthesia with the use of muscle relaxants, or under other circumstances where he is unable to breath spontaneously, mechanical ventilation is applied in order to make the organisms actual metabolic needs.

A mechanical ventilator pushes a gas mixture into the patient's lung and, usually, the patient exhales passively.

For this purpose the patient is connected to the ventilator by means of corrugated tubing. When a circle system is used, inspired gas from the ventilator and exhaled gas from the patient travel through separate tubes as the gas flow is conducted by means of one-way valves.

This way accidental inhalation of CO_2 is prevented. Under some conditions or operative procedures the presence of two tubes is a disadvantage. A co-axial system was subsequently developed in which the fresh, CO_2 free, gas is brought close to the mouth of the patient. However, this fresh gas flow is usually set, for economic reasons, at a value lower than the maximal inspiration flow rate. Thus, previous exhaled gas that is stored in and removed through the outer tubing flow momentarily back towards the patient causing some CO_2 to be reinhaled. Capnograms obtained while using such breathing systems, or Bain circuits, will show an inspired CO_2 percentage that is different from zero (see Phase I in Fig. 16.1).

To assist in the interpretation of such non-straightforward capnograms and the condition under which they exhibit a certain pattern, a computer model of the Bain circuit was devised that allows the free variation of all ventilation parameters and the introduction of a number of malfunctions.

The patient's endotracheal tube, the Bain circuit, and the connection of the Bain circuit with the ventilator can be regarded as a straight tube with different diameters for the individual components, as shown in Figure 16.3 [1].

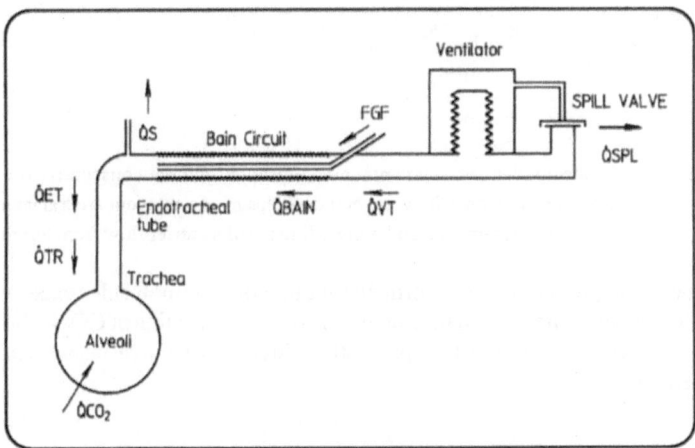

Figure 16.3.   Schematic representation of the Bain circuit (for details see text)

The alveoli compartment in Figure 16.3 represents the patient's lungs with the assumption of a uniform CO_2 distribution, followed by the trachea, the endotracheal tube, an elbow connector, the actual Bain circuit with the inner tube for fresh gas supply, the connecting tube to the ventilator, and the ventilator.

FGF indicates the fresh gas flow, QSPL, QVT, QBAIN, QS, QET and QTR are the flow rates in the spill valve, ventilator, Bain circuit, sampling tube for the sidestream monitoring, endotracheal tube and trachea, respectively. The arrows indicate positive directions of gas flow.

A mathematical formulation, which contains the relations between flow, pressure and volume in the endotracheal tube, alveolar space and ventilator, is a powerful aid for determining the CO_2 concentration, based on a description of the CO_2 transport itself.

The CO_2 transport is basically a two-dimensional problem: that is, geometrical distance along the tube as the first dimension and time dependency as the second. Therefore the mathematical formulation is based on partial differential equations. The driving functions (inputs, such as breathing patterns) are normally represented by non-analytical functions. Hence they do not have a derivative at each point, i.e. discontinuous functions. Therefore the aid of a computer is needed to solve the equations. Typically a model of this type of process is divided into a number of small, discrete units in both length and time. The passing gas exchange is time and geometry dependent; in addition the tube is segmented.

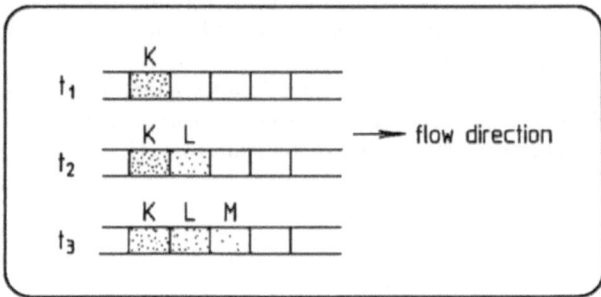

Figure 16.4.   A tube divided into segments and represented at three consecutive times -t1, t2 and t3- illustrates how several particles in one segment are transported to neighboring segments under the influence of a carrier flow from left to right.

The basic assumption is that a uniform distribution of CO_2 molecules exists within each tube segment. The method for calculating the concentration of CO_2 is shown in Figure 16.4, with t1, t2, and t3 representing three consecutive instants of time [Beneken,1985].

During t1 and t2, the volume flow of the carrier moves a certain fraction of the particles from segment K to segment L. At t2, the number of particles is reduced because of the inflow of the carrier without particles. Based on this new distribution and the known carrier volume flow, the distribution at t3 can be calculated. The calculation is repeated continually and yields not only the time course of the particle concentration in a particular segment (by looking at a segment -for example, K- as a function of time), but also the longitudinal distribution of particles along the tube at a particular instant [Beneken,1985].

When the direction of flow is reversed, particles will no longer enter segment L from K, but will enter from M instead. This change in calculating the number of particles in segment L results from the change in the flow direction of the carrier. Thus, there is a distinct cause-and-effect relationship; the carrier flow is the cause and the particle distribution is the result [Beneken,1985].

At branching points of the anesthesia circuit, where gas is added (e.g., fresh gas) or removed, the same general approach is used:

- the carrier flows are established in the different branches, according to the law of mass conservation, and
- for each subsequent instant of time, the new distribution of particles is calculated on the basis of both the previous distribution in all branches and the known carrier flows [Beneken,1985].

16.3 Mathematical Formulation of the Model

The general model principle is illustrated by a tube divided into segments. For the sake of simplicity, we assume that each segment has a volume equal to V, an assumption that does not detract from the general principle. The carrier gas volume flow through the tube from the left to the right side is equal to $\dot{Q}$. The calculation of CO_2 volume in the i-th segment at instant t is shown in equation (16.2):

$$V.F_i(t) = V.F_i(t-\Delta t) + \dot{Q}(t-\Delta t)\Delta t.F_{i-1}(t-\Delta t) - \dot{Q}(t-\Delta t)\Delta t.F_i\, t-\Delta t) \qquad (16.2)$$

where
t :   the moment in time, at which the distribution is calculated
Δt: the time interval, or difference in time, between two consecutive time instants
F_i: the fraction of CO_2 in the i-th segments
S :   the shift factor (a dimensionless factor introduced for mathematical simplicity), later defined as $\dot{Q} \cdot \Delta t/V$

The term to the left of the equal sign in equation (16.2) represents the volume of CO_2 in the i-th segment. The first term to the right of the equal sign is the CO_2 volume at time instant (t-Δt); the second term is the amount of CO_2 that entered segment i from segment i-1 during the time interval Δt, and the last term is the amount of CO_2 that leaves segment i during the interval Δt. All values to the right of the equal sign are taken at time instant $(t-\Delta t)$ and therefore lead to the new F_i value at time instant (t). Dividing by V and introducing the shift factor,

$$S(t-\Delta t) = \dot{Q}(t-\Delta t)\Delta t/V$$

gives equation (16.3)

$$F_i(t) = F_i(t-\Delta t) + S(t-\Delta t).[F_{i-1}(t-\Delta t) - F_i(t-\Delta t)] \qquad (16.3)$$

The shift factor, S, thus becomes dependent on the carrier flow rate, $\dot{Q}$, the time interval, Δt, and the segment volume, V. If the direction of the flow reverses, CO_2 is transported from the segment i+1 into segment i and from segment i into segment i-1. In that case, the CO_2 fraction in the i-th segment is calculated using equation (16.4):

$$F_i(t) = F_i(t-\Delta t) + S.[F_{i+1}(t-\Delta t) - F_i(t-\Delta t)] \qquad (16.4)$$

It should be noted that S, being dependent on the carrier flow rate, is different in equations (16.3) and (16.4), and that $F_i(t-\Delta t)$ in equation (16.3) is replaced by $F_{i+1}(t-\Delta t)$ in equation (16.4). Similar equations describe the CO_2 fractions in all segments considered. The simultaneous solution for instant, t, of all equations involved is then repeated to generate the time course of CO_2 distribution during one or more respiratory cycles. This calculation also yields quantitative information about the longitudinal distribution of CO_2 by displaying, at one instant, the different fractions F_i through F_{i+j}.
Specific data sets are necessary to solve the entire set of equations.

The patient data set comprises:
    Lung thorax compliance
    CO_2 inflow into the alveolar space (or CO_2 production)
    Respiratory quotient, accounting for inequalities of inspired versus expired volumes
    Functional residual capacity
    Dead space volume
    Airway resistance

The system data set comprises:
    Dimensions and volumes of ventilator, tube and connectors
    Spill valve pressure setting
    Resistances to flow in the different segments of the tubes
    Ventilator settings, such as I:E ratio, respiratory rate, and minute ventilation
    Fresh-gas flow rates

On the basis of these two data sets, the model structure can be defined and the number of segments for each tube section selected. A value for the time interval, Δt, should also be selected at this stage.

From the above definition of the shift factor, S, it will be clear that S may never exceed a value of 1. If it did, the product, $\dot{Q} \cdot \Delta t$, would become greater than the actual segment volume, V, and gas would be transported beyond the limit of the segment during the time interval, Δt, an occurrence that is not allowed in this type of step-by-step calculation of the CO_2 fractions. Therefore, a limit is set on the choices of segment size, V, and the time interval Δt.

Table 16.1 gives the combined volunteer and system data as they were used to initiate the model. The functional residual capacity of the volunteer (weight, 80kg; height, 178cm) who passively allowed himself to be ventilated, was measured in the pulmonary function laboratory. Tidal volume was measured using a pneumotachograph; the remaining values were estimated.

Table 16.1. Numerical Values for Volunteer and System Variables

Variable	Value
VOLUNTEER	
Lung-thorax compliance	133 ml/mm Hg
CO_2 production	210 ml/min
Respiratory quotient	1.0
Functional residual capacity	3900 ml
Dead space volume	105 ml
Tidal volume	660 ml
Airway resistance	0.0012 mm Hg/ml/min
SYSTEM	
Volume of Bain outer tube	510 ml
Volume of elbow connector	15 ml
Volume of connecting tube between Bain circuit and ventilator	530 ml
Maximum volume of ventilator	2300 ml
Resistance of Bain and connecting tube	0.003 mm Hg/ml/min
Spill valve pressure setting	2 mm Hg
Sampling flow to capnograph	205 ml/min
Quadratic pressure-flow relation through spill and exhaust valves	$Q = 12 \times \sqrt{P}$ (Q in l/min; P in mm Hg)
Fresh gas flow	10.2 and 12.2 l/min
VENTILATOR	
I:E time ratio	1:2
Respiratory rate	7.75 breaths/min
Minute volume	5 l/min

$\dot{Q}$ = carrier gas volume flow; P = pressure; I:E = ratio of inspiratory to expiratory time.

The ventilator settings and the fresh-gas flow rate were adjusted for comfortable ventilation of the volunteer. To demonstrate the model, we selected two fresh-gas flow rates, 10.2 and 12.2 l/min, for comparing the volunteer's capnograms with those generated by the model.

16.4   Limitations of the Model

The main purpose of the presented model is to demonstrate the basic functioning of the time plots of capnograms or CO_2 curves at different locations in the system based on mechanical ventilation. For that reason the model is kept as simple as possible, nevertheless, within its limitations, it is accurate and realistic. The mathematical description is based on data available from experiments or the literature. In the next paragraphs the limitations are summarized [Beneken,1985]:

- The patient's lungs are represented by one segment with a uniform distribution of CO_2 molecules; the airways by their volume and their resistance to flow.
- The ventilator volume is represented by one segment.
- The velocity profile is assumed to be flat because the corrugations in the tube are assumed to cause small eddies, which, in turn, result in good cross-sectional mixing.
- Axial diffusion is taken into account by adjusting the time intervals and the segment-size.
- Within each segment, complete mixing is assumed.
- Gas is assumed to be incompressible over the encountered pressure range.
- Compliance is neglected over the same pressure range, which creates a problem to represent small children, whose lung-thorax compliance is approximately that of the system compliance.
- Temperature is assumed to be constant.
- Uptake and elimination of anesthetic gases are not considered, since these are transient phenomena.

16.5   Operation of the Program

The operation of the models is mainly self-explanatory. We only make the following remarks in order to assist you in understanding the operation.

The case study is chosen by selecting the patient and/or system parameters in the pull-down structure. For each item, an instruction line and a helpscreen are available. The first simulation experiment deals with the time transient of computer-generated capnograms obtained from distal endotracheal segments during the first four breaths after onset of the computation using the data given in Table 16.1. At the onset of the first simulated breath, the endotracheal tube and the alveoli are assumed onset of the first simulated breath, the endotracheal tube and the alveoli are assumed to be filled with gas containing CO_2 at a partial pressure of 40mm Hg. The mouthpiece in which the sampling took place was assumed to be free of CO_2. Fresh-gas flow rates may be varied under test. Time progression is from left to right.

By changing the model parameters (you can change more than one parameter at a time), you will become acquainted with different situations in the Bain system. The program recalculates the new steady states and the new values are shown. You may also change the graphic time scale and/or the number of breaths.

16.6   Parameters Used

CLTH	constant lung thorax compliance
DT	the time interval, or difference in time, between two consecutive time instants
FAL	CO_2 fraction in the alveoli
FCT	CO_2 fraction in the last segment of the circuit closest to the ventilator
FGF	fresh gas flow
FRC	functional residual capacity
FVT	CO_2 fraction in the ventilator
PAL	pressure in the alveoli
PVENT	pressure in the ventilator
QBAIN	flow rate in the Bain circuit
QCO_2	flow rate (production) of CO_2
QET	flow rate in the endotracheal tube
QS	sampling flow rate
QSPL	flow rate through the spill valve
QTR	flow rate in the trachea
QVT	flow rate from or to the ventilator
RESIST	total resistance to flow of Bain circuit arrangement
RET	resistance in the endotracheal tube
RQ	respiratory quotient
T	the moment in time, or time instant at which the distribution is calculated
VAL	volume of the alveoli
VVT	volume of the ventilator

16.7   Conclusions

Although the model is very basic, it describes the real situation in a very instructive way. Moreover it has already proven to be accurate and realistic in explaining clinical situations that were previously misunderstood [Beneken,1987]. It can be used in many educational, research, or even clinical situations to study the basic mechanisms of the Bain circuit.

References

Beneken, J.E.W., N. Gravenstein, J.S. Gravenstein, J.J. v.d. Aa, and S. Lampotang: "Capnography and the bain circuit I: A computer model". *J. Clin. Mon.* Vol. 1, 1985, pp. 103-113.

Beneken, J.E.W., N. Gravenstein, S. Lampotang, J.J. v.d. Aa, and J.S. Gravenstein: "Capnography and the bain circuit II: Validation of a computer model". *J. Clin. Mon* Vol. 3, 1987, pp. 165-177.

17
Renal Function and Blood Pressure Stabilization

Dietmar P.F. Möller

17.1 Introduction

The model of the renal function for blood pressure stabilization as presented in this chapter is an example of long term circulatory regulation. Its structure explicitly contains the relevant biological parameters in a morphologically oriented manner for comparing simulation results with these obtained from measurements of the real system.

The model is implemented as a student controlled simulation program. The student may change parameters and add hypotheses into the program based on current research results. The computer program supplies feedback by calculating and presenting the resulting situation.

17.2 Theory

The importance of the renal body fluid compartment feedback mechanism as a long-term determinant of arterial blood pressure stabilization has been well known since Guyton's famous work [Guyton,1973]. Fig. 17.1 shows a simplified block diagram, depicting the relevant inter-relationships between venous return or cardiac output (VR or HZV), total peripheral resistance (RA), arterial blood pressure (PAS), extracellular fluid volume (VECF) and blood volume (VB).

The model shown in Fig. 17.1 is based on the following highly simplified concept: "The heart pumps whatever amount of blood flows into the right atrium (Atrium Dextrum). Hence it is assumed that the heart is capable of pumping either no blood or an infinite amount of blood depending on how much blood flows from the veins into the heart. It also assumes that an increased inflow of blood into the heart does not increase the right atrial pressure (PRA). These principles are exactly similar to those stated by the Frank-Starling law of the heart, within physiological

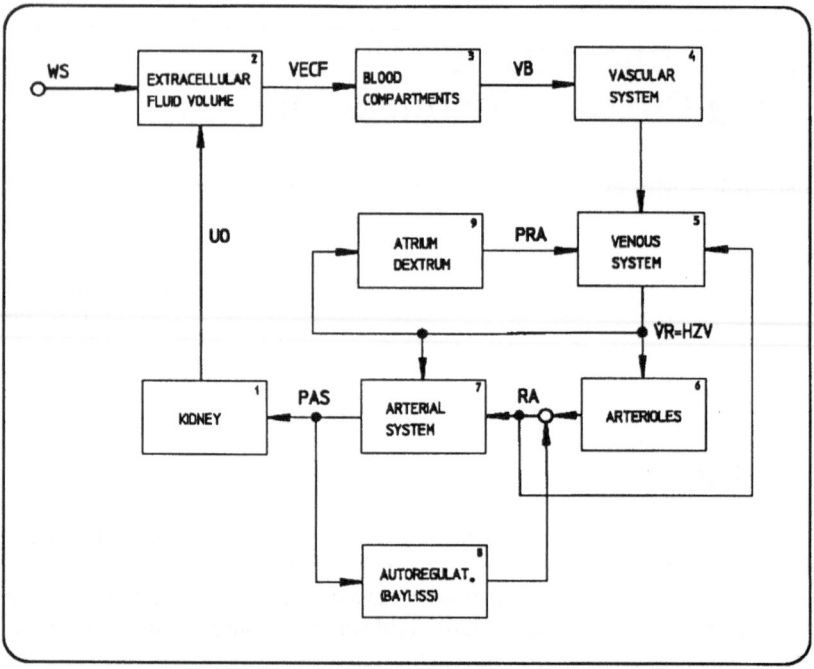

Figure 17.1. Block diagram of the renovascular system for long-term blood pressure stabilization

boundaries: "the heart pumps whatever amount of blood enters it, and does so without a significant rise in the right atrial pressure" [Guyton,1980].

Blood pressure stabilization is based on the regulation of blood volume via the pressure controlled urinary output. The urinary output (UO) increases with an elevated arterial blood pressure (PAS) [Guyton,1973], [Möller,1984]. This causes a decrease in the body fluid volume (VB), and a decrease in the mean systemic filling pressures (PMS), also the difference between the right atrial pressure (PRA) and the mean systemic filling pressure (PMS). With regard to the resistance of the venous return (RVR), the venous return (VR) decreases as does the cardiac output (CO, HZV). Taking into account a constant peripheral resistance (RA), the decrease of the arterial blood pressure (PAS) is followed by an elevated renal fluid retention. In this way the loss of blood volume is compensated by an increase in plasma volume (VP), and therefore an elevated extracellular fluid volume (VECF). This renal feedback mechanism consolidates the blood volume (VB), the right atrial pressure (PRA), the venous return (VR), the cardiac output (CO, HZV), the arterial blood pressure (PAS) and the urinary output (UO).

17.3 Aspects to Be Considered

17.3.1 Limitations of the Model

The main purpose of the model is to demonstrate the basic functioning of long-term blood pressure stabilization. It is therefore kept as simple as possible (compared to the Guyton-Coleman-Model) but is accurate and realistic within its limitations. These limitations are summarized below.

- The way the system comes to the steady state based on a renin-angiotensin-aldosterone stimulus has not yet been modeled. Moreover, extracellular potassium and protein as well as the antidiuretic hormone have been neglected.
- The model is not suited to study the effects of physical workload. Also temperature changes, eating and drug effects are not taken into account.
- Simplifications have been made, whereby autonomic nervous activity, pulmonary circulation, blood viscosity, plasma colloid/oncotic pressure and the glomerular filtration rate are neglected.

17.3.2 Outlook

Hemodynamic behavior in different kinds of hypertension can be studied using model-reference techniques to estimate non-measurable parameters (non-invasive or invasive), and this is of clinical importance in hypertension disease therapy. Also hormonal parameters will be added in future research work to test the different hypotheses explaining hypertension.

17.4 Model Description

17.4.1 Mathematical Model

To analyze circulatory function and cardiac output regulation, the block diagrammed model in Fig. 17.1 needs a representation that can be described in mathematical terms. Fig. 17.2 shows the nonlinear mathematical model of the long-term behavior of arterial blood pressure stabilization via the volume regulation mechanism [Möller,1984].

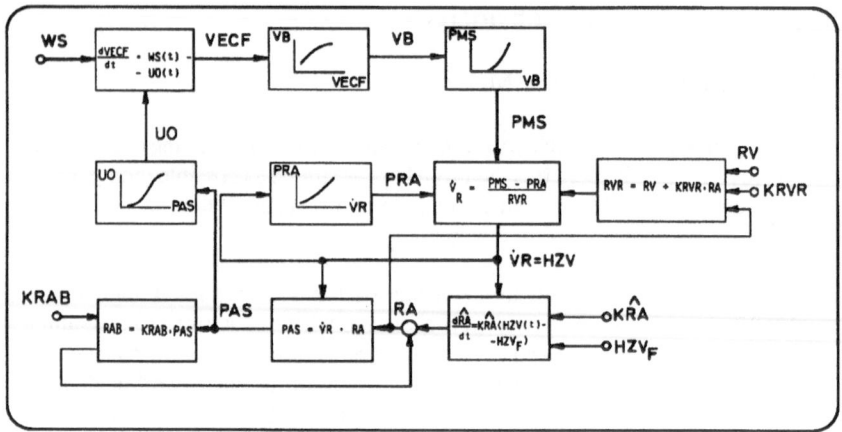

Figure 17.2. Nonlinear mathematical model of the renovascular system as shown in Fig. 17.1

17.4.2   Individual Blocks of the Model

Block 1 shows the effect of the arterial blood pressure (PAS) on urinary output (UO) by the kidney (compare Fig. 17.1 and 17.2) [Guyton,1973]. The output is expressed as output of extracellular fluid, which in terms of Guyton's notation [Guyton,1980] means output of both water and extracellular electrolytes. If the arterial blood pressure increases, there is an increased loss of extracellular fluid through the kidney and vice versa.

Block 2 subtracts the urine outflow (UO) from the oral intake of water and sodium (WS) minus losses of both of these substances through other routes besides the kidneys. The output is the actual change of extracellular fluid volume dVECF/dt.

Block 3 shows the relationship between extracellular fluid volume (VECF) and blood volume (VB) in a normal person.
It is known from Guyton's work that when VECF rises above a critical level of approximately 22 liters, VB (which by the same time has risen to about 7 liters) does not increase significantly. This is because the interstitial fluid pressure has risen from its normal sub-atmospheric value to a supra-atmospheric level at which edema fluid collects very rapidly in the interstitial spaces instead of remaining in the circulation [Guyton,1980].

Block 4 shows the relationship between blood volume (VB) and the mean systemic pressure (PMS).

Block 5 calculates the venous return (V̇R) from the mean systemic pressure (PMS), right atrial pressure (PRA) and resistance of venous return (RVR) relationship. Note the output of this block is equal to cardiac output (HZV) because the output of the heart must be equal the input to the heart over any significant period of time.

Block 6 calculates the effect of changes in tissue metabolism on the total peripheral resistance (dRA/dt).

Block 7 multiplies cardiac output times total peripheral resistance to give arterial blood pressure, which is the output quantity of this block.

Block 8 represents the myogenic influence on the total peripheral resistance (RAB), the so-called Bayliss effect.

Block 9 shows the relationship between venous return (VR) and the right atrial pressure (PRA).

17.4.3   Model Equations

Blood pressure control with the aid of the renovascular volume dependent mechanism is achieved by adjusting the urinary output. The block diagram form – shown in Fig. 17.2 – can be described in the state-space form by the nonlinear vector differential equation

$$\dot{\underline{X}} = \underline{f}\,(\underline{X},\underline{U},\, Z, \underline{Q}_s)$$

where $\underline{X}$ is the state vector

$$\underline{x} = [\text{VECF, RA}]^T$$

the components of which are the extracellular fluid volume (VECF) and the peripheral resistance (RA), and $\underline{U}$ is the control vector, the isotonic water and sodium intake. $\underline{Q}_s$ is the system parameter vector, and Z shows the disturbance influence.

From this we can conclude

$$\dot{X}_1 = U - f\,(\text{HZV } X_2)$$

$$\dot{X}_2 = \text{HZV} - \text{KHZV}_F$$

HZV represents the cardiac output and KHZV_F shows the cardiac output in the case of a fixed hypertension state. F is the function of the system equation. The complete nonlinear mathematical model of the renovascular system is given in Fig. 17.3, [Selkurit,1961], whereas Fig. 17.4 shows the linearized form.

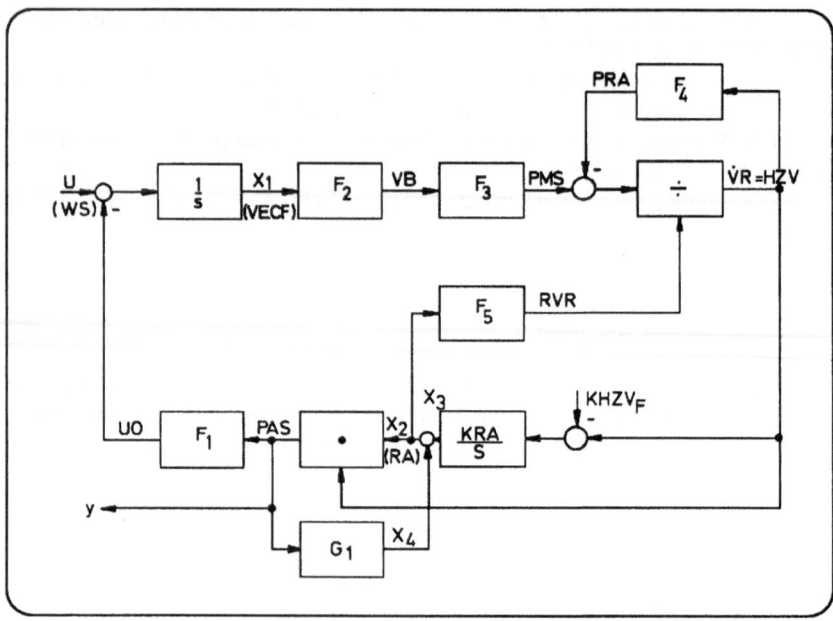

Figure 17.3. Nonlinear mathematical model of the renovascular system for long-term blood pressure stabilization

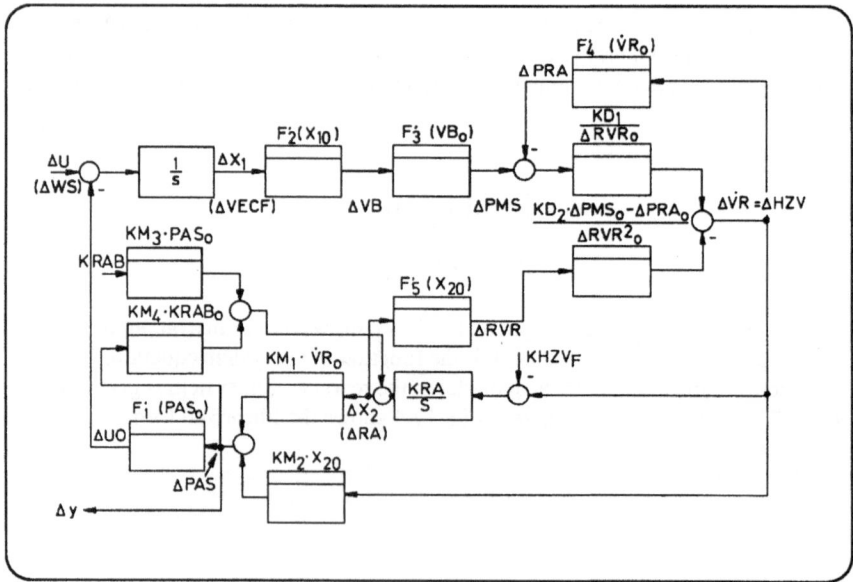

Figure 17.4. Linearized version of the nonlinear mathematical model of the renovascular system as shown in Fig. 17.3

Depending on the assumptions outlined in chapter 17.4.2, the equations of the nonlinearities are as follows:

F_1: $UO = a_{12} PAS^2 + a_{11} PAS + a10$

F_2: $VB = a_{22} VECF^2 + a_{21} VECF + a20$

F_3: $PMS = a_{32} VB^2 + a_{31} VB + a_{30}$

F_4: $PRA = a_{42} VR^2 + a_{41} VR + a_{40}$

F_5: $RVR = RV + KRVR\ RA$

$VECF = WS(t) - UO(t)$

$\overset{\wedge}{RA} = KRA.(HZV(t) - KHZV_F)$

The linearization of Fig. 17.4 is based on the linearization theory of nonlinear systems. In table 17.1 the respective correspondences are given:

Table 17.1. Correspondences between nonlinearities (left) and their linearized expressions (right)

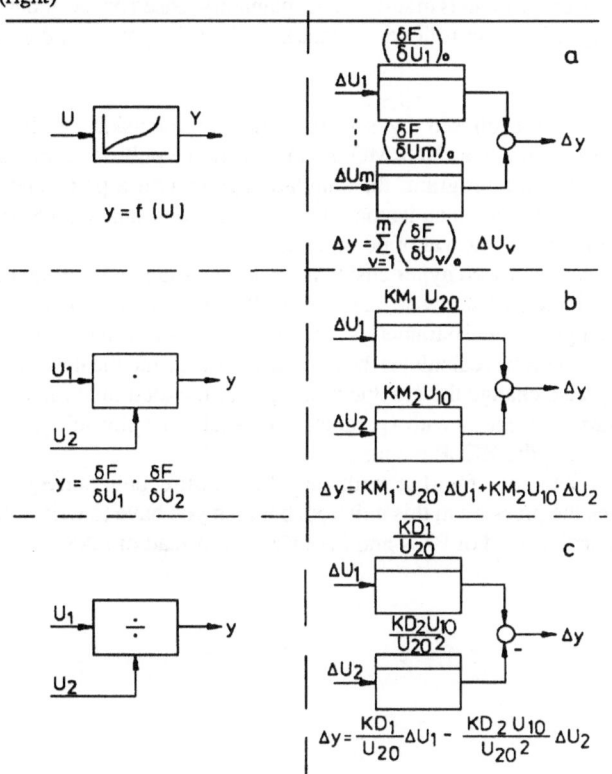

17.4.4 Parameters and Units

HZV: cardiac output in ml/min
HZVF: cardiac output in the fixed state of hypertension in ml/min
KRA: pO_2-dependent factor of the peripheral resistance in mmHg.min/ml2
KRVR: factor that considers the dynamic fluid interactions between the venous and the
         arterial compartment without dimension
PAS: mean arterial blood pressure in kPa
PMS: mean systemic filling pressure in kPa
PRA: right atrial pressure in kPa
RA: peripheral resistance in (mmHg/ml)s
RV: resistance of venous return in (mmHg/ml)s
UO: urinary output in ml/min
VB: blood volume in l
VECF: extracellular fluid volume in l
V̇R: venous return in ml/min
WS: isotonic water and sodium intake in ml/min

17.4.5 Operation of the Program

The operation of the models mainly self-explanatory, based on the general structure shown in Fig. 17.2. The following remarks will assist you in understanding the operation.

- Case studies are chosen by selecting the normotension or the Goldblatt-hypertension, shown as NORMO or HYPER in the pull-down menu structure. When starting the program, the simulation runs with a parameter set within normal range, based on an isotonic water and sodium intake (WS) of 1 ml/min and a print or plot interval of 60 minutes.
- As a first case study on hypertension the isotonic water and sodium intake (WS) should be increased stepwise, and the results obtained should be compared.
- When changing model parameters (you may change more than one parameter at a time) the program recalculates the new steady states and the new values appear. You may also change the graphic mode and/or the total time range.
- If you start the program with a parameter set outside the physiological range, you are referred to the HELP-menu.
- You may be more familiar with the old pressure unit mmHg. The model calculates the pressure in this unit, too, but then you have to write PXXMMHG for instance, instead of PXX and PASMMHG instead of PAS.

17.5   Experiments

It is generally not possible to get a closed analytical solution of the set of nonlinear differential equations describing the nonlinear mathematical model of the renovascular system as shown in Fig. 17.3.

A possible approach to a solution, and thus to understanding the dynamic properties of the mathematical model, is to use the digital simulation technique. Implementation of the model in BIOPSI is based on the parameter set, listed in the appendix for normotension and the Goldblatt hypertension form [Selkurit,1961].

17.5.1   Exercises

To help you understand the model, we will take a step-by-step approach to study the influence of increasing isotonic water and salt uptake on the most important hemodynamic parameters: mean arterial pressure (PAS), peripheral resistance (RA), cardiac output/venous return (HZV, V̇R), mean systemic pressure (PMS) and blood volume (VB) at normotension and hypertension. Therefore you will be changing the WS parameter in this exercise.

In Fig. 17.5 the stationary values of the relevant hemodynamic system variables dependent on an isotonic intake of water and salt (WS), e.g. mean arterial blood pressure (PAS), peripheral resistance (RA), venous return (HZV = V̇R), mean systemic pressure (PMS) and blood volume (VB) are presented for the orthological state (pressures in mmHg).

From Fig. 17.5 we can conclude the following: Due to the increased extracellular fluid volume (VECF) and isotonic water and sodium load, there is an increase in the blood volume (VB) which itself causes the rise of the mean systemic pressure (PMS).

As a further consequence, the difference between the right atrial pressure (PRA) and the mean systemic pressure (PMS) increases so that – taking the resistance of venous return (RVR) into account – an increased venous return can be observed, as well as a slight increase in cardiac output (HZV). As a consequence of the so-called autoregulation mechanism, the peripheral resistance (RA) increases [Selkurit,1961]. The autoregulation dependency is based on the myogenic and metabolic pathways, the result of which is an elevated mean arterial blood pressure (PAS). This is in accord with the behavior of the real biological system as shown in [Selkurit,1961].

Fig. 17.6 shows the simulation results of the pathological state of the so-called Goldblatt-hypertension (pressures in mmHg).

The Goldblatt hypertension occurs in animal experiments under a renal artery clip. In humans a renal artery stenose causes high blood pressure, called renovascular hypertension.

As a consequence the renal blood flow decreases and renin secretion increases. The conclusion is that levels of angiotensin II and aldosterone are elevated. A further

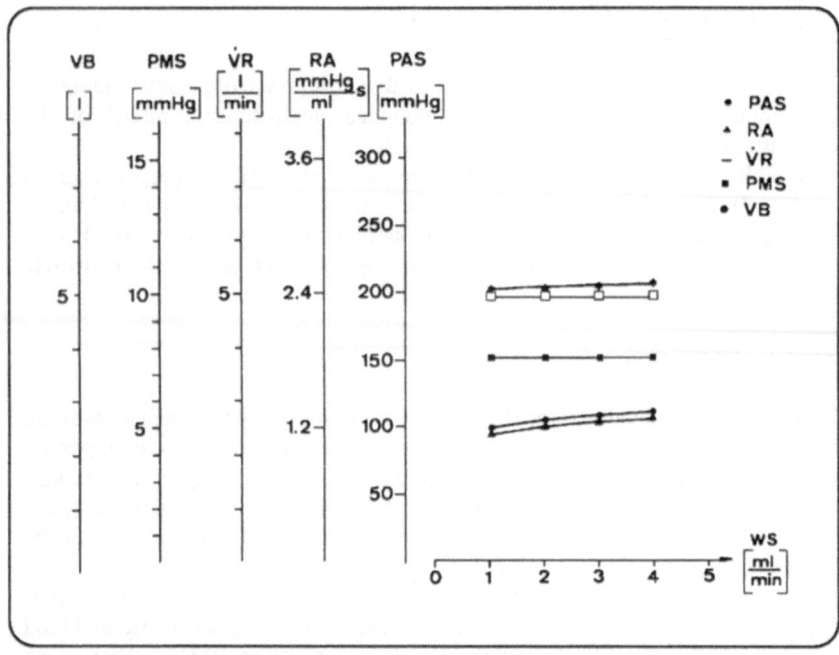

Figure 17.5. Simulation results of a stepwise increase of isotonic water and salt intake, showing the stationary values of mean arterial pressure (PAS), peripheral resistance (RA), venous return (V̇R = HZV), mean systemic pressure (PMS) and blood volume (VB) at normotension.

result of long-term renal water and sodium retention is the storage of sodium in the wall of the vessels, which causes the peripheral resistance (RA) and the mean arterial blood pressure (PAS) to increase.

Under an elevated isotonic water and sodium (WS) intake, the blood volume (VB) and the plasma volume (PV) increase. With regard to the Starling mechanism, cardiac output (CO, HZV) and venous return (VR) increase by a rise in the peripheral resistance (RA) as a result of the so-called autoregulation mechanism. There is an elevated mean arterial blood pressure (PAS), in accordance with the behavior of the real biological system [Guyton,1973].

The previous discussion and the results shown in Fiures 17.5 and 17.6 indicate the correctness of the system model and the value of the simulation results obtained from it. Built in this way, the model can be used to predict values of some system quantities as shown bythe following example [Selkurit,1961].

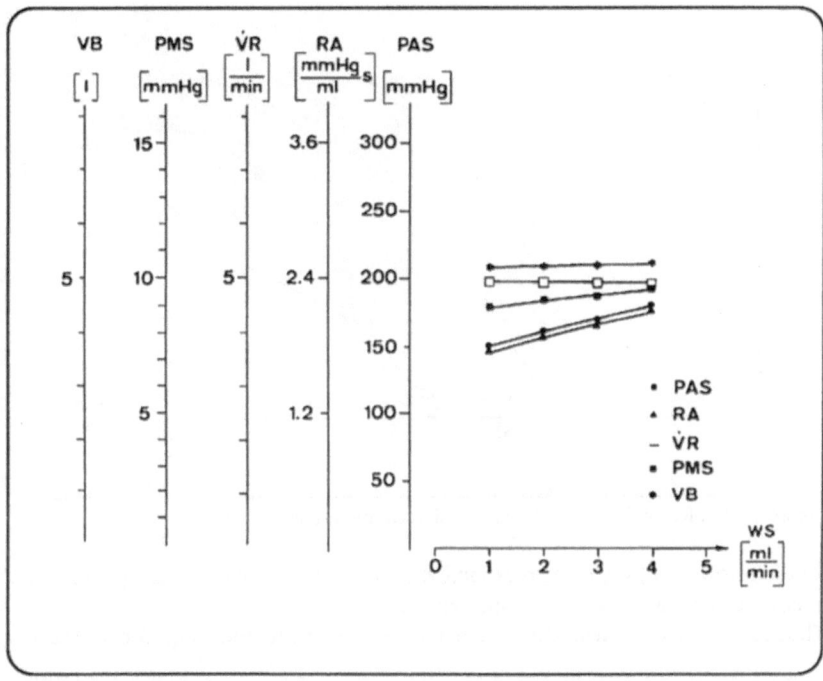

Figure 17.6. Simulation results of a stepwise increase of isotonic water and sodium intake, showing the stationary values of mean arterial blood pressure (PAS), peripheral resistance (RA), venous return (V̇R), mean systemic filling pressure (PMS), and blood volume (VB) at Goldblatt included hypertension.

17.5.2 What You Can Do on Your Own

Now that you are acquainted with the different equations that make up the model, we can start more research oriented exercises to give you more insight into the functioning of the renal system.

In Fig. 17.7 the renal function curves for the normal state (normotension) and two renal hypertension cases – the Goldblatt-Clipp ▲ and the reduction of renal mass ■ – as well as the essential hypertension ✳ are shown.

The renal function curve is a graphical curve that depicts urinary output at different arterial blood pressure levels. This curve was determined by raising the arterial blood pressure to an isolated kidney while measuring the urinary output at successive levels of blood pressure [Guyton,1973]. Fig. 17.7 illustrates average renal function curves of such experiments for different cases.

The human body and the animal body have developed extensive mechanisms for controlling the relationship between blood pressure, renal excretion of salt and water, and the control of the renal function curve itself.

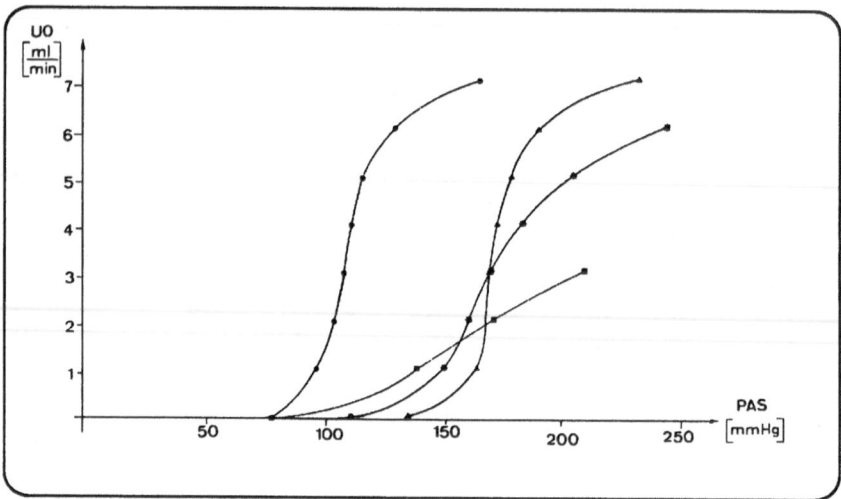

Figure 17.7.   Renal function curves for different hypertension cases.

It is therefore worthwhile to investigate these case studies of hypertension based on a reduced renal mass and essential hypertension.

The first step deals with the derivation of the respective relationships for both cases:

$$U0 = a_{12} \, PAS^2 + a_{11} \, PAS + a_{10}$$

based on the renal function curves in Fig. 17.7.

The simulation results obtained by your own study depend on the stepwise increase of isotonic water and salt intake WS.

17.6   Conclusions

Although the model is very basic and describes only the steady state, it is very instructive. It has already proven to be accurate and realistic in explaining clinical situations that were previously misunderstood.

It can be used in many educational, research, or even clinical situations to study the basic functioning of the renal mechanism in steady state.

References

Guyton, A.C., C.E. Jones and T.G. Coleman: *Circulatory Physiology: Cardiac Output and its Regulation.* W.B. Saunders Comp., Philadelphia, 1973.

Guyton, A.C.: *Arterial Pressure and Hypertension.* W.B. Saunders Comp., Philadelphia, 1980.

Möller, D.: "Modeling and Simulation of the Longterm Behaviour of Arterial Pressure Regulation with the Aid of a Block Diagrammed Interactive Simulation System". In: *Informatik Fachberichte*, Ed. W. Ameling, Vol. 71, Springer Verlag, Heidelberg, 1983 pp. 486-491.

Möller, D.: "Mathematical Modeling of the Renoprival Hypertension". *Funkt. Biol. Med.* 3, 1984, pp. 253-259.

Selkurit, E.E.: "Effect of pulse pressure and mean arterial pressure modification on renal hemodynamics and electrolyte and water excretion". *Circulation* 4, 1961, pp. 541-551.

Appendix

Parameters with default values and units for normotension and Goldblatt hypertension. Note that the isotonic water and salt uptake may vary within the range 0 to 10 ml/min.

Parameter	Normotension		Goldblatt Hypertension	Unit
a_{12}	3, 749	10^{-3}	$7,956\ 10^{-4}$	ml/(min mmHg2)
a_{11}	0, 599		-0,156	ml/(min mmHg)
a_{10}	23, 9588		6,579	ml/min
a_{22}	-0, 01			1-1
a_{21}	0, 6			without dimension
a_{20}	-1, 75			1
a_{32}	2, 38			mmHg/l2
a_{31}	-15, 47			mmHg/l
a_{30}	24, 857			mmHg
a_{42}	0, 04			mmHg/(l/min)2
a_{41}	0, 74			mmHg/(1/min)
a_{40}	-3, 5			mmHg
RV	0, 8798	10^{-3}		(mmHg/ml)s
KRVR	0, 026			without dimension
KRA	0, 5	10^{-8}		mmHg min/ml2
KHZV$_F$	5	10^{-8}		ml/min

Help-Instructions for Normo and Hyper

HELP-NORMO

IF a_{10} is greater than 25, PAS and VECF will be unstable.

IF a_{11} is greater than 0.9, PAS and HZV will be unstable.

IF a_{11} is less than 0.599, PAS and VECF as well as HZV drop to unphysiological values.

IF a_{12} is greater than 3.7 E-3, PAS shows unphysiological values.

IF a_{20} is greater than 4.5, PAS, HZV and VECF drop to unphysiological values.

IF a_{21} is greater than 0.65, PAS shows unphysiological values.

IF a_{21} is less than 0.4, PAS raises to unphysiological levels.

IF a_{22} is greater than 1.E-2 or less 5.E-3, PAS and HZV show unphysiological values.

IF a_{30} is greater than 30 or less 18, PAS and HZV rise to unphysiological levels.

IF a_{31} is greater than 17 or less 14, PAS and HZV show unrealistic values.

IF a_{32} is greater than 2.9 or less 1.6, PAS and HZV rise to unphysiological levels.

IF a_{40} is greater than 7, PAS and HZV show unrealistic values.

IF a_{41} is greater than 6.E-3, PAS, HZV and VECF are in an unphysiological range.

IF WS is greater than 50, PAS oscillates.

HELP-HYPER

IF a_{10} is greater than 9, PAS, HZV and PMS oscillate.

IF a_{11} is greater than -0.16 or less -0.13, PAS, HZV and PMS show unrealistic values.

IF a_{12} is greater than 1.E-3, PAS, HZV and PMS drop.

IF a_{20} is greater than 5, PAS increases too much.

IF a_{21} is greater than 0.7 or less 0.4, PAS, HZV and PMS show unrealistic values.

IF a_{22} is greater than 1.E-2 or less 5.E-3, PAS, HZV and PMS atlain unphysiological values.

IF a_{30} is greater than 30 or less 20, PAS, HZV and PMS show unrealistic values.

IF a_{31} is greater than 16,5 or less 13, PAS and PMS oscillate.

IF a_{32} is greater than 2.9 or less 2.2, PAS raise at unphysiological levels.

IF a_{40} is greater than 10, hemodynamic behaviour is outside physiological levels.

IF a_{41} is greater than 4.E-3, PMS is at unrealistic levels and PAS oscillates.

IF WS is greater than 50, PAS decreases dramatically.

18
Urodynamics of the Lower Urinary Tract

Willem A. van Duyl

18.1 Aim of Instruction

Urodynamics concerns the morphological, physiological, biomechanical and hydrodynamical aspects of storage and transport of urine. Since about 1950, when adequate instrumentation became available for the measurement of bladder pressure and urinary flow, clinical urodynamics has become an important subdiscipline in urology. In order to interpret the results of those measurements, basic urodynamic research became necessary. The field of urodynamics is subdivided into the urodynamics of the upper urinary tract (kidneys and ureters) and the urodynamics of the lower urinary tract (bladder and urethra). This chapter considers the urodynamics of the lower urinary tract.

The function of the lower urinary tract alternates between collection and expulsion of urine. These functions appeal to different features of the system. Clinically, the urodynamics of the lower urinary tract is mainly concerned with the unintended loss of urine (incontinence) and the retention of urine (deteriorated micturition and obstruction). A simplified model will be presented of the biomechanics of the lower urinary tract in the storage phase and in the expulsion phase. The model is based on basic research done in vitro and in vivo. Some clinical procedures are demonstrated with this model, and it may also be of help to clinicians to test a hypothesis for their clinical findings. It must be noted, however, that clinical experience is usually much more complicated than can be accounted for in the presented model, which was designed for educational purposes.

18.2 Anatomy and Function

The urinary tract is a system of tubes, valves and reservoirs that guides and propels urine from the kidneys to the outside world. Figure 18.1 is a diagram of the system.

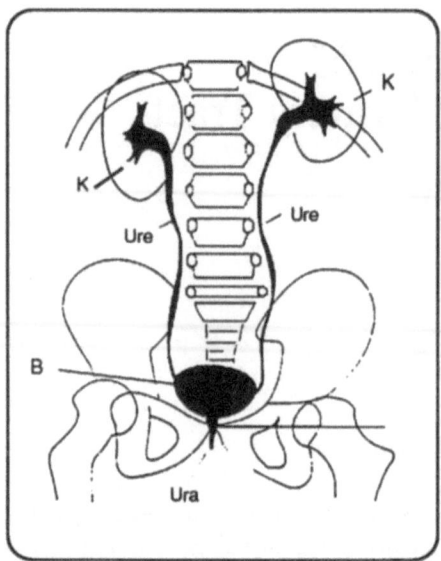

Figure 18.1. The urinary tract of a girl as seen from the front on X-ray: Ure-ureters; Ura-
urethra; S-symphysis pubis. After Griffiths 1980

Urine is formed continuously in the kidneys . Each kidney is connected to a ureter, a
long muscular tube through which the urine is propelled into the bladder by a
peristaltic wave. The urine gradually accumulates in the bladder.The urethra, the
tube that leads from the bladder to the outside world, is kept shut by muscular
contraction (sphincters). The degree of fullness of the bladder is sensed by the
nervous system. When a considerable volume has accumulated, up to 350-500 ml,
and at a socially convenient moment, micturition occurs. That is, the muscular wall
of the bladder is voluntarily made to contract so that urine is expelled through the
urethra. Simultaneously the urethra relaxes to allow the urine to pass. The rate of
outflow rises typically to about 25 ml/s. Backflow (so-called reflux) is normally
prevented by an anatomic structure acting as a valve at the junction of each ureter
with the bladder (trigones). When the bladder has been emptied, its wall relaxes and
it returns to its storage function. The urethra closes again.

The kidneys, the ureters and the bladder all lie within the abdomen. The urethra
passes through the urogenital diaphragm and so lies partly inside and outside the
abdomen. The base of the bladder is just above the urogenital diaphragm on the
pelvic floor. As a consequence of this anatomic structure, the pressure in the bladder
and the urethra consist of two components, one originating from the pressure of the
surrounding tissue (transmission pressure) and the other originating from stress in
the walls of these organs. The bladder and urethra are muscular organs coordinated
by the nervous system. The bladder wall consists of smooth muscle; the urethral wall
and its surrounding muscle tissue are both smooth and striated. Stress in the walls of

these organs is due to contraction activity of muscle tissue and/or to passive properties of the tissue. More accounts of the anatomy of the lower urinary tract are given by Gosling [1979]. Some physical features of the lower urinary tract that have been implemented in the model are described in more detail by Griffiths [1980].

18.2.1 The Bladder as a Spherical Organ

The bladder is a hollow organ that has an almost spherical shape when filled with urine. In the storage phase its contents increase from almost zero to typically 500 ml, or even up to 1000 ml. When the bladder capacity is reached, the urge to micturition is felt. At the end of a normal micturition the bladder is empty. In abnormal situations the bladder cannot be emptied completely and a so-called residual volume (Vres) of urine is retained. The bladder can be filled up to a certain amount without straining the bladder wall. This volume is called the rest volume (V0). In the clinical procedure called cystometry, the bladder is filled slowly (e.g. 1 ml/s) with saline via a catheter from outside. The registration of the relation between infused volume and measured bladder pressure, the cystometrogram, has the typical course as schematized in Figure 18.2.

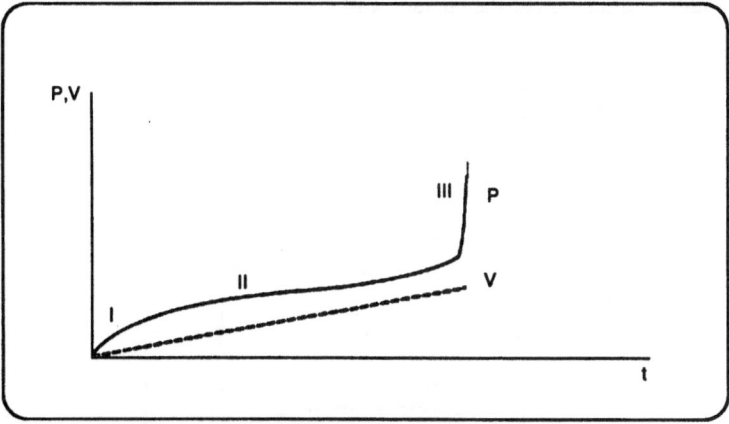

Figure 18.2. Schematized cystometrogram:
　　　　Segment I -　initial pressure rise from zero to the first inflection.
　　　　Segment II -　initial limb, begins at the first inflection and ends at micturition contraction or, in the absence of micturition, continues into segment III.
　　　　Segment III- ascending limb during micturition or, in the absence of micturition, during terminal pressure rise.

In the very first part of the cystometrogram, the pressure increases with the infused volume only as a consequence of hydrostatic pressure and of the transmission of abdominal pressure. When the rest volume is passed, pressure will increase mainly as a consequence of stress building up in the bladder wall. Usually it is impossible to observe this transition in the cystometrogram and hence to derive a reliable figure of the rest volume. The steep increase of pressure in the beginning is followed by a course with a downward curvature. Within the main volume range the cystometrogram is almost linear with a relatively small increase in bladder pressure. When bladder capacity is reached, the increase in pressure is steep again.

   Our first step in modeling concerns the relation between bladder pressure and stress in the bladder wall. We note that this step is necessary to simulate the bladder function both in the storage phase and in the expulsion phase.

In Figure 18.3 the spherical bladder is represented as two hemispheres that are in equilibrium. The equilibrium demands

$$F = \pi R^2 p \tag{18.1}$$

where F   - the force that keeps both hemispheres together
      P   - pressure in the bladder as it is built up only by the stress in the bladder
             wall (detrusor pressure)
      R   - the (mean-)radius of the bladder as defined in Figure 18.3
      2ΔR - thickness of the bladder wall

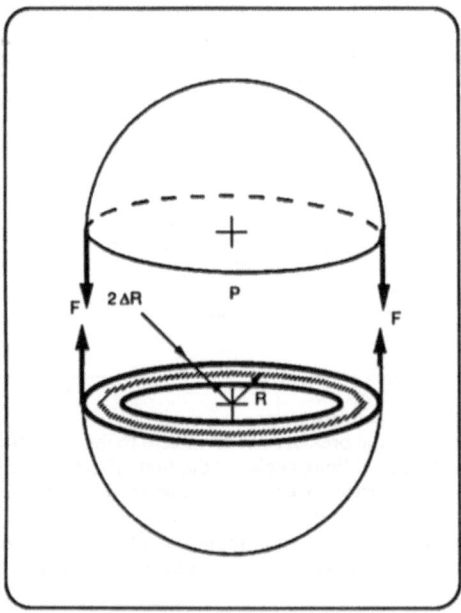

Figure 18.3. Illustration of the equilibrium of the bladder as a spherical organ

The circumference ℓ of the bladder is defined as follows:

$$\ell = 2\pi R \qquad (18.2)$$

From (18.1) and (18.2) it follows that

$$F = \frac{\ell^2}{4\pi}\, p \qquad (18.3)$$

Force F is related to the (mean-)stress (σ) in the wall according to

$$\sigma = \frac{F}{S} \qquad (18.4)$$

where S is the cross-sectional area of the wall.
From (18.3) and (18.4) we derive that

$$\sigma_S = \frac{1}{4\pi}\, \ell^2 p \qquad (18.5)$$

For the tissue volume of the bladder wall (V_t) we derive

$$V_t = \frac{4}{3}\left[(R + \Delta R)^3 - (R - \Delta R)^3 \right] \cong 8\pi\, R^2\, \Delta R \qquad (18.6)$$

For the cross-sectional area of the wall (S) we write

$$S = \pi \left[(R + \Delta R)^2 - (R - \Delta R)^2 \right] \cong 4\pi\, R\, \Delta R \qquad (18.7)$$

Combination of (18.6) and (18.7) yields

$$S = \frac{V_t}{2R} = \pi\, \frac{V_t}{\ell} \qquad (18.8)$$

By combining (18.8) with (18.5) we find

$$p = 4\pi^2\, V_t\, \frac{\sigma}{\ell^3} \qquad (18.9)$$

Tissue volume of the bladder is taken to be a constant for a particular bladder, typically $V_t = 20$ ml.
Circumference (ℓ) is related to bladder volume (V) as follows

$$V = \frac{4}{3}\pi R^3 = \frac{1}{6\pi^2}\, \ell^2 \qquad (18.10)$$

Substitution of (18.10) in (18.9) yields the useful relation

$$p = 4\pi^2 \, V_t \frac{1}{6\pi^2 \, V} \, \sigma = \frac{2}{3} \, V_t \frac{\sigma}{V} \qquad (18.11)$$

This equation expresses the important fact that a certain stress in the bladder wall is in equilibrium with a value of the bladder pressure that is inversely proportional to bladder volume. This is a consequence of the decreasing thickness of the wall with increasing bladder volume. From this we conclude that the increase of bladder pressure with increasing volume implies an increase in stress that is more than necessary to compensate for decreasing wall thickness.

18.2.2 Passive Stress-Strain Properties

For a strip of elastic material that is strained in one direction, we may write

$$\sigma = \varepsilon \, E \qquad (18.12)$$

where $\varepsilon = \dfrac{\ell \cdot \ell_0}{\ell_0}$ = strain

ℓ_0 = the initial rest length
E = the elastic modulus

From in vitro studies on bladder strips, it has been concluded that the elastic modulus depends on the strain according to the following equation (van Mastrigt et al., 1978):

$$E = E_0 \, e^{\beta\varepsilon} \qquad (18.13)$$

where β is the stiffness factor.

In order to account for the fact that a strip of tissue in the intact bladder is strained in all directions, we need to use a correction factor

$$\frac{1}{1 - \mu} \qquad (18.14)$$

where μ is the Poisson ratio.

For the case that V_t is supposed to be constant, it can be shown that $\mu = 0.5$. Hence we find for the elasticity of the bladder wall

$$\sigma = 2\varepsilon \, E_0 \, e^{\beta\varepsilon} \qquad (18.15)$$

Combining (18.11) and (18.15) yields:

$$p = \frac{2}{3} \, V_t \frac{1}{V} \, 2 \, \varepsilon \, E_0 \, e^{\beta\varepsilon} = V_t \frac{4}{3} \, \varepsilon \, E_0 e^{\beta\varepsilon} \frac{1}{V} \qquad (18.16)$$

The pressure component developed by the elastic properties of the tissue will be denoted by P_{pe}.

In equation (18.16), ε is the strain relative to the rest length ℓ_0, which is related to the rest volume V_0 by

$$V_0 \frac{1}{6\pi^2} \ell_0^3 \tag{18.17}$$

In the model, ε is derived from the variable V according the relation

$$\varepsilon = \frac{\sqrt[3]{V} - \sqrt[3]{V_0}}{\sqrt[3]{V}} \tag{18.18}$$

In practice the cystometrogram appears to be dependent on the infusion rate: the higher this rate, the higher is the pressure measured at a certain infused volume. This observation has been related to visco-elastic properties of bladder tissue that was studied on bladder strips in vitro.

When a strip of bladder tissue is elongated step-wise, the force across the strip also increases step-wise. But afterwards this force decays to a certain value. This behaviour is known as stress-relaxation. It can be simulated by a mechanical model of a spring in series with a dashpot (see Figure 18.4).

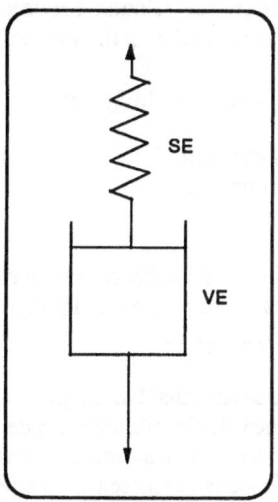

Figure 18.4. Mechanical visco-elastic model:
SE-series-elastic element, VE-viscosity element

Similar pressure decay is obtained after an almost stepwise filling of the bladder. It turned out that the decaying pressure component (P_{ve}) can be described by a mathematical model consisting of three exponential terms:

$$P_{ve}(t) = A_1\, e^{-t/\tau_1} + A_2\, e^{-t/\tau_2} + A_3\, e^{-t/\tau_3} \qquad (18.19)$$

where A_1, A_2, A_3 are coefficients representing the magnitude of each exponential term and τ_1, τ_2, τ_3 are the time constants of each exponential term.
Experiments have shown that the time constants are not dependent on bladder volume.

From these experimental results we derive the theoretical pressure response to a pulsewise infusion of 1 ml fluid. To account for the visco-elastic behaviour of bladder tissue in the model, we implemented a system that is characterized by a unit-pulse response according to the following formula:

$$h_{ve}(t) = 0.8\delta(t) - \frac{0.04}{37}\, e^{-t/37} - \frac{0.04}{370}\, e^{-t/370} \qquad (18.20)$$

The infused volume (V_u) is the input of this system. The response of this system (pve) is added to the pressure component developed by the elastic properties of the tissue (ppe). This yields the total passive intravesical pressure (pp).

18.2.3 Active Stress-Strain Properties

For a stimulated striated muscle, the relation between the initial velocity of shortening v and the load F is described with the well-known Hill's equation:

$$(F+a)(v+b) = b(F_0 + a) \qquad (18.21)$$

where   a   = physiological constant
          b   = physiological constant
          F_0 = isometric force

The isometric force is the maximum force that can be developed and depends on the degree of stimulation of the muscle. According to (18.21) the relation between v and F is hyperbolic, as shown in Figure 18.5.

It has been shown with in vitro studies that Hill's equation is also valid for bladder tissue (Griffiths et al., 1979). The velocity of shortening depends on the length of the contracting muscle; it becomes zero when a certain contracted length lr is reached. In Hill's equation this dependence is expressed by the length dependence of the isometric force F_0. During micturition bladder muscle shortens considerably. For the simulation of micturition it is necessary to account for the dependence of F_0 on the decreasing circumference of the bladder. In the model we implemented the following expression:

$$F_0 = l_v \cdot ffn \qquad (18.22)$$

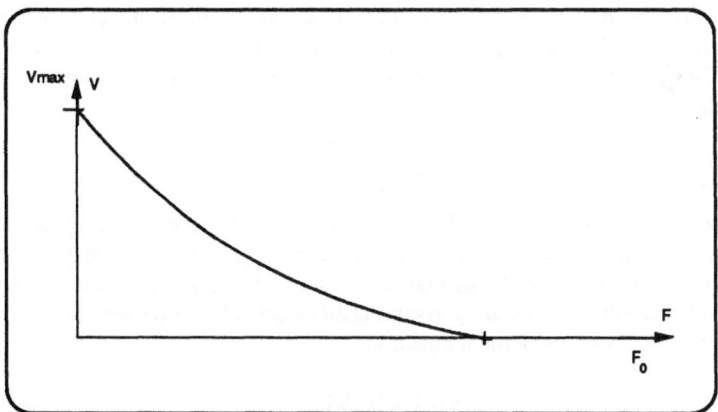

Figure 18.5. Force-velocity curve according Hill's equation

where ffn = a measure of nervous stimulation
       lv = contractile part of circumference of the bladder

It appears that a good fit of experimental force-velocity curves is obtained when it is assumed that

$$\frac{a}{F_0} = 0.25 \qquad\qquad (18.23)$$

a relation that also holds for striated muscle. Equation (18.21) characterizes the contractile tissue in the bladder wall. When stimulated muscle is limited in shortening, force is built up by straining elastic tissue components. In the model we represent contractile and elastic tissue by separate elements in series as shown in Figure 18.6.

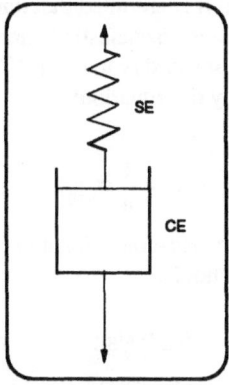

Figure 18.6. Mechanical contractile-elastic model:
          SE-series-elastic element, CE-contractile element

The contractile element CE is governed by (18.21) while the elastic element SE is characterized by the elastic properties of the tissue in series with the contractile tissue. The velocity of shortening (vcon) is simulated by implementation of the equation, derived from (18.21)

$$v = \frac{b\,F_0 - bF}{F + a} \qquad (18.24)$$

At the end of micturition, when lv becomes small and because of (18.22) F_0 also decreases, the value of parameter a decreases because of (18.23). Then the denominator in (18.24) becomes so small that unrealistic velocities are simulated in our model. This problem is solved by assuming an offset value of parameter a, so that instead of (18.23) we implement equation

$$a = 1 + 0.25\,F_0 \qquad (18.25)$$

In our model, F is derived from the active pressure component (P_a) in the bladder and the circumference of the bladder by using (18.3). The active intravesical pressure (P_a) is developed by elongation of the series elasticity. The stress-strain relation of the series elasticity is similar to that of the parallel elasticity as given in (18.15), i.e:

$$\sigma_{SE} = 2\varepsilon\,E_0\,e^{\beta\varepsilon} \qquad (18.26)$$

For the elongation of SE, ΔSE holds

ΔSE = bladder circumference (ℓ_u) - initial rest length (ℓ_0) - part of circumference concerning contractile tissue (ℓ_v)

When the bladder is filled physiologically or artificially by infusion, the contractile unstimulated tissue is passively elongated. Elongated contractile tissue will restore its shorter length by stimulation. It is assumed that during passive elongation the series arrangement CE-SE displays viscoelastic behaviour similar to the parallel tissue components previously discussed [van Duyl, 1985]. This means that during the elongation CE is governed by the equation:

$$v_{vis} = \frac{1}{n}\,\sigma_{vis} \qquad (18.27)$$

where (v_{vis}) is the velocity of viscous elongation and (n) is the viscosity constant. Because CE is in series with SE holds,

$$v_{vis} = \sigma_{SE} \qquad (18.28)$$

In other words, the CE-SE arrangement in the unstimulated situation of our model contributes to the passive properties of the bladder tissue with a third viscoelastic component.

In our model the elongation velocity of CE (v_{vis}), is integrated. This yields the length of contractile tissue (ℓ_v) being part of the circumference of the bladder (ℓ_u). The value of (ℓ_u) is derived from the volume of the bladder (volu). When the difference (ℓ_u)-(ℓ_v) is larger than the rest length (ℓ_0), this difference equals the elongation of SE. Consequently a stress component is generated. During slow (physiologic) filling, this stress component is negligible.

When CE is stimulated, then the simulated velocity of shortening (v_{con}) is integrated. This integrated value is subtracted from the initial value of ℓ_v set in the integrator, which corresponds to the initial volume of the bladder. If micturition is prevented by a closed urethra, the pressure will increase to a certain level, depending on the bladder volume. This is called an isovolumetric contraction. Such an isovolumetric contraction is generated by initiation of a micturition during the period before the urethra has opened. The registration of the pressure during this prephase of a micturition has been proposed as a clinical procedure to assess the contractility of the bladder [van Duyl et al., 1978]. It turned out that the phase plot of dp/dt versus p during the prephase of a micturition is of particular clinical value.

18.2.4 The Urethra

The urethra is an active organ that, in principle, relaxes during voiding to lower its resistance to flow. If relaxation is complete, reproducible pressure-flow plots are obtained (passive urethral resistance relation). In rest, the urethra is a collapsed tube. A certain threshold pressure level, the urethral opening pressure, must be surpassed to open the tube. In the opened situation the pressure-flow relation is determined by the cross-sectional area of the distensible and contractile urethral tube. This relation is complex and has a considerable influence on the course of a micturition. Some attempts for a theoretical analysis are given in the literature, and in clinical practice it has been attempted to derive this relation experimentally from pressure-flow studies. In our model we implemented a simplified pressure-flow relation based on such clinical studies (Griffiths et al., 1988), where

$$P = P_{uo} + \frac{Q^2}{c} \qquad (18.29)$$

where P_{uo} = urethral opening/closing pressure
  Q = flow rate
  c = constant related to effective cross-sectional area

It appears that in man P_{uo} and $1/c$ are correlated so that in practice the urethra can be characterized by one parameter only. Figure 18.7 gives a typical, normal pressure-flow relation with P_{uo} = 20 cm H_2O and $1/c = 0.08$.

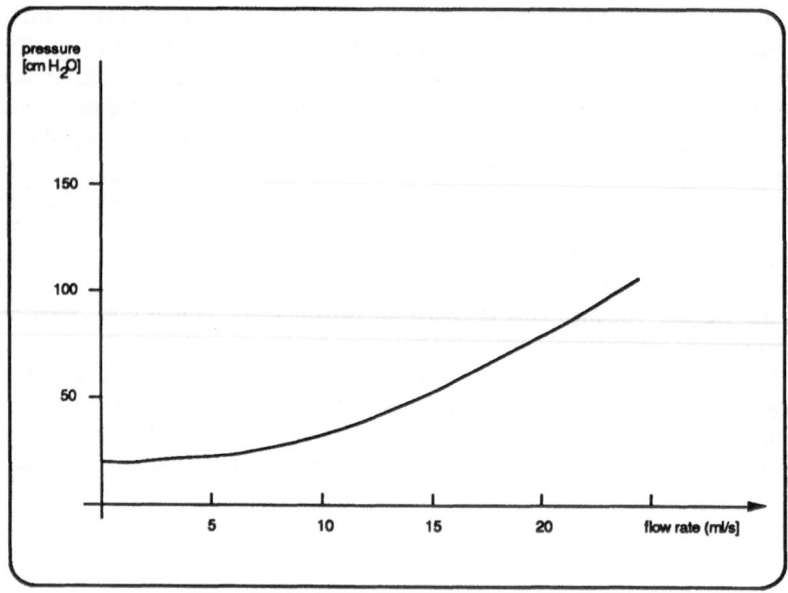

Figure 18.7. Pressure-flow relation of the urethra

18.3   Simulations

18.3.1 Model Description

Figure 18.8 shows the blockdiagram of the total model of the lower urinary tract based on the relations derived in the previous section.

In this scheme the following sub-blocks are distinguished:

sub-block -PVE-
The output of this sub-block is the passive intravesical pressure (P_p), which is composed, via a sum-block, of the elastic component (P_{pe}) and the viscoelastic component (P_{ve}). The elastic component is simulated according to (18.16) with the following default values:
    $V_t = 20$ ml $V_0 = 25$ ml $E_0 = 20$ g/cm^2 $\beta = 0.7$
With these values the implemented relation is:

$$P = 534e^{0.7\varepsilon}\,[0.34V^{1/3} - 1] \tag{18.30}$$

By means of a DSP-block it has been realized that the elastic component is zero when the intravesical volume (volu) is less than the restvolume (v_0).

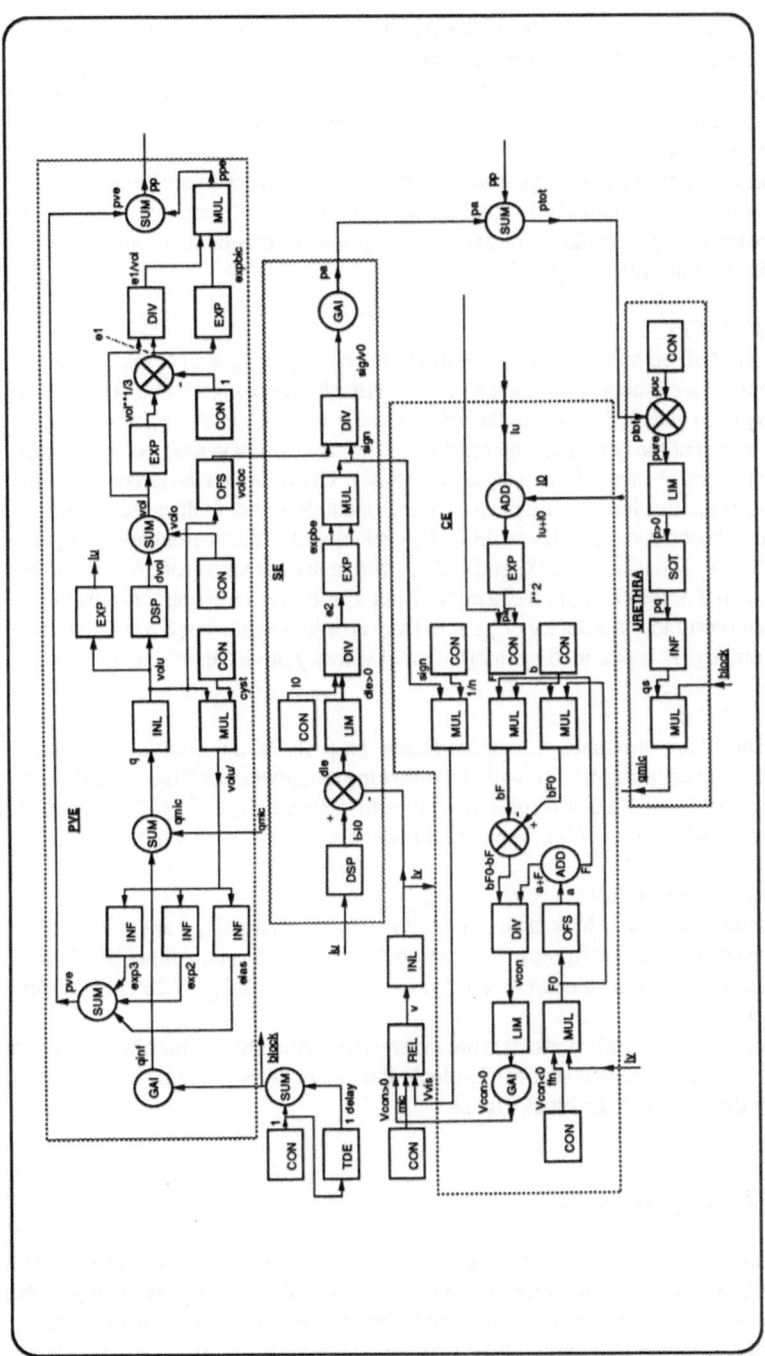

Figure 18.8. Block-diagram of the simulation model of the lower urinary tract

The viscoelastic component is simulated according to eq. (18.20).The two exponential terms of the unit-pulse response are implemented by using two INF blocks (exp2) and (exp3). The pulse term is also simulated with an INF block (elas) with a relative small value of the time constant namely 0.1 s.

The input of this sub-block is the flow (q) via a SUM block. For the simulation of a micturition it is set to be equal to the flow of micturition (qmic) while for the simulation of a cystometry it is set to be equal to the infusion rate (qinf). The intravesical volume (volu) is derived from (q) via an integrator. From (volu) the bladder circumference (ℓ_u) is derived.

sub-block -CE-

The output of this sub-block is the contraction velocity ($v_{con} < 0$) of the contractile element. This element is stimulated to contraction by choosing a value of the neural activity (ffn), which determines the value of the isometric tension (F_0) proportional to the length of the contractile element (ℓ_v). The contraction velocity is determined by the tension (F) according to equation (18.24). The value of F is derived from the value of the active intravesical pressure component (P_a), as default values have been initially chosen b = 1 and ffn = 1000. The velocity of viscous elongation (V_{vis}) is simulated to equations (18.28) and (18.29) with a default value for 1/n = 0.01 cm-H_2O/s.cm. For the simulation of a micturition, ℓ_v is derived by subtraction of the intragrated contraction velocity ($v_{con} < 0$) from an initial value. For the simulation of a cystometry, ℓ_v is obtained by integration of velocity of viscous elongation (V_{vis}).

sub-block -SE-

This sub-block simulates the series-elastic part SE of the contractile element according to equation (18.26) with the following default values: $2E_0 = 1000$ g/cm^2 and $\beta = 0.6$. The input of this sub-block is the difference in ($\ell_u > \ell_0$) and the length of the contractile element (ℓ_v). The output is the active pressure component (P_a).

sub-block -URETHRA-

The input of this sub-block is the total intravesical pressure (P_{tot}) and the output is the flow of micturition (qmic). The implemented relation is equation (18.30) with the following default values, corresponding to figure 18.11: $P_{uo} = 20$ cm H_2O and 1/c = 0.08.

Furthermore a first order system is put in series to account for the inertia of the urine in the urethra. By means of a MUL-block, the model of the urethra can be opened during the period of the block-signal.

18.3.2   Experiments

Simulations of different urodynamic procedures will be done with the total model of Fig. 18.8 using different settings of some parameter. Table 18.1 is the listing of the model with the parameter set for the simulation of a cystometry with an infusion rate of 1 ml/s. In Table 18.2 the parameter settings are given for some other experiments with the values chosen in the default demonstrations.

18.3.2.1 Cystometry

In this demonstration the bladder is filled by infusion at the rate of 1 ml/s, which corresponds to a slow filling cystometry. Defaults shows the bladder volume (volu) and the intravesical pressure (P_{tot}) during 400 seconds (Fig. 18.9). By using the XY-command, with x = 1 and y = -2, the cystometrogram is obtained as it is recorded in the clinic. The infusion rate is so slow that during infusion the stress developed across the SE-element is negligible compared to the stress developed across the parallel elastic components.

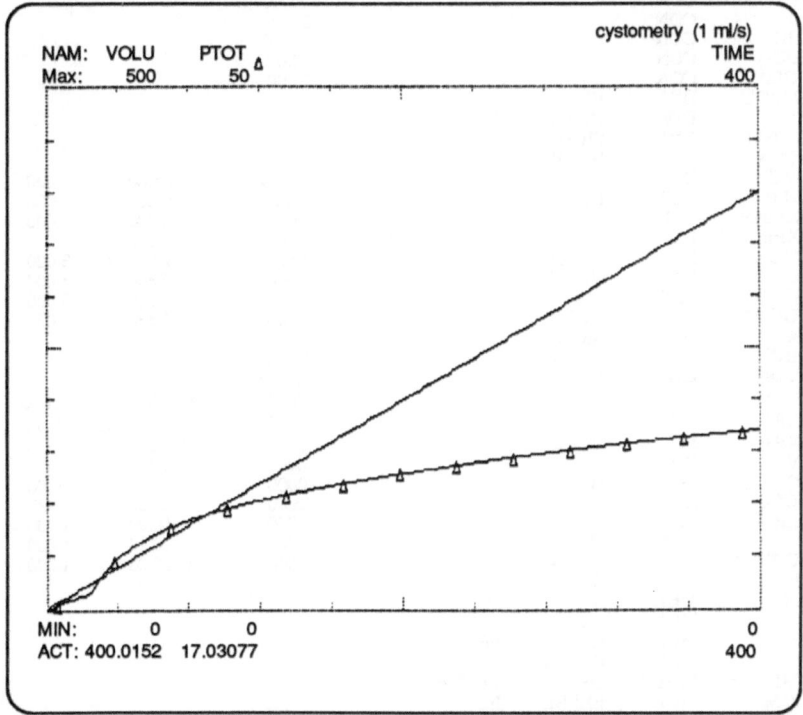

Figure 18.9.   Output of demonstration udyn10: simulation of an infusion with an infusion rate of 1 ml/s, i.e. slow-filling cystometry

Table 18.1. Listing of simulation model of the lower urinary tract with parameter settings for the simulation of an infusion of 1 ml/s.

			- - *STRUCTURE AND PARAMETERS UDYN10.PS6* - -				
Block	Type	Input1	Input2	Input3	Par1	Par2	Par3
A+F	ADD	F	A				
LU+L0	ADD	LU	L0				
1	CON				1.000		
1/N	CON				1.0000E-02		
B	CON				1.000		
CYST	CON				1.000		
FFN	CON				.0000		
L0	CON				5.000		
MIC	CON				.0000		
PUC	CON				20.00		
VOL0	CON				25.00		
E1/VOL	DIV	E1	VOL				
E2	DIV	DLE>0	L0				
SIG/VO	DIV	SIGM	VOL0C				
VCON	DIV	BF0-BF	A+F				
DVOL	DSP	VOLU			-10.00	25.00	1.000
L>L0	DSP	LU			-10.00	5.000	1.000
EXPB1E	EXP	E1			534.0	.7000	1.000
EXPBE	EXP	E2		1.0000E+04	.6000		1.000
L**2	EXP	LU+L0			7.9600E-02	2.000	3.000
LU	EXP	VOLU			3.900	.3330	3.000
LU**2	EXP	LU			7.9600E-02	2.000	3.000
VOL**1/3	EXP	VOL			.3400	.3330	3.000
PA	GAI	SIG/VO			11.11		
QINF	GAI	BLOCK			1.000		
VCON<0	GAI	VCON>0			-1.000		
ELAST	INF	VOLU/			.0000	8.0000E-02	.1000
EXP2	INF	VOLU/			.0000	-4.0000E-02	37.00
EXP3	INF	VOLU/			.0000	-4.0000E-02	370.0
LV	INL	V			.0000	.0000	100.0
QS	INF	PQ			.0000	.0000	.5000
VOLU	INL	Q			.0000	.0000	1000.
DLE>0	LIM	DLE			.0000	10.00	1.000
P>0	LIM	PURE			.0000	100.0	1.000
VCON>0	LIM	VCON			.0000	20.00	1.000
BF	MUL	B	F				
BF0	MUL	B	F0				
F	MUL	PA	L**2				
F0	MUL	LV	FFN				
PPE	MUL	E1/VOL	EXPB1E				
QMIC	MUL	QS	BLOCK				
SIGM	MUL	EXPBE	E2				
VOLU/	MUL	VOLU	CYST				
VVIS	MUL	SIGM	1/N				
A	OFS	F0			.2500	10.00	
VOL0C	OFS	VOLU			1.000	2.000	
V	REL	MIC	VVIS	VCON<0			
PQ	SQT	P>0				3.500	
BF0-BF	SUB	BF0	BF				
BLOCK	SUM	1	1DELAY	1.000	.0000		
DLE	SUB	L>L0	LV				
E1	SUB	VOL**1/3				1	
PP	SUM	PPE	PVE		1.000	1.000	
PTOT	SUM	PA	PP		1.000	1.000	
PURE	SUB	PTOT	PUC				
PVE	SUM	EXP2	EXP3	ELAST	1.000	1.000	1.000
Q	SUM	QINF	QMIC		1.000	-1.000	
VOL	SUM	DVOL	VOL0		1.000	1.000	
1DELAY	TDE	1		.0000	36.00	1.000	

Table 18.2.   Parameter setting of the model in the demonstrations

Exercise Block	infusion 1 ml/s			infusion 10 ml/s			isovol.-contr.			micturition			stoptest		
cyst.	1	-	-	1	-	-	0	-	-	0	-	-	0	-	-
qinf.	1	-	-	10	-	-	0	-	-	0	-	-	0	-	-
mic	0	-	-	0	-	-	1	-	-	1	-	-	1	-	-
ffn	0	-	-	0	-	-	500	-	-	1000	-	-	1000	-	-
ldelay	0	36	1	0	36	1	0	36	1	0	50	1	0	5	1
ptot	1	1	-	1	1	-	1	1	-	1	1	-	1	1	-
qs	0	0	0.5	0	0	0.5	0	0	0.5	0	1	0.5	0	1	0.5
pp	1	1	-	1	1	-	1	0	-	1	0	-	1	0	-
block	1	0	-	1	1	-	1	1	-	1	1	-	1	1	-

18.3.2.2   Fast Filling Cystometry

In this demonstration the bladder is filled by infusion at a rate of 10 ml/s, which corresponds with a so-called fast filling cystometry or 'stepwise' cystometry. The bladder is filled for 10 seconds after which the bladder volume kept at 360 ml. Volume and pressure are recorded during 500 seconds (Fig. 18.10). When the

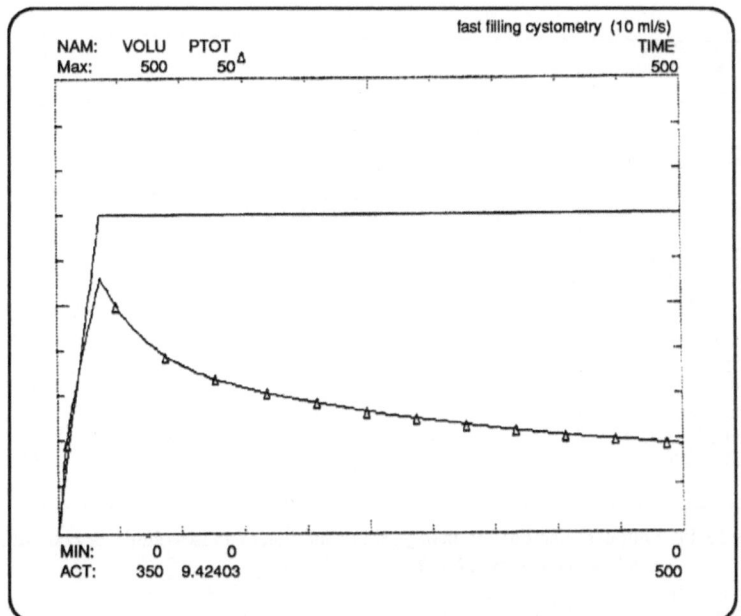

Figure 18.10.   Output of demonstration udyn20: simulation of an infusion with and infusion rate of 10 ml/s, i.e. fast-filling cystometry, during 10 seconds, followed by isovolumetric pressure decay

infusion is stopped the pressure decays because of the simulated viscoelastic properties of bladder tissue. Stepwise cystometry has been proposed as an alternative to the classic, (slow-filling) cystometry in order to determine the elastic as well as the viscoelastic properties of the bladder. By choosing different infusion rates (qinf) the effect of the infusion rate over the course of the cystometrogram can be demonstrated.

18.3.2.3   Isovolumetric Contraction

In this demonstration an isovolumetric contraction is simulated at a bladder volume of 360 ml. The starting conditions for this demonstration are obtained after a simulation of an infusion at a rate of 1 ml/s for 360 seconds followed by a DIC-command. Figure 18.11 shows the results of this simulation. Similar clinical registrations are used to derive parameters concerning the CE-element as explained in the text.

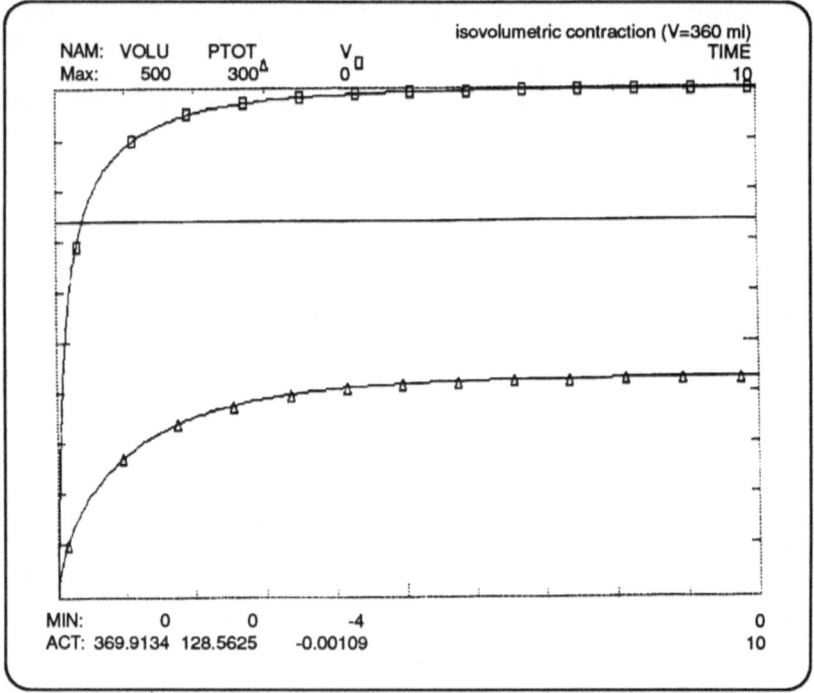

Figure 18.11. Output of demonstration udyn30: simulation of an isovolumetric contraction at a bladder volume of 360 ml

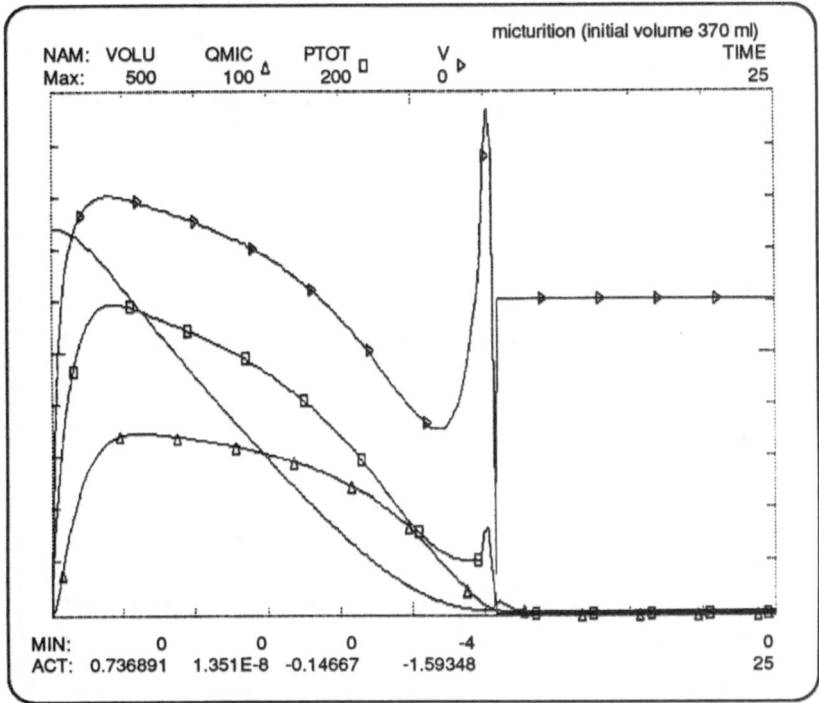

Figure 18.12. Output of demonstration udyn40: simulation of a micturition starting at a bladder volume of 370 ml

18.3.2.4 Micturition

The micturition simulated in this demonstration starts at initial bladder volume of 370 ml. The corresponding conditions of the integrators can be obtained by simulating an infusion followed by a DIC-command. The output is shown in Figure 18.12.

The simulated flow curve (qmic versus t) has the characteristic course of graphs obtained in uroflowmetry, i.e. a steep initial increase of the flow followed by a period of almost constant flow (plateau) and finally a less steep decrease of the flow to zero. An attractive alternative presentation of flow data is the plot of the flow versus the bladder volume [Rollema et al., 1977]. Such a graph can be simulated by using the XY-command. Figure 18.12 also shows the bladder contraction velocity during a micturition. The course of this signal corresponds well with that of the graph obtained from uroflow data. The peak in the velocity curve at the end of the micturition is very characteristic. When micturition has stopped the simulation shows a constant value of v. Of course this is a simulation error as a consequence of maintaining a value of ffn.

The course of other variables during micturition can be easily studied with this model.

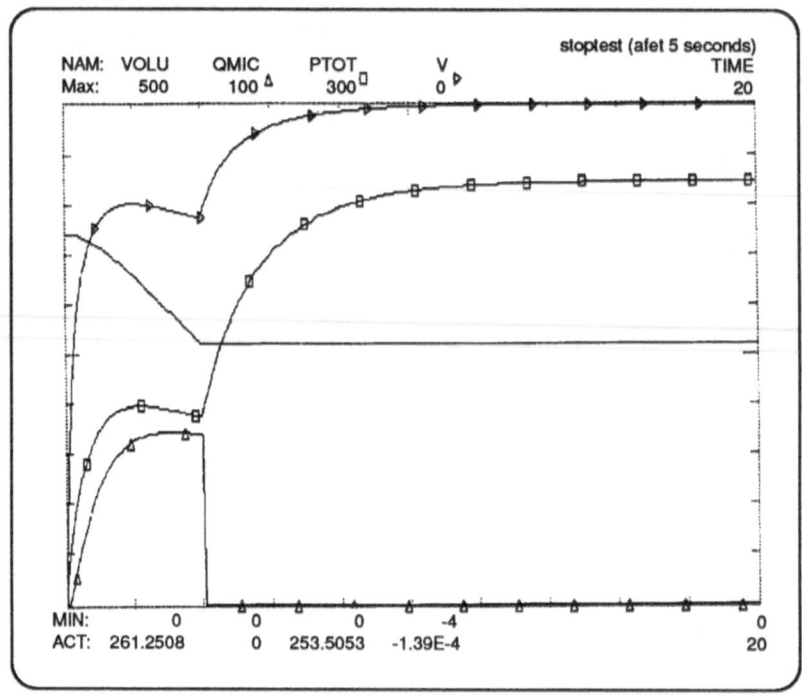

Figure 18.13. Output of demonstration udyn50 (the stoptest): simulation of a micturition starting at a bladder volume of 370 ml and stopped after 5 seconds by closing the urethra

18.3.2.5 Stop Test

The stop test has been proposed as a clinical alternative to the isovolumetric pressure recording in the preflow phase of a micturition for the assessment of bladder contractility. In the stop test the micturition is interrupted by closing the urethra by means of an inflated balloon. The stop test is simulated in udyn50. Micturition starts at an initial volume of 370 ml and is stopped after 5 seconds. The output is shown in Figure 18.13.

Concluding Remarks

With this model, simulations corresponding to various clinical situations can be made by changing the numerical values of different parameters. The passive properties found by cystometry can be adjusted by changing the values of the elastic and viscoelastic parameters. Reduced bladder contractility can be simulated by changing the parameter values in Hill's equation. Another relevant change concerns the resistance of the urethra and thus the simulation of an obstructed flow. Investigation is going on to look for parameters to be derived from flow data and isovolumetric pressure data enabling the clinician to do a differential diagnosis in case of abnormal micturition caused by reduced bladder contractility or an obstructed urethra. By using a model, those parameters can be evaluated in relation to the theoretical concept of the urodynamics of the lower urinary tract.

The model can be extended in order to adapt better to clinical and experimental observations by introducing a modulation of the parameter representing nervous stimulation (ffn), e.g. making it dependent on bladder pressure. Also the urethral resistance can be made dependent on nervous stimulation. In this rather complicated model, unstable micturition can be simulated. Another observation not accounted for in our model concerns the spontaneous bladder contraction activity. However, too little is known to build a useful model of the urodynamics of the lower urinary tract that includes all these phenomena.

References

Griffiths D.J.: Medical Physics Handbooks 4: Urodynamics. The mechanics and hydrodynamics of the lower Urinary tract. Adam Hilger Ltd, Bristol, 1980.

Mastrigt R. van, Coolsaet B.L.R.A. and Duyl W.A. van: "Passive properties of the urinary bladder in the collection phase". Med. and Biol. Eng. and Comp. 16, 1978 pp. 471-482.

Griffiths D.J., Mastrigt R. van, Duyl W.A. van and Coolsaet B.L.R.A.: "Active mechanical properties of the smooth muscle of the urinary bladder". Med. Biol. Eng. and Comput. 17, 1979 pp. 281-290.

Mastrigt R. van, Coolsaet B.L.R.A. and Duyl W.A. van: "First results of stepwise straining of the human urinary bladder and bladder strips". Invest. U., 19, 1, 1981 pp. 58-61.

Rollema H.J., Griffiths D.J., Duyl W.A. van, Berg J. van der, and Haan R. van: "Flowrate versus bladdervolume". Urol. Int, 32, 1977, pp. 401-412.

Duyl W.A. van: "A model for both the passive and active properties of urinary bladder tissue related to bladder function". Neurorol. Urodynam, .

Griffiths D.J., Mastrigt R. van and Bosch R.: "Quantification of urethral resistance of bladder-function during voiding with special reference to the effect of prostate size reduction on urethral obstruction due to BPH". Neurol. and Urodynam. 8, 1, 1989 pp.17-27.

Gosling J.A., Dixon J.S. and Humpherson J.R.: Functional anatomy of the urinary tract. Gower Medical Publishing, London, New York, 1983.

19
Fluid Volumes:
The Program "FLUIDS"

Fredericus B.M. Min

Abstract

This chapter describes the program FLUIDS. The mathematical model underlying this program contains over 200 variables and describes control mechanisms of body fluid volumes and electrolytes as well as respiratory control mechanisms. This model allows a variety of simulations for conditions such as thirst, fluid loss, exaggerated drinking, carbon dioxide inhalation, severe physical exercise, etc. Students can also infuse fluids of different compositions or give a diuretic, etc. The basic physiology of respiratory and metabolic acidosis and alkalosis can also be studied with this model.

19.1 Aim of Instruction

The computer simulation program (CS-program) FLUIDS enables students to experiment with a model of the human fluid and electrolyte system, as well as the regulation of respiration and its underlying (basic) physiology and pathophysiology. The simulation consists of the following: heart and cardiovascular system, lungs, intra- and extracellular fluid compartments, nerve reflexes, the kidney, and a number of hormonal systems with an influence on the kidney function. In some ways this simulation coincides with the model of the CS-program CARDIO, but it is much more extensive in a number of important ways. FLUIDS simulates a kind of experimental laboratory environment, so that students can do research on water and salt regulation and the regulation of respiration under all sorts of circumstances.

Students have at their disposal "a testee", so to speak, with which they can do a series of experiments. Thus it is possible to imitate excessive thirst, sweating, or loss of water and salt. A change in the oxygen and carbon dioxide ratio during breathing can also be simulated, as well as disorders in the acid-base balance and the function of the kidney.

Besides interventions that could make a patient ill, there are therapeutic interventions possible, such as administering a diurecticum or applying infusions of various compositions in order to supply extra oxygen. It is also possible to do the glucose tolerance test (GTT). Despite the complexity of the model, students easily find their way in this CS-program. An emphasis is given to free investigation while working with the simulation, so students develope an overview of the way in which the human body works as a complex physiological dynamic system.

19.2 Model of the Water and Electrolyte Regulation

The model of the CS-program FLUIDS is a model of the water and electrolyte regulation of a healthy male of 55 kg. The model of the regulation of the body fluids was designed by Ikeda, Marumo, Shiritako, and Sato in 1979 and is based on by, among others, [Guyton, 1972] and [Blaine, 1972]. [Ikeda et al., 1979].

The model by Ikeda and his colleagues consists of the following sub-systems: circulation, respiration, kidney function, and the intra- and extracellular fluid compartments. Overall, the model consists of 30 integral equations plus other algebraic equations, totalling more than 270 variables. Water and salt regulation varies greatly among individuals and also between the sexes. In this model of a young man of 55 kg, the body fluids are divided into three major groups: the intracellular (20.0 l), the interstitial (8.8 l) and the intravascular compartment (2.2 l). The relatively small amount of transcellular water (brain fluid, fluid in the stomach and intestinal canal, eye compartment fluid and such) is not taken into account and only plays a small role in the total weight. The extracellular volume is 11.0 l and the blood volume 4.0 l. The composition of the blood plasma, fluid and extracellular fluid, as well as that of the dissolved ionised matter and the molecular dissolved matter of gasses, is very specific.

In the CS-program FLUIDS the normal values of ion concentrations and others in the extracellular and intracellular fluid compartments are:

Extracellular concentrations:			Intracellular concentrations:	
Sodium Na^+	140	mEq/l	10	mEq/l
Chlorine Cl^-	104	mEq/l	4	mEq/l
Potassium K^+	4.5	mEq/l	140	mEq/l
Calcium Ca^{++}	5	mEq/l	$\ll 10^{-3}$	mEq/l
Magnesium Mg^{++}	3	mEq/l	58	mEq/l
Bicarbonate HCO_3^-	24	mEq/l	10	mEq/l
Phosphate PO_4^{--}-etc.	1.1	mEq/l	75	mEq/l
Sulphate SO_4^{--}	1	mEq/l	2	mEq/l
Glucose	6	mosmol/l	0	mosmol/l
Urea	2.5	mosmol/l	2.5	mosmol/l

There are many regulatory mechanisms to keep changes in the concentrations in the extra- and intra-cellular fluid compartments as small as possible. The kidney has an important role in this, by regulating the volume and the colloid osmotic pressure of the fluid compartments. In addition there are regulatory mechanisms for the degree of acid, for the stabilisation of the blood gas pressure of O_2 and CO_2, for the stabilisation of the ionic composition, and for the stabilisation of the glucose percentage. The model includes seven distinguishable parts, between which there exist the relations shown in Figure 19.1. This figure also indicates some interventions which are possible in this model. The different parts of the model are:

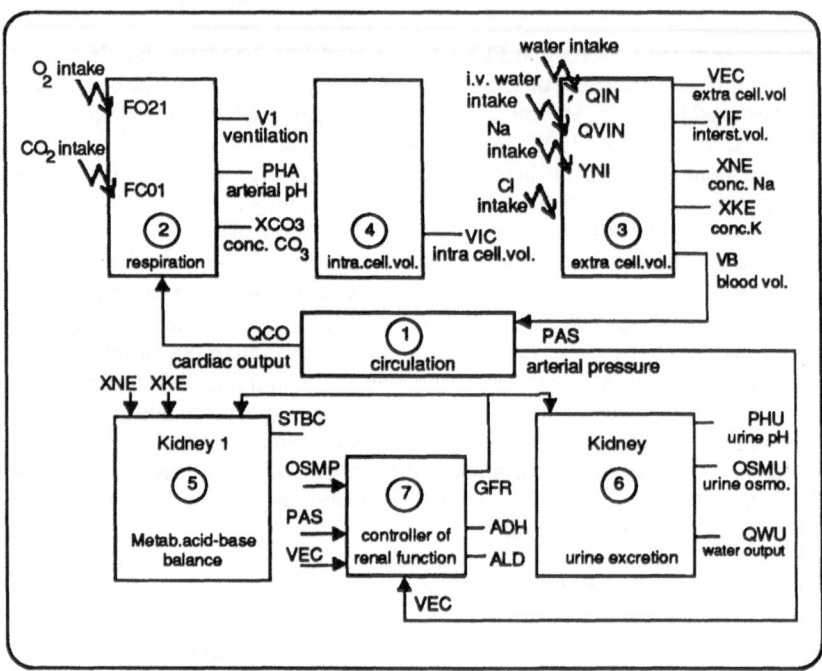

Figure 19.1. Block outline of the model by Ikeda et al. (1979) of the water and electrolyte regulation. The arrows going into the blocks are the possibilities of interventions; the arrows touching the edges of the blocks are input variables, and arrows leaving the blocks are output variables

19.2.1 Cardiovascular System

The cardiovascular system has been minimalized to a functional unit for the cardiac output and the mean arterial pressure, which in this model depend only on the blood volume, the elasticity of the vascular system, and the peripheral resistance. Because the cardiovascular system has been so minimalized, simulation of cardiovascular affliction cannot easily be done with this model.

19.2.2 Respiratory Regulation System

The regulating system of the respiration is in this model a functional unit in which ventilation depends on the pH, the CO_2-, and the O_2-pressure in the arterial blood. The pH of the blood is determined by the percentage of freely dissolved CO_2 and the percentage of HCO_3^-. Here the equation of Henderson-Hassalbalch is valid. The hemoglobin buffer system can keep the pH of the blood constant by intake or release of H^+-ions, although there are situations when a large or fast increase or decrease of H^+-ions takes place. The pH in the blood is for an important part determined by animo acids, composed of proteins. This buffer system of H^+-ions regulates the pH around 7.4 (iso-hydric point). Another buffer system for the pH is the bicarbonate buffer. The proper functioning of this system requires that the CO_2 surplus is adequately removed through the lungs.

19.2.3 Extracellular Fluid Compartment

This part of the model simulates the regulation of the fluid volumina. The dissolved ion concentration in the extra-cellular fluid compartment is closely connected to the quantity of water taken in and excreted by the body. The fluid volume in the extra-cellular compartment is determined by the quantity of water which is administered orally or intravenously and by the loss of fluid through urine and sweat. The intake of concentrations of the dissolved matter held in this system usually comes via the food from the stomach and intestinal canal. Also intravenous administration through an infuse is possible in this model. All intakes been normalized on an average of 24 hours with certain referential values per minute.

19.2.4 Intracellular Fluid Compartment and Electrolytes

In this part of the model Ikeda and his collegues have classified the intracellular fluid volume, the osmotic active substances in the intra- and extra-cellular compartments and the intra-cellular acid-base balance, among others sodium, potassium, chlorine, glucose, ureum and mannitol. The change in the extra cellular sodium quantity (ZNE) is represented in the model as follows:
d(ZNE)/dt=intake by Na(YNIN) - excretion of Na(YNU) + increase of Na, in exchange for H+-ions in the cell and the change of the quantity of potassium in the extra-cellular compartment (ZKE) as follows:
d(ZKE)/dt=intake of K(YKIN) - excretion of K(YKU) + increase of K, in exchange

for H^+-ions in the cell - K which goes into the cell in connection with glucose metabolism and insulin secretion.

The ureum reabsorption mechanism in this model is passive while it is assumed that about 60% of the filtered quantity will eventually by excreted. The plasma osmolality (OSMP) is in this model dependent on the sodium, potassium, chlorine, glucose, ureum and mannitol concentrations, as well as a constant factor for the other osmotic active substances.

19.2.5 The Kidney

The renal excretion for bicarbonate, calcium, magnesium, phosphate, and organic acids are in this model functions of the glomerular filtration velocity and the concentrations of these ions. A quantity of fluid from the plasma is filtered through the glomerulus. The blood pressure is in this model the driving force. The largest part of the filtrate returns then, selectively, into the plasma in the peritubuluar capillaries. The transport of potassium and sodium is active, going against the concentration gradient as well as the electric gradient. The negative ions, like chlorine, diffund passively, with the exception of the bicarbonate. The transport of water only happens if there is an osmotic driving power. The model attempts to keep the osmolality of the plasma constant. Chemoreceptors play an important role in this. The active transport of ions and the transport of water is influenced by hormones. In this model the active transport of sodium is aided by aldosteron which holds (reabsorbs) sodium and excretes potassium. The excretion of water can be checked by the antidiuretic hormone.

19.3 CONSTRUCTION OF THE MODEL

The following describes the formulation of the most essential parts of the various blocks.

19.3.1 The Circulation

The cardiac output (QCO) is only dependent on the blood volume (BV) and determines together with the total peripheral resistance (RTOT) the mean arterial pressure (PAS).

PAS: = QCO0 * RTOT + 20;

19.3.2 The Fluid Volumunia

The plasma volume (VP) is determined by the quantity of water which is administered intravenously (QVIN), the quantity which is taken in by the body orally (QIN), the quantity of water which leave the body via urine (QWU) and, among others, also by the quantity of water which leaves the body through sweating (QIWL) and that which is formed metabolocally (QMWP).

VP: = VP + (VIN/10 + QVIN + QWMP - QIWL - QWU) . dt

VIF: = VIF + (QCFR - QLF - QIC) . dt

The blood volume (VB) and the extracellular fluid volume (VEC) are ultimately determined by the plasma volume. The interstitial fluid volume (VIF) is calculated in a similar way.

19.3.3 The Na, K and Cl Concentrations

The sodium concentration in the extracellular fluid (XNE) depends on the quantity of water in the extracellular compartment (VEC) and the absolute quantity of sodium (ZNE). The absolute quantity of sodium is, among others, a function of the mean average daily sodium intake (YNIN) and the quantity which leaves the body via urine (YNU). The same is true for potassium (XKE, ZKE, YKIN, YKU) and chlorine (XCLE, ZCLE, YCLI, YCLU). The equations are:

ZNE: = ZNE + (YNIN - YNU + YHI + CION/1440) . dt

ZKE: = ZKE + (YKIN - YKU - DZI) . dt

XCLE: = ZCLE + (YCLI - YCLU + CION/1440) . dt

The model is also built thus for calcium, magnesium, glucose, bicarbonate, phosphate, sulphate, urea, mannitol, other organic acids, and proteins.

The standard bicarbonate concentration at pH=7.4 (STBC) is defined in this model as the difference between the concentration of cations and anions (in the ECF) different from that of bicarbonate.

19.3.4 Kidney and Urine

The osmolality of the urine (OSMU) is determined by the urine output (QWU), the glomerular filtration velocity (GFR) and the antidiuretic hormone (ADH).

OSMU: = (1.86 . (YNU + YKU) + YGLU + YURU + YMNU) . QWU

The pH of the urine (PHU) is calculated from, among others, the renal clarification velocity of organic salts (YORG) and the renal clarification velocity of phosphate (YPO4).

PHU: = - ^{10}log (f(HU))

19.3.5 The Glomural Filtration Velocity

The glomural filtration velocity (GFR) is here only a function of the mean arterial pressure (PAS) and the extracellular fluid volume (VEC).

    GFR: = GFR0 . GFR1 . VEC/11.0

19.3.6 The Respiration

The ventilation (VI) is determined by a function (F) of the CO_2 pressure (PCOA), the O2 pressure in the alveoli (PO2A) and the pH of the arterial blood (PHA). The pH in the alveoli (PHA) is made out through the Henderson-Hasselbalch equation:

$$pH = pK + {}^{10}\log ([HCO_3^-]/S.PCO2)$$

Here the S.PCO2 is the freely dissolved percentage CO2 in the alveoli (PCOA) and [HCO_3^-] the percentage HCO_3^- in the alveoli (XCO3). The equations are:

    VI: = F(PHA, PO2A,PCOA)
    PHA: = 6.1 + ${}^{10}\log$(XCO3/(0.03 . PCOA))

19.4 Results

The CS-program FLUIDS, built with the RLCS system of the University of Limburg, measures and registers eight important variables in its starting position.

    QWU: Urine output 0.0015 l/min
    VEC: Extracellular volume 11 litres
    OSMP: Osmolality in the plasma 287 mOSM/l
    PAS: Mean arterial pressure 100 mmHg
    VP: Plasma volume 2.2 litres
    VIC: Intracellular volume 11.0 litres
    VIF: Interstitial volume 8.8 litres
    STBC: Standard bicarbonate (at pH=7.4) 24 mEq/l

These values are only valid when no interventions have taken place in the model and one can speak of a steady state. In this rationale with regard to thirsting, sweating, drinking and infusions, other arbitrary variables often play an important role and so it is possible to measure and register them. If some of these variables go above or below a critical value, the students will be warned by a message.

The symptoms appearing in the CS-program FLUIDS are determined by a range of values, above or below which a message is given to the student:

    6.95 < PHA < 7.05   Your patient has convulsions
    PHA < 6.95   Your patient goes into a coma

PHA < 6.9	Your patient does not breathe anymore
PHA < 6.8	Your patient just died ...,
	Do you have a good lawyer?
XKE > 9.5	"Doctor, my heart"
	... (fibrillate, hyperkalemy) ...
XKE > 7.0	This patient has an abnormal EEG
XKE < 2.5	"Doctor, I'm becoming terribly weak"
XKE < 2	Your patient becomes too weak to be able to breathe
HT < 35	Your patient starts to look very pale
HT < 30	"Doctor, I have been so tired these last few weeks"
PAS < 85	"Doctor, I feel so dizzy"

19.4.1 Thirsting

An important simulation is to withhold water from the testee (the model) over an extended period. For that purpose the water intake (QIN), which is normally 0.0015 litre/min. (equal to 90 ml/hour), can be reduced to zero.

In Figure 19.2 it can be seen that the plasma volume (VP) and the extracellular volume (VEC) slowly decrease. The potassium and sodium concentrations become constant by an increase of the antidiuretic hormone (ADH), enabling the kidney to keep its fluid while the urine production (QWU) decreases and the osmolality of the urine (OSMU) increases.

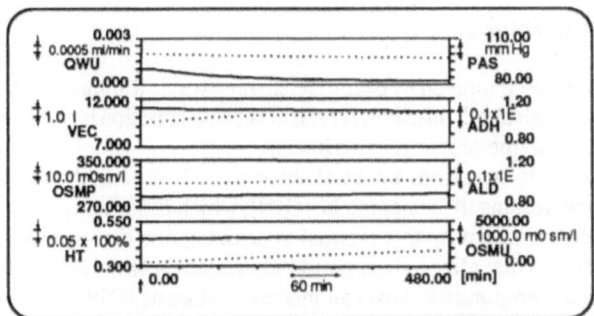

Figure 19.2. Thirsting during 8 hours (QIN = 0 l/min). Adjustment of the scale: t-axis from 0 to 9 hours. The variables mentioned on the left along the y-axes belong to the uninterrupted registered graphs and the variables mentioned on the right, to the dotted registered graphs (RLCS-system, Maastricht)

19.4.2 Thirsting, Sweating and Loss of Salt

When there is also loss of fluid (QIWI) and much loss of salt (via YNIN and YCLI) together with the thirsting, the results are as can be seen in Figure 19.3. During the simulation time the water intake is QIN = 0 l/min and the velocity with which fluid is withdrawn (QWIL) from the body has increased from 0.007 l/min to 0.008 l/min (which is equal to 0.5 l/hour). The sodium and chlorine intake (YNIN and YCLI), which are normally 0.12 respectively 0.133 mEq/min, are both fixed on -1.2 mEq/min. It can be clearly seen that the ADH increases and the urine excretion reduces sharply. The blood pressure (PAS) falls, in contrast with the simulation in Figure 19.2 in which the osmolality of the plasma and urine increase slightly.

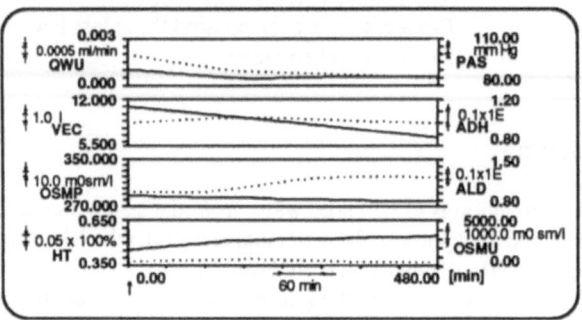

Figure 19.3. Thirsting , sweating and salt loss (QWIL = 0.008 l/min, YNIN = -1.2 mEq/min and YCLI = -1.2 mEq/min). The intake of water is 0 (QIN = 0,0 l/min).

19.4.3 Water Intake

Water intake can be simulated by the model in three ways: drinking, hypotonal infusion or a physiological salt infusion. A typical starting situation is one in which there is desiccation as in the previous experiments.

It is also possible to start from a normal situation. When normal water drinking is simulated by increasing the water intake (QIN), which is normally 0.0015 l/min, to 0.2 l/min over five minutes, this is equal to drinking one litre of water.

Figure 19.4 shows that after a short period the variables have their normal values again. The body temporarily shows an increased diuresis (QWU increased).

Figure 19.5 shows the consequences of intravenous injection of one litre of water. This is the same as increasing QVIN to 0.2 ml/min over five minutes, and is equal to an infusion of one litre of pure water. Following a raised diuresis the variables reach their normal values again after a short time.

Figure 19.6 shows the consequences of intravenous injection of one litre of physiological salt in five minutes, while the sodium intake (YNIN) and the chlorine intake are both 31 mEq/min. This is equal to a quantity of sodium and chlorine of 154 mEq. In Figure 19.6 it can be seen that the model reacts with a rise in blood pressure (PAS). Here also the kidney will react to the rise of the blood pressure by excreting salt. This lasts for several hours, during which time water is also excreted along with the salt.

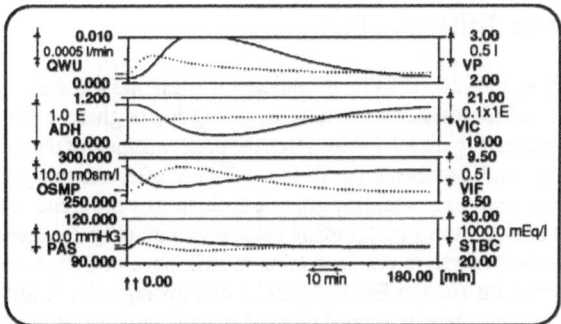

Figure 19.4. Drinking 1 litre water (QIN = 0.2 l/min during 5 minutes). All variables return to their normal values within 3 hours. The diuresis is increased during these hours.

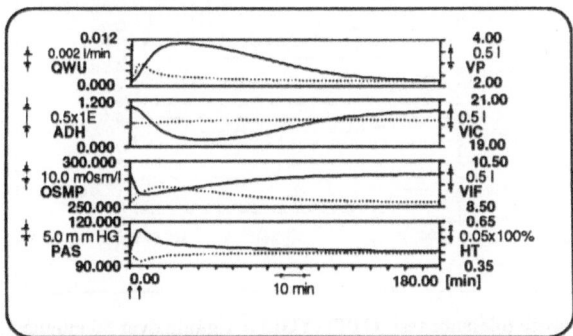

Figure 19.5. Water infusion of 1 litre (QVIN = 0.2 l/min during 5 minutes). After 180 minutes (3 hours) all variables return to their values at time = 0.

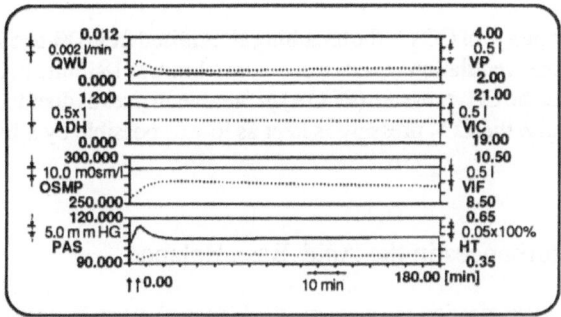

Figure 19.6. Infusion with physiological salt (QVIN = 0.2 l/min, YNIN = 31 and YCLI is 31 mEq/min over 5 minutes).

19.4.4 Glucose Tolerance Test

The glucose tolerance test (GTT) can indicate if someone has a normal glucose-insulin regulation. This test can be simulated, giving 50g of glucose through an infusion over 50 minutes. Figure 19.7 shows that the concentration of the glucose (in the ECFV) increases and decreases again immediately after stopping the infusion. Besides that it can also be seen that the glucose is cleared by the kidney (YGLU). The glucose level is back on its normal starting value rather quickly. Furthermore the fall in the extracellular potassium concentration is striking. For a diabetes patient the glucose concentration (in the ECFV) would remain high for a long time after stopping the glucose infuse. In a mild form of diabetes mellitus the insulin will be able to lower the glucose level, but it will go slowly.

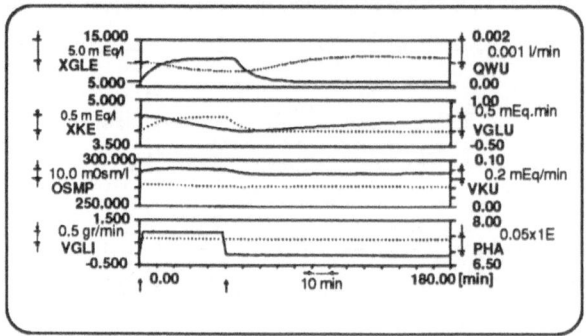

Figure 19.7. Glucose tolerance test (GTT) (YGLI = 1 g/min over 50 minutes).

19.4.5 Increased CO_2 Intake

When the percentage of CO_2 in the breathing air is raised from 0% to 5% for 50 minutes, the alveolar ventilation will increase from 5 l/min to 18 l/min. In Figure 19.8 it can be seen that the CO_2 pressure (PCOA) in the alveoli rises as does the O_2 pressure (PO2A). Clearly, the CO_2 pressure is kept as low as possible by a fast regulating mechanism.

19.4.6 Disturbances in the Acid-Base Balance

In the body's compensation mechanism, the kidney plays an important role during disturbances in the acid-base balance. The 'long-term' conduct of this model during respiratory acidosis and alkalosis is put by Ikeda et al. in the "pH-[HCO3]"-plane, as is shown in Figure 19.9. Curve OA corresponds to the response to an experiment in which the breathing air contains 10% CO_2 over an extended period. Curve OB

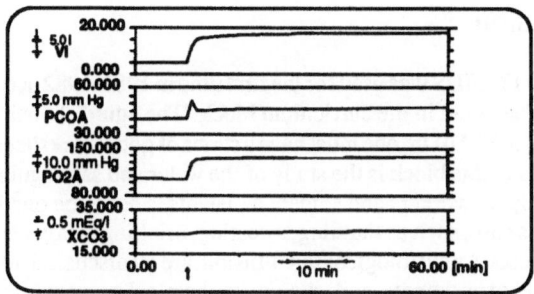

Figure 19.8. The respiration during raised percentage of carbon dioxide in the breathing air to 5% CO_2, i.e. FCOI = 0.05.

corresponds to the response to hyperventilation when the ventilation (VI) is three times higher than normal over an extended period. In the first two hours of experiments A and B there is compensation through intracellular buffering; after that there is mainly renal compensation (see Figure 19.9). Ikeda et al indicate that the results of this model correspond of the results found by other authors.

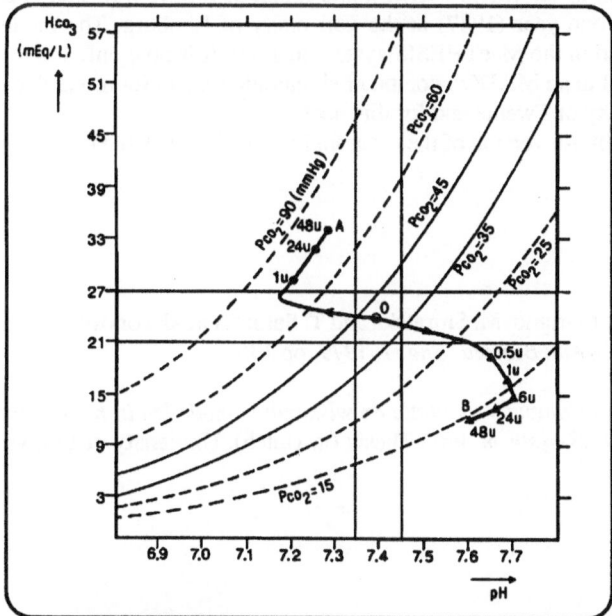

Figure 19.9. Two exercises with the values for the arterial blood pH, HCO_3-concentration and CO_2-pressure (Ikeda et al , 1979)
   A. Hyperventilation when VI=15 (after 48 hours) partially compensated respiratory acidosis).
   B. Inhalation 10% CO_2 during 48 hours (partially compensated respiratory alkalosis).

19.5 Discussion

The CS-program FLUIDS was used for the first time in 1981-1982 at the University of Limburg at Maastricht in the curriculum block "The adult". In using a complex model such as used FLUIDS, one must have a series of good worksheets. One of the educational goals in this block is the study of the water and salt regulation, and the case developed by co-workers and student assistants is based on one in this block. Aspects from this case, such as thirsting, sweating, drinking (sea)water, and giving the dessicated person a physiological salt infusion, are all discussed. Many students have done these experiments and shown in their subsequent discussions the willingness to tackle more problems in the teaching group and to study them more closely in the literature.

Working with the CS-program FLUIDS and the complex model of water and electrolyte regulation demands that the teacher be well-versed in the relationships within the model and that students have adequate workbooks or worksheets to guide them. The results obtained with this model are illuminating and give good cause for further study.

As mentioned in the Chapter 10, the CS-programs FLUIDS, CARDIO (both implemented in the RLCS-system), and MacDOPE have now had 3000 student sessions in one year (1987) at the University of Limburg. The model will be implemented in the MacTHESIS system in 1989. It is now only implemented in TurboPascal in an MS.DOS computer simulation system for research purposes at the University of Twente and for this book.

A MacIntosh version of the program FLUIDS is available from the author on request.

References

Ikeda, N., F. Marumo, M. Shirataka and T. Sato: "A model of overall regulation of body fluids". *Ann Biomed. Eng.* 7, 1979, pp.135.

Min, F.B.M.: *Computersimulatie en wiskundige modellen in het medisch onderwijs: Het RLCS-systeem.* PhD Thesis (in Dutch), University of Limburg, Maastricht, 1982.

---

*Acknowledgement - The author gratefully acknowledges the contribution of Dr. Ikeda and his co-workers (Kitasato University, Japan) to the CS-program FLUIDS.

Appendix

Starting values in the CS-program FLUIDS:

```
T:=0.0; {minutes}          QMWP:=0.0002;
PAS:=100.0;                QIWL:=0.0007;
PVS:=3.000;                PC:=17.0;
PAP:=20.000;               D57:=0.0;
PVP:=4.000;                XNE:=140.0;
RTOT:=20.000;              XKE:=4.5;
RTOP:=3.000;               YHI:=0.0;
KR:=0.300;                 YGLU:=0.0;
KL:=0.200;                 YMNU:=0.0;
PVP0:=0.000;               YURU:=0.15;
PVS0:=0.000;               VTW:=31.0;
QC0:=5.000;                OSMP:=287.0;
PCOA:=40.000               QIC:=0.0;
PHA:=7.4;                  XGLE:=6.0;
XCO3:=24.0;                YGLS:=0.0;
DCLA:=0.0;                 XKI:=140.0;
VI:=5.0;                   VIC:=20.0;
UCOV:=0.60753;             ZNE:=1540.0;
UO2V:=0.15148;             ZKE:=49.5;
FCOA:=0.0561;              ZKI:=2800.0;
FO2A:=0.1473;              ZHI:=100.0;
PO2A:=105.0;               ZGLE:=66.0;
MRCO:=0.2318;              ZMNE:=0.0;
MRO2:=0.2591;              ZURE:=77.5;
FCOI:=0.0;                 YNIN:=0.12;
FO2I:=0.21;                YKIN:=0.047;
PBA:=760.0;                YGLI:=0.0;
XHB:=15.0;                 YMNI:=0.0;
KVI:=0.0;                  YURI:=0.15;
KLF:=1.0;                  YINS:=0.0;
UHBO:=0.2;                 XGL0:=108.0;
TRSP:=2.0;                 CSM:=0.0003;
VB:=4.0;                   CKEI:=0.0007
VEC:=11.0;                 CGL1:=1.0;
PPCO:=28.0;                CGL2:=1.0;
XPP:=70.0;                 CGL3:=0.03;
HT:=0.45;                  PHI:=7.0;
VP:=2.2;                   YINT:=0.0;
VIF:=8.8;                  YCO3:=0.015;
PIF:=-6.3;                 YPO4:=0.025;
QLF:=0.002;                YORG:=0.01;
QCFR:=0.002;               STBC:=24.0;
ZPP:=154.0;                YCLU:=0.1328;
ZPIF:=176.0;               XCLA:=104.0;
ZPG:=20.0;                 ZCAE:=55.0;
ZPLG:=70.0;                ZMGE:=33.0;
VIN:=0.015;                ZSOE:=11.0;
XPIF:=20.0;                ZPOE:=22.0;
```

```
CFC:=0.007;
CRAV:=5.93;
VFBC:=1.8;
QIN:=0.0015;
QVIN:=0.0;
YPOI:=0.025;
YOGI:=0.01;
YCLI:=0.1328;
CION:=0.0;
D111:=0.0;
QWU:=0.001;
YNU:=0.12;
YKU:=0.047;
YNH:=1.4;
YNH4:=0.024;
STPG:=0.03823;
OSMU:=461.0;
PHA1:=7.4;
PHU1:=6.0;
PHU:=6.0;
YNH0:=0.024;
YTA0:=0.00677;
CPRX:=0.2;
YTA:=0.01677;
PHU2:=6.0;
CHEI:=5.0;
CBFI:=1.0E-9;
D129:=0.0;
ADH:=1.0;
ALD:=1.0;
THDF:=1.0;
GFR:=0.1;
ADH0:=0.0;
ALD0:=0.0;
GFR0:=0.1;
CPAD:=1.0;
COAD:=0.5;
ACTH:=1.0;
CKAL:=0.5;
CPVL:=0.1;
CNAL:=0.1;
CPAL:=0.01;
ALD1:=0.0;
DVWE:=0.0;
DVWI:=0.0;
DZN:=0.0;
DZK:=0.0;

ZOGE:=66.0;
ZCLE:=1144.0;
YCAI:=0.007;
YMGI:=0.008;
YSOI:=0.02;
DZCL:=0.0;
XCLE:=104.0;
D151:=0.0;
D152:=0.0;
TLOOP:=32000.0;
DVW:=0.0;
T0:=0.0;
TMAX:=60.0;
TMA:=180;
DTG:=0.2;
DTAS:=10.0;
DT:=0.199;
ZFLAG2:=0.0;
ZFLAG3:=0.0;
ZFLAG4:=0.0;
REQ:=0.0;
ERR:=0.0;
TUUR:=0.0;
TDAG:=0.0;
D270:=0.0;
REQ1:=0.0;
REQ2:=0.0;
REQ3:=0.0;
REQ4:=0.0;
REQ5:=0.0;
REQ6:=0.0;
REQ7:=0.0;
REQ8:=0.0;
REQ9:=0.0;
REQ10:=0.0;
REQ11:=0.0;
REQ12:=0.0;
REQ13:=0.0;
REQ14:=0.0;
QIPI:=2.45E-4;
QIL:=2.45E-4;
QIF:=7.0E-5;
QIZ:=1.4E-4;
YCA:=0.007;
YMG:=0.008;
YSO4:=0.02;
TMI:=0.0;
```

Model of FLUIDS in THESIS (MS.DOS version) and MacTHESIS.*)

N. Ikeda, F. Marumo, M. Shirataka and T. Sato
(Kitasato University, Japan).

Version F.B.M. Min & R.H.A.W. ter Hedde**)
Thanks to P. van Schaick Zillesen & H. van Kan

```
ERR:=0;
T := T + DT; (* IN MINUTES! *)
T_in_cursor := T;
```

┌─────────────────────────────┐
│ (1) HEART DYNAMICS   │
└─────────────────────────────┘

```
QCO0:=VB;
QCO:=QCO0+1;
PAS:=QCO0*RTOT+20;
PVS:=QCO0/KR-10.33;
PAP:=QCO0*RTOP+8.0;
PVP:=QCO0/KL-16.0;
IF PVS < 0.0
    THEN PVS:= 0.0;
IF PVP < 0.0
    THEN PVP:= 0.0;
IF ( PAS < 85.0 ) AND ( REQ14 <> -14.0 )
    THEN REQ14:= 14.0;
IF PAS < 0.0
    THEN ERR:= -5.0;
IF PAS > 250.0
    THEN ERR:= 5.0;
```

┌─────────────────────────────┐
│ (2)+(3) BODY FLUIDS*)   │
└─────────────────────────────┘

```
DV:=QIN-VIN/10.0;
VIN:=VIN+DV*DT;
DV:=VIN/10+QVIN+QMWP-QIWL-QWU-QCFR+QLF;
VP:=VP+DV*DT;
IF VP < 0.1
    THEN VP:= 0.1;
VB:=VP+VRBC;
VEC:=VP+VIF;
IF VEC <= 0.0
    THEN ERR:= -4.0;
IF VEC > 30.0
    THEN ERR:=4.0;
HT:=VRBC/VB;
DV:=QCFR-QLF-QIC;
VIF:=VIF+DV*DT;
X:=VIF/8.8;
IF X < 0.9
```

```
      THEN PIF:= -15.0;
 IF ( X >= 0.9 ) AND ( X < 1.0 )
      THEN PIF:= 87*X-93.3;
 IF ( X >= 1.0 ) AND (X < 2.0 )
      THEN PIF:=-6.3*EXP(10*LN(ABS(2-X)));
 IF (2-X) = 0.0
      THEN ERR:= 6.0;
 IF X >= 2.0
      THEN PIF:= X-2;
 QLF:= 0.002 * 24 / (1 + EXP(-0.4977 * PIF));
 PC:= (PAS + PVS * CRAV) / (1.0 + CRAV);
 PPCO:= XPP * 0.4;
 PICO:= XPIF * 0.25;
 QCFR:= (PC - PIF - PPCO + PICO) * CFC;
 YPLC:= ( XPP - XPIF ) * 2.768E-6 * SQR(PC);
 YPLF:= XPIF * QLF;
 YPLV:= XPP * 0.00047 - 0.0329;
 YPLG:= ( XPP-ZPLG ) * 0.00023;
 DZ:= YPLF - YPLV - YPLG - YPLC;
 ZPP:= ZPP + DZ * DT;
 XPP:= ZPP / VP;
 YPG:= ( XPIF - ZPG ) * 0.0057;
 DZ:= YPLC - YPLF - YPG;
 ZPIF:= ZPIF + DZ * DT;
 XPIF:= ZPIF / VIF;
 DZ:= ( XPIF - ZPG ) / 150.0;
 ZPG:= ZPG + DZ * DT;
 DZ:= ( XPP - ZPLG ) / 24.0;
 ZPLG:= ZPLG + DZ * DT;
 IF ( HT <= 35.0 ) AND ( REQ9 <> -9.0 )
      THEN REQ9:= 9.0;
 IF ( HT <= 30.0 ) AND ( REQ10 <> -10.0 )
      THEN REQ10:= 10.0;
```

```
     (4)  ELECTROLYTES
```

```
 DZ:= YNIN - YNU + YHI + CION / 1440.0;
 ZNE:= ZNE + DZ * DT;
 XNE:= ZNE / VEC;
 QIC:= ( XKI + 10.5 - XNE - XKE - XGLE ) * CSM;
 VIC:= VIC + QIC * DT;
 F41A:= XKE / (56.744 - 7.06 * PHA );
 IF F41A <= 0.0
      THEN ERR:= 1.0;
 F41:=1.0+0.5*(LN(F41A)/NL10);
 DZI:=(F41*2800.0-ZKI)*CKEI+CGL3*YGLS;
 ZKI:=ZKI+DZI*DT;
 DZ:=YKIN-YKU-DZI;
 ZKE:=ZKE+DZ*DT;
 XKE:=ZKE/VEC;
 IF (XKE >= 9.5) AND (REQ5 <> -5.0)
```

```
      THEN REQ5:= 5.0;
IF (XKE > 7.0) AND (REQ6 <> -6.0)
      THEN REQ6:= 6.0;
IF (XKE <= 2.5) AND (REQ7 <> -7.0)
      THEN REQ7:= 7.0;
IF (XKE < 2.0) AND (REQ8 <> -8.0)
      THEN REQ8:= 8.0;
XKI:=ZKI/VIC;
YHI:=CHEI*(PHI-PHA+0.4);
ZHI:=ZHI+YHI*DT;
PHIA:=ZHI*CBFI;
IF PHIA <= 0.0
      THEN ERR:= 2.0;
PHI:= (LN(PHIA)/NL10);
DY:=XGLE-XGL0/18.0-YINT;
YINT:=YINT+DY*DT;
YGLS:=CGL1*YINT+CGL2*YINS;
DZ:=YGLI/0.18-YGLU-YGLS;
ZGLE:=ZGLE+DZ*DT;
XGLE:=ZGLE/VEC;
YGLU:=FNTML(XGLE*GFR, 0.65);
DZ:=YMNI-YMNU;
ZMNE:=ZMNE+DZ*DT;
XMNE:=ZMNE/VEC;
YMNU:=XMNE*GFR;
DZ:=YURI-YURU;
ZURE:=ZURE+DZ*DT;
XURE:=ZURE/VTW;
YURU:=XURE*GFR*0.6;
VTW:=VIC+VEC;
OSMP1=86*(XNE+XKE)+XGLE+XURE+XMNE+9.73;
```

```
┌─────────────────────────────────┐
│      (5) ION EXCRETION           │
└─────────────────────────────────┘
```

```
F50:=PCOA/120.0+1.33333;
X:=XCO3*GFR*F50;
IF X < 2.0
      THEN YCO3:= 0.0;
IF (X-2) = 0.0
      THEN ERR:= 7.0;
IF (X >= 2.0) AND (X < 4.0)
      THEN YCO3:= 0.1638*EXP(2.61*LN(X-2));
IF X >= 4.0
      THEN YCO3:= X-3;
DZ:=YCAI-YCA;
ZCAE:=ZCAE+DZ*DT;
DZ:=YMGI-YMG;
ZMGE:=ZMGE+DZ*DT;
DZ:=YSOI-YSO4;
ZSOE:=ZSOE+DZ*DT;
DZ:=YPOI-YPO4;
```

```
ZPOE:=ZPOE+DZ*DT;
DZ:=YOGI-YORG;
ZOGE:=ZOGE+DZ*DT;
DZ:=YCLI-YCLU+CION/1440.0;
ZCLE:=ZCLE+DZ*DT;
XCAE:=ZCAE/VEC;
XMGE:=ZMGE/VEC;
XSO4:=ZSOE/VEC;
XPO4:=ZPOE/VEC;
XOGE:=ZOGE/VEC;
XCLE:=ZCLE/VEC;
YCA:=FNTML(XCAE*GFR,0.493);
YMG:=FNTML(XMGE*GFR,0.292);
YSO4:=FNTML(XSO4*GFR,0.08);
X:=XPO4*GFR;
IF X < 0.2
    THEN YPO4:=X/8;
IF X >= 0.2
    THEN YPO4:=X-0.175;
X:=XOGE*GFR;
IF X < 0.6
    THEN YORG:=X/60;
IF X >= 0.6
    THEN YORG:=X/3-0.19;
XCLA:=XCLE-DCLA;
YCLU:=STPG+YNU+YKU+YNH4-YCO3+YCA+YMG-YSO4;
IF YCLU < 0.0
    THEN YCLU:=0.0;
STBC:=
-XCLE-XOGE-XPO4-

XSO4+XMGE+XCAE+XNE+XKE-0.2214*XPP;
```

┌─────────────────────────────────────┐
│ (6) KIDNEY AND URINE BUFFER   │
└─────────────────────────────────────┘

```
GFR1:=GFR*CPRX*THDF;
YKD:=XKE*GFR1*0.5*0.9+0.0187*ALD*XKE;
YNH:=XNE*GFR1*0.5;
YND:=YNH*0.9-0.09*ALD;
QWD:=1.86*(YND+YKD)+YGLU+YURU+YMNU+0.32)/OSMP;
QWU:=QWD*(1-0.9*ADH);
OSMU:=1.86*(YNU+YKU)+YGLU+YURU+YMNU)/QWU;
F61:=-0.5*PHU1+4;
F62:=-2.5*PHA1+19.5;
IF PHU2 < 4.0
    THEN F63:=0.0;
IF (PHU2 >= 4.0) AND (PHU2 < 5.0 )
    THEN F63:=PHU2-4;
IF PHU2 >=5.0
    THEN F63:=1.0;
```

```
YNH4:=YNH0*F61;
YTA1:=YTA0*F62*F63;
YNDD:=YCO3+YNH4+YTA1;
IF YNDD < 0.0
    THEN YNDD:= 0.0;
YNU:=YND*0.116-YNDD;
IF YNU < 0.0
    THEN YNU:= 0.0;
YKU:=YKD*0.39;
DP:=(PHA-PHA1)/200;
PHA1:=PHA1+DP*DT;
DP:=(PHU-PHU1)/300;
PHU1:=PHU1+DP*DT;
DP:=(PHU-PHU2)/20;
PHU2:=PHU2+DP*DT;
YTA:= YTA1+0.001*ALD+0.009;
STPO:=YPO4*(1+1/(1+EXP((6.8-PHA)*LN(10))));
STPG:=YTA+STPO+YORG;
IF STPG < 0.0
    THEN STPG:= 0.0;


(*-----PHU CALCULATION-----*)


A:=STPG-YPO4;
BA=(CPO4+CORG)-(CPO4*YPO4+CORG*YORG);
C:=CPO4*CORG*(A-YPO4-YORG);
D:=B*B-4*A*C;
IF D <= 0
    THEN  D:= 0.00001;
HU:=0.5*(-B+SQRT(D))/A;
IF HU >1.0E-4
    THEN HU:= 1.0E-4;
IF HU < 1.0E-8
    THEN HU:= 1.0E-8;
PHU:=-(LN(HU)/NL10);
```

```
    (7)  KIDNEY CONTROL
```

```
DA:= (-CPAD*(PVP-4)+COAD*(OSMP-287)-ADH0)/30;
ADH0:=ADH0+DA*DT;
ADH:=1.1/(1+EXP(-0.5*(ADH0+4.605)));
DL(ACTH-1)+CKAL*(XKE-4.5)-CPVL*(PVP-4)
    -CNAL*(YNH-1.4)-CPAL*(PAS-100);

(*----------------BUFFER----------------*)
(* 100 MIN DELAY:  IN:DL
OUT:ALD1 (DELAYED with DT minutes) *)

DELAY(DL,ALD1,100,DT,BUF);
```

```
DA:=(ALD1-ALD0)/30;
ALD0:=ALD0+DA*DT;
ALD:=10/(1+EXP(-0.4934*(ALD0-5)));
X:=PPCO/28;
THDF:=-5*(X-1)+1;
IF X > 1.0
    THEN THDF:=1.0;
X:=PAS;
IF X < 40.0
    THEN GFR1:=0.0;
IF (X >= 40.0) AND (X < 80.0)
    THENGFR1:=0.02*X-0.8;
IF (X >= 80.0) AND (X < 100.0)
    THENGFR1:=-0.0005*(X-100)*(X-100)+1;
IF X >= 100.0
    THEN GFR1:=1.0;
GFR:=GFR0*GFR1*VEC/11.0;
DVWE:=VEC-11;
DVWI:=VIC-20;
DZN:=ZNE-1540;
DZK:=ZKE+ZKI-2849.5;
DZCL:=ZCLE-1144;
DVW:=DVWE+DVWI;
```

```
     (9) RESPIRATION AND PH
```

```
UCOA:=PCOA*6.732E-4+XCO3*0.02226;
DU:=(UCOA-UCOV)*QCO+MRCO)/VTW;
UCOV:=UCOV+DU*DT;
PBL:=PBA-47;
CRUF:=QCO*863/PBL;
DF:=(UCOV-UCOA)*CRUF+(FCOI-FCOA)*VI)/3;
FCOA:=FCOA+DF*DT;
PCOA:=FCOA*PBL;
IF PO2A < 33
    THEN PO2A:=33;
VR:=0.22*MACHT(10,(9-PHA))+0.262*32-18.238+0.2125*
    (1+17/(PO2A-32))*(PCOA-32);
IF VR <= 0.0
    THEN VR:=0.02;
DV:=(5.0*VR*KLF-VI)/TRSP;
VI:=VI+DV*DT;
IF KVI > 0.0
    THEN VI:=KVI;
PHAA:=XCO3/(0.03*PCOA);
IF PHAA <= 0.0
    THEN ERR:=3.0;
PHA:=6.1+(LN(PHAA)/NL10);
X:=PHA;
Y:=0.454+X*(0.44921+X*(-0.10098+X*0.0066815));
Z:=PO2A*Y;
```

```
UHB:=XHB/75;
UHBO:=UHB*(1-EXP(-Z))*(1-EXP(-Z));
XCO0:=STBC-(0.527*XHB+3.7)*(PHA-7.4)+

(UHB-UHBO)*0.375/0.02226;
DX:=(XCO0-XCO3)/TRSP;
XCO3:=XCO3+DX*DT;
DCLA:=XCO3-STBC;
UO2A:=UHBO+3.168E-5*PO2A;
DU:=((UO2A-UO2V)*QCO-MRO2)/VTW;
UO2V:=UO2V+DU*DT;
DF:=((UO2V-UO2A)*CRUF+(FO2I-FO2A)*VI)/3;
FO2A:=FO2A+DF*DT;
PO2A:=FO2A*PBL;

(* MESSAGES OR REQUESTS:


 *)
(* TEST IF VARIABLES ARE OUT OF RANGES:  *)

IF (PHA >6.95) AND (PHA <7.05) AND
    (REQ1 <> 1.0)
    THEN REQ1:= -1.0;
IF (PHA < 6.95) AND (REQ2 <> -2.0)
    THEN REQ2:= 2.0;
IF (PHA < 6.9) AND (REQ3 <> -3.0)
    THEN REQ3:= 3.0;
IF (PHA < 6.8) AND (REQ4 <>-4.0)
    THEN REQ4:= 4.0;
```

---

*)FLUIDS is available on MS.DOS (implemented in THESIS), on VAX (in the RLCS-system) and on Macintosh (in colour, in MacTHESIS version 4.0).

**)For more information about the model: Write to F.B.M. Min, University of Twente, P.O. Box 217, 7500 AE ENSCHEDE, The Netherlands.

20
Pharmacokinetics

Dietmar P.F. Möller

20.1 Introduction

There are a number of different ways in which a drug can be administered. The most obvious way is oral medication. A second way is injection directly into a vein, or (seldom) into an artery or into body tissue, such as a muscle.

Since it is desirable to provide a dosage regimen that is individually tailored to the patient, it is important to understand the pharmacokinetics of the drugs being used. In the case of drugs being administered orally, doses and the timings of their intake can be determined by using compartment models, enabling the student to develop therapies that are in certain ways optimal.

The subject of pharmacokinetics has been defined [Gibaldi and Perrier,1975] as "the study of the time course of drug and metabolic levels in different fluids, tissues and excreta of the body, and of the mathematical relationships required to develop models to interpret such data". The main point of modeling physiological kinetic problems is to separate the human body into a finite number of component parts, which interact through the exchange of matter. These components are called compartments.

20.2 Theory of Pharmacokinetic Compartment Models

A compartment is defined [Brown,1980] as "a vessel which contains a single distinct form of matter (or energy), and to which the law of conservation of matter (or energy) applies. Often the quantity of matter in the compartment is deduced from analysis of small samples (concentration measurements), for which purpose it is necessary to specify a volume for the compartment and to postulate that the compartmental contents are uniformly distributed (well mixed). In an interconnected system of compartments, there is the possibility of bidirectional flow between any one compartment and any other. Sometimes, different compartments contain different species, in which case intercompartmental flows refer not to transport of matter, but to biotransformation of matter (with the possibility that different species may occupy the same physical space)".

Models of compartment systems are easy to understand and well accepted by biomedical scientists who are not experts on mathematics. The basic equations of compartmental analysis are a set of differential equations that determine the exchange of matter among various compartments and environments. By "analysis" is meant the determination of the form of the compartmental responses to external perturbation of one or more compartments, given the rate constants. Such analysis is a prerequisite for tackling the important problem of identifiability - that is, whether, given some or all of the responses, the unknown rate constants can be found" [Gibaldi,1985].

Hence the pharmacokinetics of a drug administered non-orally can be characterized by three different transport processes: absorption, distribution and elimination. In case of oral drug administration, a pre-absorption process must be added, describing the dissolution of the drug in the stomach and its transportation to the site of absorption.

Moreover, all drugs have a therapeutic concentration level between one that is too low, rendering the drug ineffective, and one that is too high, which leads to toxicity. Therefore an optimal administration of drugs at a therapeutic level is the main point of every therapy.

20.2.1 Linear Pharmacokinetic Models

For many drug applications, a model such as shown in Fig. 20.1, is proper to use.

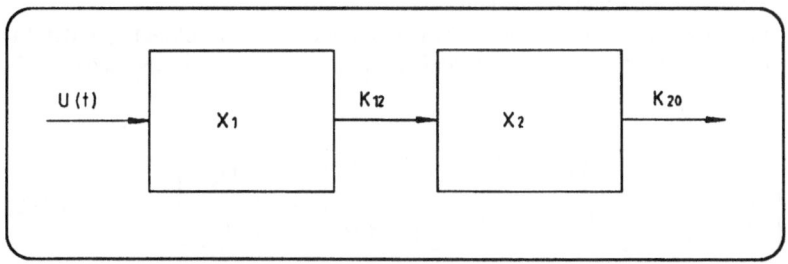

Figure 20.1. Two compartment model

Supposing that the drug is taken orally, it enters the gastrointestinal tract, is absorbed into the circulation and distributed throughout the body to be metabolized and finally eliminated. Compartment X_1 describes the gastrointestinal tract and the gastrointestinal vascular bed (circulation); compartment X_2 stands for the bloodstream (between the distribution and elimination process), whereas K_{12} and

K_{20} represent the distribution and elimination, respectively. Let us start at time zero, and let $X_1(t)$ denote the mass of drug in compartment 1 and $X_2(t)$ be the mass of drug in compartment 2. If the ingestion rate of the drug $U(t) > 0$, we find the plausible assumption for the model [Godfrey,1983]

$$\dot{X}_1(t) = U(t) - \text{drug distribution rate compartment1 to 2} \qquad (20.1)$$

Equation (20.1) is commonly a mass balance equation. In the case of first-order kinetics, drug distribution rate from compartment 1 to 2 is assumed to be proportional to the mass (or concentration) of the drug in the first compartment. If $K_{12} > 0$ is the corresponding proportionality constant, then (20.1) becomes

$$\dot{X}_1(t) = U(t) - K_{12}.X_1(t) \qquad (20.2)$$

Compartment 2 is described by a flow rate equation that balances the inflow and outflow rates described by equation (20.2)

$$\dot{X}_2(t) = \text{inflow rate - outflow rate} \qquad (20.3)$$

where $K_{12}.X_1$ is the inflow rate of drug distribution from the first compartment. With respect to first order kinetics, the outflow rate of compartment 2 is proportional to X_2. Thus (20.3) becomes

$$\dot{X}_2(t) = K_{12}.X_1(1) - K_{20}.X_2(t) \qquad (20.4)$$

K_{20} being the elimination constant.

Equations (20.2) and (20.4) constitute the linear pharmacokinetic model. In a matrix-vector format the constant coefficient linear differential equation model is $(t > 0)$

$$\begin{bmatrix} \dot{X}_1(t) \\ \dot{X}_2(t) \end{bmatrix} = \begin{bmatrix} -K_{12} & 0 \\ K_{12} & -K_{20} \end{bmatrix} \begin{bmatrix} X_1(t) \\ X_2(t) \end{bmatrix} + \begin{bmatrix} U(t) \\ 0 \end{bmatrix} \qquad (20.5)$$

The pharmacokinetic model described in equation (20.5) is a second-order linear model. The first differential equation is uncoupled from the second differential equation, the second differential equation is coupled with the first differential equation. It should be noted that this observation is important, since mathematical models should not be excessively difficult for analytical studies. Otherwise there is no closed analytical solution.

The primary interest of studying compartment models as shown in Fig. 20.1 is to govern how input ingestion rate and/or the initial mass (concentration) of the drug in the body affects the person. The main interested variable $X_2(t)$ (and hence compartment X_2) is of great interest, because it is accessible to analysis by taking blood samples.

20.2.2 Nonlinear Pharmacokinetic Models

In contrast to the metabolic system kinetics considered previously in the linear compartmental models, nonlinear compartmental models arise in enzyme kinetics [Godfrey,1983]. The enzyme molecule E combines with the substrate molecule S (input of the compartment) to form a molecule complex C. When bound to the enzyme molecule in this complex as shown in [Godfrey,1983], the substrate is somehow more likely to form the product P of the reaction (output of the compartment). When this takes place, the complex breaks into the product molecule P and the enzyme molecule E. It is proper to follow the reaction schematically.

$$S + E \underset{K_{-1}}{\overset{K_1}{\rightleftharpoons}} C \xrightarrow{K_2} P + E \tag{20.6}$$

which will lead to the nonlinear compartmental model in state space notation

$$\begin{bmatrix} \dot{S} \\ \dot{C} \end{bmatrix} = \begin{bmatrix} G_{11} & G_{12} \\ G_{21} & G_{22} \end{bmatrix} \cdot \begin{bmatrix} S \\ C \end{bmatrix} \tag{20.7}$$

where $G_{11} = K_1 . E, G_{12} = K_1.S + K_{-1}, G_{21} = K_1 E, G_{22} = -(K_1 S + K_{-1} + K_2)$ and $E = E(t) + C(t)$, which turns out to be constant for the entire time t.

20.3 Aspects to Be Considered

20.3.1 Limitations of the Model

The main purpose of the model is to demonstrate the basic functioning of the time course of drugs in different fluids, tissues and/or organs and excreta of the body. For that reason the model is kept as simple as possible while remaining accurate and realistic. The mathematical description is based on data available from experiments or the literature. In the next paragraphs the limitations are summarized.

- Since simplifications are made in the formulation of the model, feedback influences are neglected.

- As in real biological systems, the model does not have setpoints. We cannot instruct the model to adjust the output variables to certain values of the state variables. We can only change the parameters. A brief introduction into the model reference adaptive approach technique will be given in section 20.6.
- The model is not suitable for studying the effects of physical workload because the mechanisms that maintain the oxygen utilization during workload are neglected.

20.3.2 Outlook

The time courses and the parameter sensitivity in different diseases (diabetes, hydrocephalus) can be studied with second-order nonlinear compartment models. Moreover the effects of drugs can be studied as predictive dose-response compartment models and predictive compartment models for therapeutic patient management or diagnostics.

Using identification techniques (see section 20.6) as a tool to model and simulate compartmental processes, the system structure can be identified by measured data, or internal non-measurable parameters can be estimated.

20.4 Model Description

20.4.1 Mathematical Model

The physical foundations and equations used in the model have been presented in section 20.2.1.
To simulate the metabolism of a drug in a given individual, the equation that describes the ingestion must be adapted to the particular drug.

20.4.2 Individual Blocks of the Model

Fig. 20.2 shows the block diagram realization of the model of Fig. 20.1 in BIOPSI:

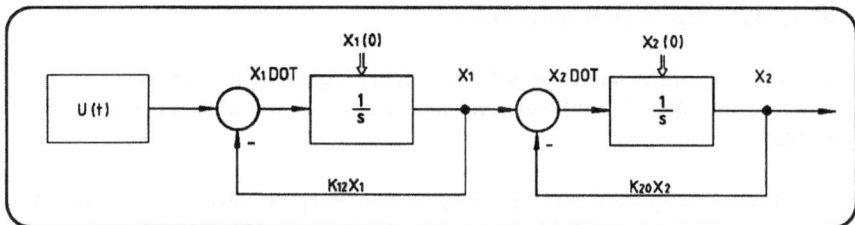

Figure 20.2. BIOPSI-realization of the two compartment model of Fig. 20.1

The implementation in BIOPSI is shown in Table 20.1.

Table 20.1. Two compartment model

Block	Type	Input 1	Input 2	Input3	Par1	Par2	Par3
U	CON				.0000		
X1	INT	X1DOT			100.0	1.000	
X2	INT	X2DOT			100.0	.5000	
X1DOT	SUM	U	X1		1.000	-1.000	
X2DOT	SUM	X1	X2		1.000	-1.000	

From Table 20.1 it can be concluded that parameter Par1 of blocks X_1 and X_2 describe the initial drug concentrations $X_1(0)$ and $X_2(0)$, respectively, in mg/ml. Parameter Par2 of X_1 and X_2 are the respective rate constants K_{12} and K_{20} in h^{-1}. The parameters Par1 and Par2 of blocks X1DOT and X2DOT describe the weighting and the sign of the summarizers.

20.4.3 Parameters and Units

U: ingestion rate in mg/min, ml/min
X_1:   mass (or concentration) of drug in the first compartment in mg (mg/ml)
X_2:   mass (or concentration) of drug in the second compartment in mg (mg/ml)
K_{12}:   the transfer rate constant from compartment 1 to 2 in min^{-1}, h^{-1}
K_{20}:   the transfer rate constant from compartment 2 to excretia or another compartment in min^{-1}, h^{-1}
$X_1(0)$: the initial amount (concentration) of drug in compartment 1 in mg (mg/ml)
$X_2(0)$: the initial amount (concentration) of drug in compartment 2 in mg (mg/ml)

20.4.4 Operation of the Program

The operation of the models is self-explanatory, based on the general structure shown in Fig. 20.1. The following remarks will assist you in understanding the operation.

- The case study number must be chosen from the pull-down menu. Case 1 is case study 1, it runs with $U = 0$, case 2 is case study 2 with $U = 0$. When starting the program, the simulation runs with a parameter set within normal range, shown as case number 1 in Table 20.2, and a print or plot interval of 1 hour.
- As a first case study we try to understand how different values of the transfer rate constants K_{12} and K_{20} and the initial drug masses $X_1(0)$, $X_2(0)$ affect the amounts of drug in the respective compartments, e.g. gastrointestinal tract, bloodstream.
- When changing model parameters the program recalculates the new transients and the new values appear.
- If you start the program with a parameter set outside physiological range, you are referred to the Help-menu.

20.5 Experiments

Implementation of the model shown in Fig. 20.1 in BIOPSI is based on the parameter set listed in Table 20.1 as case 2 of study 1, listed in Table 20.2.

20.5.1 Exercises

Some typical time responses are shown as case examples in Table 20.2 to demonstrate how different values of $X_1(0)$, $X_2(0)$, K_{12} and K_{20} affect the amounts of drug in the compartments X_1 or X_2.

Case	K_i / $X_i(0)$	$X_1(t)$ / $X_2(t)$									
No.	t in (h)	0	1	2	3	4	5	6	7	8	9
1	$K_{12} = 0.5$ $K_{20} = 0.25$ $X_1(0) = 100$	100	60,7	36,8	22,3	13,5	8,2	4,9	3,0	1,8	1,1
	$X_2(0) = 100$	100	95,1	84,5	72,2	60,0	49,1	39,6	31,7	25,2	19,9
2	$K_{12} = 1.0$ $K_{20} = 0.5$ $X_1(0) = 100$	100	36,8	13,6	5,0	1,8	0,7	0,2	-	-	-
	$X_2(0) = 100$	100	84,5	60,0	39,7	25,2	15,8	9,7	6,0	3,6	2,2
3	$K_{12} = 2.0$ $K_{20} = 1.0$ $X_1(0) = 100$	100	13,7	1,9	0,3	-	-	-	-	-	-
	$X_2(0) = 100$	100	60,0	25,2	9,8	3,7	1,4	0,5	0,2	-	-
4	$K_{12} = 1.0$ $K_{20} = 0.5$ $X_1(0) = 100$	100	36,8	13,6	5,0	1,8	0,7	0,2	-	-	-
	$X_2(0) = 0$	0,0	23,8	23,2	17,3	11,7	7,5	4,7	2,9	1,8	1,1
5	$K_{12} = 0.25$ $K_{20} = 5.0$ $X_1(0) = 200$	200	155,8	121,3	94,5	73,6	57,3	44,6	34,8	27,1	21,1
	$X_2(0) = 100$	100	163	127,7	99,5	77,5	60,3	47	36,6	28,5	22,2

Table 20.2. Simulation examples of the two compartment models of Fig. 20.1, for case study 1

How can we understand Table 20.2 ? K_i and $X_i(0)$ are the respective rate constants and initial conditions of the case examples 1 to 5. The state variables $X_1(t)$ and $X_2(t)$ belong to their discrete time points in h. Compared to case 2, we find a reduction of drug concentration for case 3 in both compartments.

The transient behaviour of case number 2 in Table 20.2 is shown in Fig. 20.3.

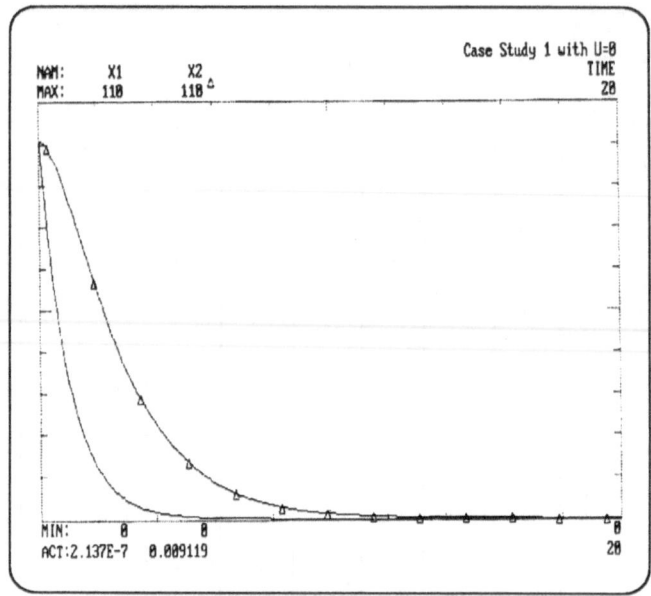

Figure 20.3. Transient behaviour of the concentration X_2 in compartment two

Increasing the initial dosage of compartment one, with the assumptions that distribution and elimination rates are $K_1 = 0.25\ h^{-1}$ and $K_2 = 5\ h^{-1}$, we get the results in case number 5: an overshoot in X_2 and the transient of X_2 in Fig. 20.4. Ordinate setting is changed with DSA 0,220.

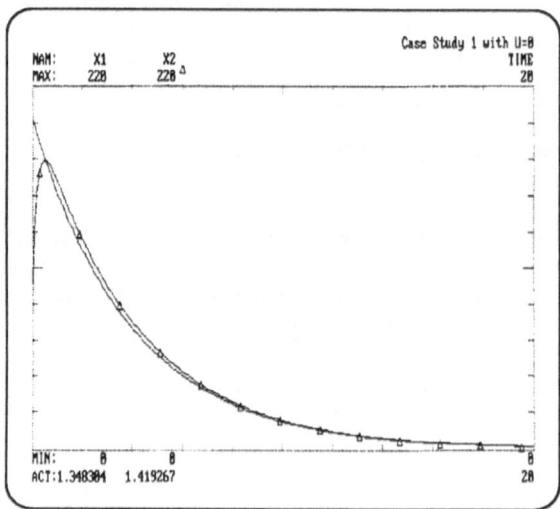

Figure 20.4. Transient behaviour of X_2 for case number 5

The influence of the input ingestion rate on the dynamics of the compartments can be investigated by introducing the input
as a rapid injection of a known amount of tracer (mass, concentration). Hence we can write

$$U(t) = A \cdot \exp(-K_{31} \cdot t) \tag{20.8}$$

The realization in BIOPSI of case study 2 is shown in Table 20.3:

Table 20.3. Two compartment model with input ingestion rate U(t)

Block	Type	Input 1	Input 2	Input 3	Par1	Par2	Par3
A	CON				200.0		
B	CON				1.000		
K3	CON				-.2000		
Y	EXP	E1			1.000	1.000	1.000
X1	INT	X1DOT			.0000	1.000	
X2	INT	X2DOT			.0000	.5000	
X3	INT L	K3	X3				
U	MUL	A	Y				
X1DOT	SUM	U	X1		1.000	-1.000	
X2DOT	SUM						

The realization of equation (20.8) requires the blocks A, B, K3, Y, X3, E1 and U for the simulation, time t has to be generated by t = X3 = INT B. The simulation results for U, X1 and X2 are shown in Fig. 20.5. Since $K_{31} < K_{20} < K_{12}$, it takes a much longer time for the drug in the bloodstream ($X_2(t)$) to reach its maximum value and to decay than it does in the gastrointestinal tract ($X_1(t)$).

The programmes in Tables 20.1 and 20.2 are particularly simple and should be easy to follow. The notation can be used for individual case study examples. Analysis of such cases could be done using the analytical framework that has been developed in the equations above.

20.5.2 What You Can Do on Your Own

Now that you are acquainted with the different equations that make up the model, we can start more research-oriented exercises that will give you more insight into the functioning of ingestion, distribution and elimination of a drug.

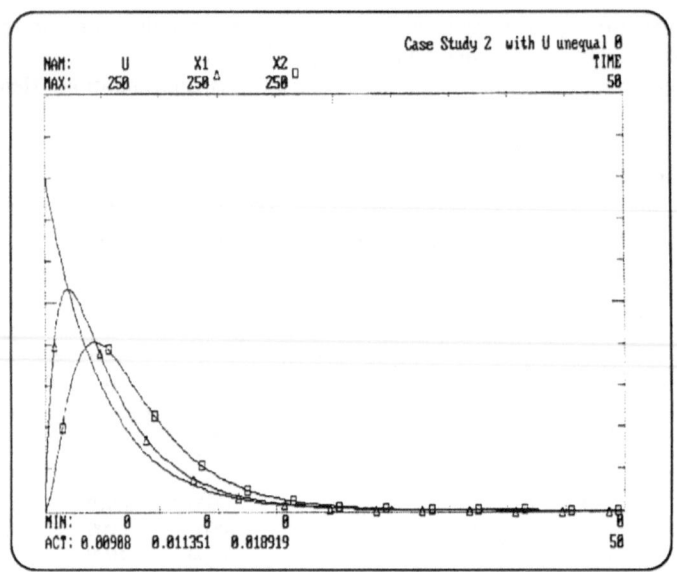

Figure 20.5. Time response for the ingestion rate U(t), the drug concentration (g/ml) in compartment 1 (X_1) and compartment 2 (X_2)

Case A: There was a premedication of X_{10} = 100 mg/ml. In the bloodstream X_{20} = 0, and the ingestion rate is U(t) = 0.
The dynamic behaviour of $X_1(t)$ and $X_2(t)$ should be simulated. What are the maximum values for each of the drug levels and at what time do these maximums occur when changing the values of K_{12} and K_{20}.

Case B: There was no premedication, $X_{10}=X_{20}=0$. The drug ingestion rate into the gastrointestinal tract is based upon equation (20.8). The dynamic behaviour of $X_1(t)$ and $X_2(t)$ should be shown via simulation for

$$U(t) = A.exp(-K_{31}.t)$$

with $0 \le t \le 3h, t \ge 3h$.
The maxima and the times of these maxima should be discussed.

20.6 Optimization of Compartmental Models

A comparison is made between experimental data and data obtained from the model outputs. Then the parameters of the model can be estimated by minimizing a least squares criterion function with an adaptive optimization algorithm.

The criterion function normally chosen is

$$J = \sum_{i=1}^{n} W_i \, [\, Y_{meas} \, (t_i) - Y_{mod} \, (t_i \, , Q) \,]^2 \tag{20.9}$$

where W_i is the weighting matrix assigned to the ith point; $Y_{meas}(t_i)$ is the ith experimental measurement; $Y_{mod} \, (t_i)$ is the model output at $t = t_i$, which is a function of the model parameters and n is the number of experimental data.

J is a non-linear function of the model parameters. For the minimization of J, the algorithm of Powell, Rosenbrock, Davidon-Fletcher-Powell has been employed [Möller,1983]. The identification task is to adjust the parameter vector Q of an identification model having the same structure as the true model, in such a way that the output $<Y_{meas} \, (t_i, Q)>$ coincides with the output $<Y_{mod} \, (t_i, Q)>$ of the true model [Möller,1985].

Since $<Y_{mod} \, (t_i, Q)>$ is not known, it is at most possible to compare $<Y_{mod} \, (t_i, Q)>$ with $<Y_{meas} \, (t_i, Q)>$.

If the difference

$$V_K(Q) = Y_{meas} \, (t_i, Q)_K - Y_{mod} \, (t_i, Q)_K \tag{20.10}$$

is interpreted as an estimate of VK, with

$$V_K = Y_{meas} \, (t_i, Q)_K - Y_{mod} \, (t_i, Q)_K,$$

the identification task is to adjust 0 in such a way as to impress on the sequence $<V_K(0)>$ some known statistical properties of $<V_K>$, e.g. its mean and its variance. Assuming that $<V_K>$ is white and stationary with mean zero and variance V2, this task can be done by minimizing the well-known output error least squares function

$$J(Q) = \sum_{K=1}^{n} \hat{V}_K(Q)^2 = \sum_{K=1}^{n} (\hat{Y}_{measK} - Y_{modK} \, (\hat{Q}))^2 \tag{20.11}$$

the minimizing argument in which, Q_{min} is a consistent estimate of the true parameter vector Q_s. The schematic diagram is shown in Fig. 20.6.

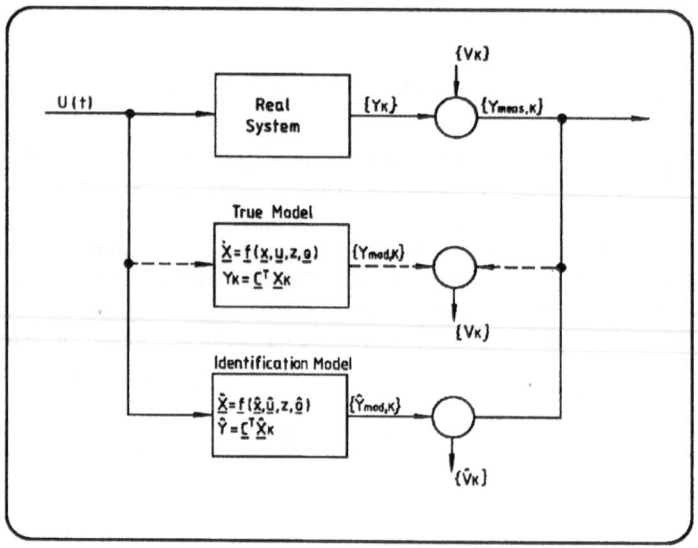

Figure 20.6.   Relationships between the real system, the true model, and its identification
model

20.7   Conclusions

Although the model is very basic and only describes the time course, it is very instructive. Moreover it has already proven to be accurate and realistic in explaining
clinical situations or pharmacokinetical dose-response regimens.

It can be used in education, research such as studies for optimal drugs design, or
even clinical applications to study the basic mechanisms of first order kinetics
under different boundaries.

References

Anderson, D.H.: Compartment Modeling and Tracer Kinetics.
Springer Verlag, Berlin-Heidelberg, 1983.

Brown, B.F.: "Compartment System Analysis : State of the Art".
IEEE-BME-27, 1, 1980, pp. 1-11.

Gibaldi, M. and D. Perrier: Pharmacokinetics. Marcel Dekker, New York, 1975.

Godfrey, K.: Compartment Models and their Application. Academic Press,
London, 1983.

Möller, D.P.F., D. Popovic, and G. Thiele: Modeling, Simulation and Parameter-
Estimation of the Human Cardiovascular System.
Vieweg-Verlag, Wiesbaden, 1983.

Möller, D.P.F., D. Popovic, and G. Thiele: "Reliability of Parameter Estimation
Methods applied to the Identification of Biomedical Multi-Compartment Systems'.
In: Identification and System Parameter Estimation, H.A. Borker and P.C. Young,
(Eds.), Vol.1, Pergamon Press, Oxford, 1985, pp. 1385-1390.

21
Optimal Experiment Design in Pharmacokinetics

Jiří Potůček

21.1   Aim of Instruction

The problem of "Improved Experimental Design" is very important part of the whole proces of model building. We can find in the literature many articles concerning the parameter estimation problem. But rarely is there discussion concerning optimal experimental design, i.e., measures that improve the reliability range of the paramter vector and minimize the costs necessary to perform the experiment. If the parameter estimates are judged to be insufficiently accurate for their intended purpose, it may be necessary to redesign the experiment that supplied the model idenfication data.

A project called EXPDES makes part of software for pharmacokinetic studies. We present this program to fill in the gap in pharmacological software equipment, i.e. to find an effective tool for experiment design of pharmacological kinetics studies in order to obtain more reliable parameter estimations (more reliable kinetics constants).

A major benefit of this approach is that it can lower the costs of experiments in a pharmacological study. In the EXPDES program, the basic calculation method is based on differential equations describing the compartmental model. The calculation of parameter reliabilities is based on sensitivity functions derived from a set of differential equations. The program is quite general, but is was designed with phamacokinetic applications in mind.

The current version of the EXPDES program makes it possible to define the optimal time points for measurement of one variable (usually concentration in blood). For pharmacological practice, special types of procedures have been prepared for standard drug administrations and pharmacological models.

21.2   Theory

21.2.1   Optimal Experiment Problem

An optimal experiment problem is illustrated in Fig. 21.1 [J.V. Beck, K.J. Arnold, 1977]:

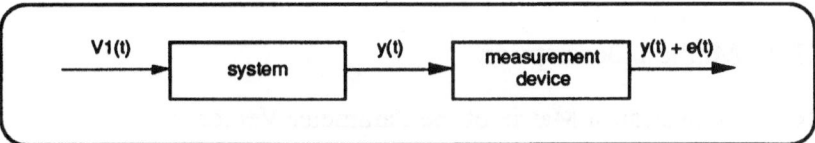

Figure 21.1.   Parameter estimation problem

In finding the optimal experiment one seeks to determine the conditions under which the estimation of unknown parameters is made within a minimal reliability range. The first step in this procedure is to find the covariance matrix of the parameter vector; the second step is to define an appropriate criterion in order to minimize variance of the parameters.
Finally, a strategy is developed for minimizing the defined criterion. This minimum defines the optimal conditions necessary to carry out the experiment.

21.3   Aspects to be Considered

21.3.1   Limitations of the Method

Before starting the experimental design, a preliminary hypothesis about the model structure and parameter values (according to experimental data from the preliminary experiment) is required. Moreover, the model must be theoretically identifiable [Carson, Labelli, 1983].
The method is correct only if the hypothesis about the model structure is correct. In the case of systematic deviation in the experimental data, it is necessary to change the model structure and start again. In the case of an experimental data error, there are only five possible ranges in the time scale. The optimal time schedule can be calculated using a special algorithm [Di Stefano, 1982]. For practical applications, this method is recommended. Special types of procedures for standard input dosages and pharmacological models have been developed.

21.3.2 Outlook

There are only a very limited number of publications on this subject, and most of them deal with bioengineering applications [Carson, Cobelli, 1983]. As to software, a similar package has not yet been found. In the future it is intended to extend this method to metabolic studies, ecological studies, etc.

21.4 Method Description

21.4.1 Covariance Matrix of the Parameter Vector

Assume a general model whose mathematical description in the state space has the following form:

$$\dot{x} = f(x,p,u).x(t_0) = x_0 \tag{21.1}$$

$$y = q(x,p)$$

$p \in R^k$ - parameter vector, $u(t) \in R^r$ - input vector
$x(t) \in R^n$ - state $\leftrightarrow$ vector, $y(t) \in R^m$ - output vector

The system (21.1) for a linear biological model has the form:

$$\dot{x} = Ax + Bu \tag{21.2}$$

$$y = Cx$$

where A is the system matrix
       B is the input matrix
       C is the output matrix (definition of relation beween output and state vector).

Let x (t), y (t) and u (t) be Laplace transformable functions. Let the transfer function matrix of the dynamical system (21.2) be expressed by means of the Laplace transform:

$$G(s) = C(sE - A)^{-1} . B \tag{21.3}$$

Provided there are independent parameters of the model in the transfer function matrix (21.3), then m of these parameters can be determined unambiguously on the basis of the experiment for a given model's structure. Further, assume that functions f and q in relation (21.1) are continuous and have continuous first partial derivatives with respect to vector components x, u, p.
Let the nominal state x*, the nominal input u*, the nominal output y* and the nominal parameter vector p* be denoted according to Pean's principle [Hartman, 1964].

$$\lambda = \frac{\partial x}{\partial p} / * \qquad (21.4)$$

and solution of the linear differential equation

$$\dot{\lambda} = A_{11}(t)\,\lambda + A_{12}(t)\,.\,\lambda(t_0) = 0 \qquad (21.5)$$

where

$$A_{11}(t) = \frac{\partial f}{\partial x}\bigg|_* ,\; A_{12}(t) = \frac{\partial f}{\partial p}\bigg|_*$$

From the output equation of the system (21.1), it can be

$$\eta = C_1(t)\lambda + C_2(t) \qquad (21.6)$$

where matrix η

$$\eta = \begin{bmatrix} \dfrac{\partial y_1,}{\partial p_1} & \cdots & \dfrac{,\partial y_1}{,\partial p_k} \\[2ex] \cdot & & \cdot \\[1ex] \cdot & & \cdot \\[2ex] \dfrac{\partial y_m,}{\partial p_1} & \cdots & \dfrac{,\partial y_m}{\partial p_k} \end{bmatrix} = \begin{bmatrix} y_i \\ \overline{} \\ \partial p_j \end{bmatrix} ,\; i < m \atop 1 < k \qquad (21.7)$$

will be termed the sensitivity function matrix with respect to the parameter vector p of the model described in system (21.1), and the equations (21.5).
(21.6) will be called the sensitivity equations of the biological model (21.1).

For the linear biological model equation (21.2), sensitivity equations in the form

$$\lambda = A\lambda + Hx + Zu$$
$$\eta = C\lambda + Vx \qquad (21.8)$$

are obtained, where $H = \dfrac{\partial A}{\partial p}$, $v = \dfrac{\partial C}{\partial p}$, $z = \dfrac{\partial B}{\partial p}$ are constant matrices.

Assuming that vector y of the output variables of the biological model (21.1) is coupled with additive noise, the vector of observations $z(t)$ becomes

$$z(t) = y(t) + v(t) \tag{21.9}$$

Assuming that measurements are made during the discrete time intervals $t_1, \ldots, t_s$, we have the vector of observation $z(tl), \ldots, (ts)$.

Let the discretized vector of observation z, the discretized output vector Y and the discretized error vector V be introduced:

$$Z = \begin{bmatrix} z(t_1) \\ . \\ . \\ . \\ z(ts) \end{bmatrix}, y = \begin{bmatrix} y(t_1) \\ . \\ . \\ . \\ y(ts) \end{bmatrix}, V = \begin{bmatrix} v(t_1) \\ . \\ v(ts) \end{bmatrix} \quad (Z, Y, v \in R^{ms}) \tag{21.10}$$

Let the error vector V have a zero mean value and a positive definitive matrix of covariance R : $e(V) = O$, $E(VV') = R$.

The problem of optimizing of the parameter vector p of model (21.1), i.e., to obtain maximum agreement between observations and calculations in the sense of maximal probability, leads to minimization of the criterion:

$$J(p) = (Z - Y)^T . R^{-1}(Z - Y) \tag{21.11}$$

If one denotes

$$M = \begin{bmatrix} \eta(t_1) \\ . \\ . \\ . \\ \eta(ts) \end{bmatrix} \tag{21.12}$$

then the covariance matrix of the parameter vector W equals

$$W = (M^T . R^{-1} . M)^+ \tag{21.13}$$

where + denotes pseudoinversion.

21.4.2 Some Properties of the Covariance Matrix of the Parameter Vector

The covariance matrix W is symmetric and positively semidefinitive if the number of measurements is equal to or greater than the number of parameters. If k random quantities in the random vector p identically equal zero or are a linear combination of the remaining random quantities, the matrix W is positively semidefinitive and has a nullity k.

Assume that the random vector v has a normal density distribution. If it is presumed that the difference between signal and observation is caused by noise, one can make $y = v$, and therefore:

$$V(t) = \eta(t) . \Delta p$$

With respect to the equations (21.10 and 21.12) one can write:

$$V = M . \Delta p \qquad (21.14)$$

Because the linear transformation (21.14) maintains a normal distribution, the random vector p can be described as

$$f(p) = \frac{1}{(2\pi)^{r/2} \sqrt{\det W}} . \exp\left(-\frac{1}{2} p^T W^{-1} \Delta p\right) \qquad (21.15)$$

where r is the number of parameters. This relation holds only if W is a regular matrix. If $f(p) = $ const, the expression in the exponent of equation (21.15) is a constant and

$$\Delta p^T . W^{-1} . \Delta p = \rho^2 \qquad (21.16)$$

We obtain in the space of the parameter vector p a hypersphere corresponding to the constant density of probability.

Equations (21.16) define a hyperellipsoid in the space of parameters p. Let an orthogonal matrix whose columns represent the orthonormalized system of vectors of matrix W be denoted. Then:

$$Q^T . W^{-1} . Q = \Lambda^{-1} \qquad (21.17)$$

where Λ is a diagonal matrix formed from the eigenvalues of matrix W. Since matrix Q is orthogonal, it holds that

$$Q^T . W . Q = \Lambda \qquad (21.18)$$

Thus matrix W has an identical system of orthonormalized eigenvectors and its eigenvalues are reciprocal values of the eigenvalues of matrix W^{-1}. Writing the vector $\Delta p = Q . j$, where $j = [j_1,...,j_r]^T$ and substituting these relations in (21.16), we obtain

$$j^T . Q^T . W^{-1} . Q . j = \rho^2 \tag{21.19}$$

If we use relation (21.17), we obtain

$$j^T . \Lambda^{-1} . j = \rho^2 \tag{21.20}$$

Equation (21.20) can be written in the form

$$\frac{j_1^2}{\lambda_1} + \frac{j_2^2}{\lambda_2} + \dots + \frac{j_r^2}{\lambda_r} = \rho^2 \tag{21.21}$$

Relation (21.21) is the equation of a hyperellipsoid in the coordinate system consisting of an orthonormalized system of the eigenvectors of matrix W. The axes of the hyperellipsoid are identical with those of the new base, and the length of the semiaxes is $\rho . \sqrt{\lambda_i}$

The above findings can be summarized as follows:

1. Equation (21.12) defines the hyperellipsoid system.
2. Axes of the hyperellipsoids form the eigenvectors of matrix W.
3. The length of individual semiaxes is $\rho . \sqrt{\lambda_i}$, where λ_i is the eigenvalue of matrix W, $i = 1, \dots, r$.
4. For calculation, the eigenvectors and eigenvalues of matrix W can be used because the eigenvectors are equal and the eigenvalues are reciprocal to those of matrix W.

The random vector Δp is a vector with a zero mean value, a covariance matrix W, and a normal density with r degrees of freedom.

If the given vector parameters are to be (with some degree of probability) an element of the set of vectors with the covariance matrix W, the value must not exceed the limits that can be found in statistical tables. All the above mentioned considerations are irrelevant if we consider the estimation of variance (resulting reliability range of individual parameters in the direction of their original axes).

The variance of parameter p_i equals the diagonal element of the covariance matrix W. However, the hyperellipsoid may be elongated in some direction. Thus, even though individual parameters are relatively well defined, there is a direction in which they may far exceed their original reliability interval. For two parameters, this situation is represented in Fig. 21.2.

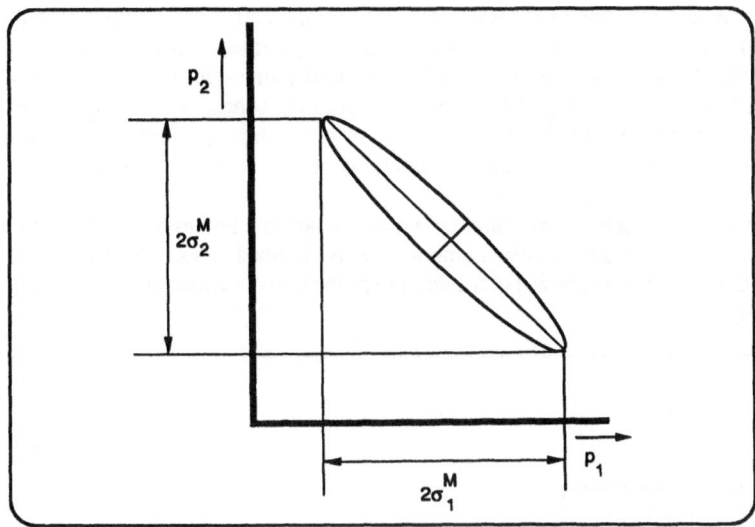

Figure 21.2. Hyperellipsoid for two parameters

The markedly elongated hyperellipsoid indicates a considerable degree of linear dependence between the parameters. An index of this dependence is the $\lambda_{min}/\lambda_{max}$ ratio. If the ratio is l, a hypersphere is involved. The lesser the value, the greater the linear dependence of individual parameters.

21.4.3   Criterion for Finding the Minimal Variance of Parameters

The literature [Mehra, 1974] offers some standard design criteria such as:
a.   A-optimality minimization of the trace of W.
b.   E-optimality maximization of W's smallest eigenvalue
c.   D-optimality: This is equivalent to minimizing the volume of the ellipsoid.
We use D-optimality in our program. An important advantage of D-optimality is that it is invariant under scale changes in the parameters and linear transformations of output, whereas A-optimality and E-optimality are affected by these transformations.

21.4.4   Strategy for Minimization of Criterion

Now it is necessary to study the possibility of improving the estimation of the parameter vector, i.e. how to diminish its reliability range. It follows from [Himmelblau, 1972] that, for a parameter vector once estimated and hence known, the sensitivity functions are time functions only. Consequently, the covariance matrix W is a function of measurement times, input and R matrix (measurement errors). Since it can be assumed that an improved parameter vector will not differ significantly from the proposed one, the sensitivity functions can be taken to be equal. Therefore the proposal of new measurement times, new input and best-measurement errors permits the calculation, from the already calculated sensitivity functions, of the matrix of sensitivity functions as well as W, which will generate a more convenient hyperellipsoid.

In this way it is ascertained if and how the estimate can be improved. If it can be improved, an estimate for measurement times is obtained and a new experiment is carried out. During the new experiment the procedure described above is repeated (see Fig. 21.3).

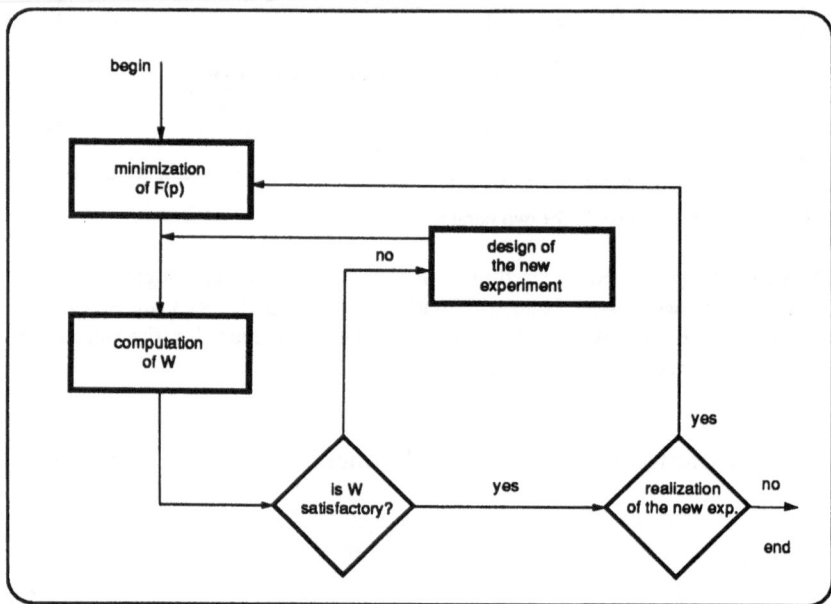

Figure 21.3. Flow diagram of procedure

21.5   Experiments

21.5.1   The Default Demo

In this chapter you can learn the whole procedure step by step using help commands. The following basic steps are presented in detail:

1.   Simulation output from the predefined simulation model, including sensitivity functions.
2.   Proposal of sampling (i.e. introductory analysis of the problem) makes it possible to have the first rough information about the most important sampling time intervals
3.   Design of sampling - serves for final, detailed, time intervals design including reliability range computation.
4.   Time schedule - serves to optimize the cost of performing the necessary experiment.

21.5.2   Different Types of Administration

In this exercise you can learn how the body responds to drug concentration in the blood after the following administrations:
a.   Injection
b.   Infusion
c.   Oral or intramuscular
d.   Combination of the doses

21.5.3   Introductory Analysis

In this exercise we can learn the preliminary rough experimental design. We have the following input variables for this purpose:
-   measurement errors derived max. in 5 time intervals on your time scale
-   level of interest dependent on your interest in each parameter

21.5.4   Design of Sampling

-   In this exercise you can learn the detailed design of sampling. The reliability range calculation follows after each individual design. You can repeat this procedure in order to improve the reliability range of your parameter vector. The best design is automaticaly stored. You also have to realize the practical aspects of the experiment, i.e. daily clinical practice for precise design. In order to improve the reliability of the selected parameter, the plot of sensitivity function is available.

The time intervals where the sensitivity functions reach their peak values are recommended.

21.5.5 Minimization of Costs

The costs of each sample are dependent on the time of day in the clinical department. You can define your day-time breakpoints (max. 5) and the corresponding cost. The optimal injection moment of the experiment (with minimal cost) is also calculated on demand. If you would like to improve this time in relation to clinical practice, you can be more liberal with the moment of beginning. Finally you can print the output protocol including beginning, duration and cost for laboratory purposes.

21.6 Free Design

In this exercise you can make your own design according to the procedure presented above.

References

Beck, J.W. and Arnold J.K.: *Parameter Estimation in Engineering and Science*. J. Wiley, New York 1977.

Di Stefano, J.J.: "Experience with sequential optimal sampling schedule designs for pharmacokinetic and physiologic experiments". *10th IMACS World Congress on System Simulation and Scientific Computation*, Montreal 1982, pp. 203-205.

Carson, E.R. and Cobelli C., Finkelstein L.: *The Mathematical Modelling of Metabolic and Endocrine Systems*. John Wiley, New York, 1983.

Hartman P.: *Ordinary Differential Equations*. John Wiley, New York, 1964.

Himmelblau D.M.: *Applied Nonlinear Programming*. New York, McGrawHill, 1972.

Mehra R.K.: "Optimal Input Signals for Parameter Estimation in Dynamic Systems - Survey and new results'. *IEEE Transactions on Automatic Control*, Vol. AC-19, No 6, 1974, pp. 753-768.

Nahi N.E. and Wallis D.E.: "Optimal Inputs for Parameter Estimation in Dynamic Systems with White Observation Noise". *Proc. JACC. Boulder, Colorado 1969*, paper IX - A5.

Parakyriazis P.A.: "Optimal Experimental Design in Econometrics". *Journal of Econometrics 7*, 1978, pp. 351-372.

Potucek J., Hajek M., and Brodan V.: "Methods of Determining the Reliability Range of Parameters of a Biological Model using a Hybrid Computer'. *Proc. 8th AICA Congress*, Delft, The Netherlands, 1976.

Potucek J., Brodan V. and Sechser T.: "Proposal of Optimum Drug Dosage by means of Simulation Pharmacokinetics Models". *Proc. 10th IMACS Congress on System Simulation and Scientific Computation*, Montreal 1982, pp. 244-246.

Whittle P.: *Some general Points in the Theory of Optimal Experimental Design*. PhD Thesis, University of Cambridge, 1972.

22
The Cerebrospinal Fluid Circulation Model

Lourens J. van Briemen, John H.M. van Eijndhoven, and Dirk A. de Jong

22.1 Introduction

The four ventricles and the subarachnoid space, formed by the cerebral meninges, contain a watery fluid (Fig. 22.1).
As this fluid also appears in the spinal compartments, it is called the cerebrospinal fluid (CSF) or liquor cerebrospinalis.
This fluid serves as a buffer that reduces the effect of skull impact on the brain. The CSF also serves as a heat buffer and takes care of the disposal of certain waste products.

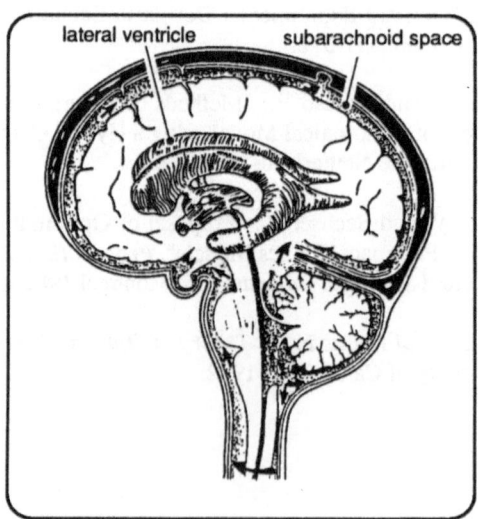

lateral ventricle   subarachnoid space

Figure 22.1. The cerobrospinal fluid system

Secretion from the choroid plexus, on the walls of the first and second ventricle, is the main source of CSF formation. CSF is also formed in the third and fourth ventricle and in the spinal cord.

From the ventricular system the flow is directed towards the cranial and spinal subarachnoid space. The main portion of the liquor is absorbed from the cranial subarachnoid space into the venous system within the dural sinuses.

Under normal conditions, when there is free communication of CSF between the cranial and spinal compartments, the processes of CSF formation and CSF absorption are in equilibrium. Then the pressure is equal in all compartments.

When this equilibrium is disturbed, the craniospinal system must compensate for changes in volume.

22.2 The CSF Circulation Model

The compensatory capacities of the CSF space and the mechanisms involved in the circulation of the CSF were studied by numerous authors [Cserr, 1971; Cushing, 1910; Davson, 1967; Welch, 1975]. Many of these studies were originally aimed at obtaining a better understanding of the pathogenesis and management of hydrocephalus. Since the early seventies, physiological studies of the CSF circulation have been supported by bio-physical and mathematical models, describing the interaction between formation, storage and absorption of CSF [Agarwal,1969; Benabid, 1970; Bloch, 1976; Guinave, 1972; Marmaron, 1973]. Although these models were primarily developed to describe the process involved in the origin of hydrocephalus, they can also be applied to situations in which there is elevated ICP due to other lesions of a space-occupying nature, such as tumours, hematomas, etc.

The model parameters provide information on the rate of CSF formation and absorption as well as on the storage capacities of the system. The absolute values of these parameters, however, depend largely on the assumptions made in the mathematical relationships. In most models, it is assumed that the parameters are not affected by the rate at which volume changes occur, although the results of a few studies contradict this assumption [Borgesen, 1979; Sullivan, 1979].

At present the Marmarou model is the most widely used CSF circulation model [Marmaron, 1973]. The model presented here deviates from the Marmarou model, since the storage capacity of the system was derived from an exponential volume-pressure function extended by a constant term (P_0) [Avezaat, 1969].

22.3 Model Equations

The basis of the CSF circulation model is given by the equilibrium existing between CSF formation, CSF absorption, and changes in total craniospinal volume. This equilibrium exists under all circumstances within the craniospinal system:

$$F_i - F_o - F_{cs} = 0 \tag{22.1}$$

where: F_i =   CSF formation rate (ml/hr)
        F_o =   CSF absorption rate (ml/hr)
        F_{cs} =   rate of changes in CSF volume (ml/hr)

22.3.1 CSF Formation

The exact relationship between CSF formation and the pressure gradient across the choroid plexus is not known. We shall therefore develop the model for two different situations:

-   CSF formation takes place at a constant rate, as assumed in most models

$$F_i = F_1 \tag{22.2a}$$

-   CSF formation is linearly dependent on the pressure gradient between the arteries in the choroid plexus and the cerebral ventricles

$$F_i = (P_a - P) / R_i \tag{22.2b}$$

where: P_a =   arterial pressure in the choroid plexus (mmHg)
        P =   pressure in the cerebral ventricles (mmHg)
        P =   ICP
        R_i =   resistance to formation of CSF

The resistance to formation of CSF (Ri) is assumed to be constant.
Since the arterial pressure in the choroid plexus is difficult to measure, it is often replaced by the systemic arterial pressure (SAP).

22.3.2 CSF Absorption

The CSF absorption rate is generally accepted to be linearly related to the pressure difference between the subarachnoid space and the dural sinuses, if and so long as the ICP exceeds the dural sinus pressure:

$$F_o = (P - P_d) / R_o \qquad \text{for } P >= P_d \tag{22.3}$$
$$F_o = 0 \qquad \text{for } P < P_d$$

where: P_d =   sinus pressure (mmHg)
        R_o =   resistance to absorption of CSF
             or outflow resistance

In the following model equations we shall assume that the ICP always exceeds the dural sinus pressure.

22.3.3 CSF Storage

The storage capacities of the system are deduced from the volume-pressure relationship [Avezaat, 1984]:

$$P = P_0 + P_1 \exp(E_1 V_e) \qquad (22.4)$$

where: P = ICP (mmHg)
P_1, P_0 = constant pressure terms (mmHg)
E_1 = elastance coefficient
V_e = elastic volume, change in total craniospinal volume with respect to equilibrium volume

Boundary conditions: $E_1 > 0$ so $P > 0$
and $P_0 < P$

The term P_0 is primarily introduced for mathematical reasons without a physiological concept. Changes in P_0 will cause a shift of the volume-pressure curve as a whole along the pressure axis without affecting its shape. For example, the ICP is always measured with respect to a chosen reference level, at which level the pressure transducer is placed.

Changes in the height of the reference level will thus automatically affect the magnitude of the ICP measured, but they will not affect the elastic properties of the craniospinal system.

Assessment of the volume-pressure relationship with a model that does not take into account P_0 or, more correctly, that assumes that P_0 equals zero, will produce different results according to the chosen reference level of the ICP. This error is prevented by the introduction of P_0.

The same applies to changes in ICP due to postural changes. The volume-pressure relationships are not likely to be influenced in this case either. With a model including P_0 the same volume-pressure relationships will be found irrespective of body position. Another clue with regard to the physiological meaning of the constant term can be found in a study by Löfgren et al. [Löfgren, 1973a; Löfgren, 1973b; Löfgren, 1973c].

The volume-pressure curve was recorded in dogs during variations in central venous pressure that were produced by applying positive or negative pressure to the outlet tube of the respirator. The curve was found to be displaced upwards in a parallel manner along the pressure axis with positive central venous pressures and in the reverse direction with negative pressures.

These findings suggest that P_0 may be related to the pressure in the intra- and extradural venous system, constituting the basic hydrostatic pressure level of the system.

From (22.4) we derive

$$dP/dV = E_1 \cdot (P - P_0) \tag{22.5}$$

and

$$dP/dt = (dV/dt) \cdot (dP/dV)$$
$$= F_{cs} \cdot E_1 \cdot (P - P_0) \tag{22.6}$$

22.3.4 External Signals

The model can be tested and parameters calculated by adding a known external flow to the physiological flows.

The 'continuous liquor infusion test' will be implemented in this model.

Artificial liquor with a temperature of 37°C is injected into the ventricles or into the cisterna magna, at a constant rate.

The infusion rate is chosen so that a new equilibrium pressure is reached within a few minutes. The continuous infusion can be regarded as a step function on the formation rate:

$$F_{ex} = const \tag{22.7}$$

22.4 Implentation

A block diagram of the interrelationships between formation, absorption and storage of cerebrospinal fluid is shown in Figure 22.2. This diagram will serve as a basis for the implementation of the circulation model in PSI.

22.4.1 Pressure-Independent CSF Formation

Figure 22.3 shows the PSI implementation of the model where constant liquor formation is assumed (model: LIQUOR1).

The parameters used in this example are given in Table 22.1A.

The initial value of the integrator, the baseline ICP, is calculated from the balance between CSF production and absorption, when there is no change in the volume of liquor stored.

From equations 22.1, 22.2a and 22.3 we derive

$$P_b = R_o F_1 + P_d \tag{22.8}$$

Where P_b is the baseline ICP. $P_b = P$, from eq. 22.4, as the elastic volume V_e is assumed to be zero under equilibrium conditions. Consequently, P_b is always greater then P_0.

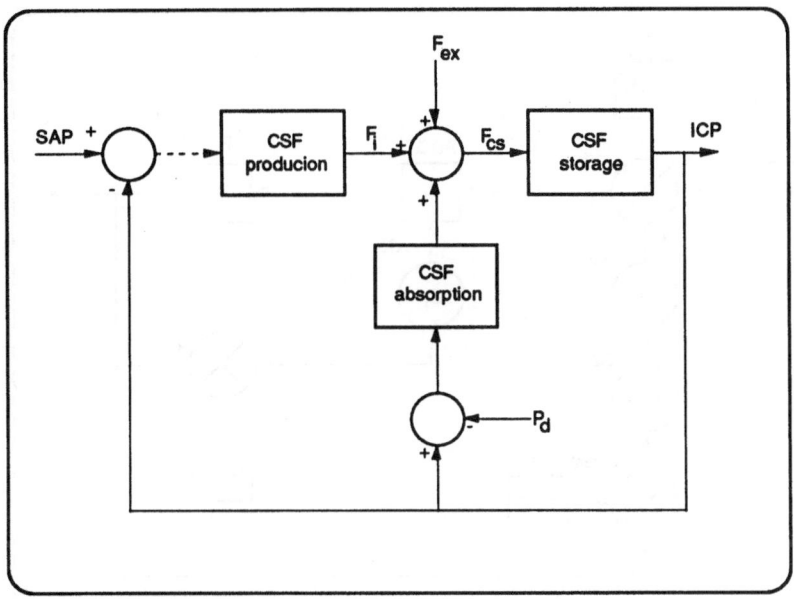

Figure 22.2.   Block diagram of the CSF circulation model

When the CSF formation rate is increased, by infusing artificial CSF at a constant rate, the new equilibrium can be calculated from

$$F_1 + F_{ex} - F_o = 0 \tag{22.9}$$

or
$$F_1 + F_{ex} - (P - P_d) / R_o = 0$$

so
$$P_{eq} = R_o F_{ex} + R_o F_1 + P_d \tag{22.10}$$

$$= R_o F_{ex} + P_b \tag{22.11}$$

Where P_{eq} is the new equilibrium ICP.

22.4.2 Pressure-Dependent CSF Formation

Figure 22.4 shows the PSI implementation of the model where the CSF formation rate is assumed to be linearly related to the pressure gradient between the choroidal arterial system and the ventricular system (model: LIQUOR2).

The parameters used in this example are given in Table 22.1b. Equations 22.8 and 22.10, describing the equilibrium conditions before and during steady state infusion, must be modified by substituting $(P_a - P) / R_i$ for F_1 (equation 22.2b).

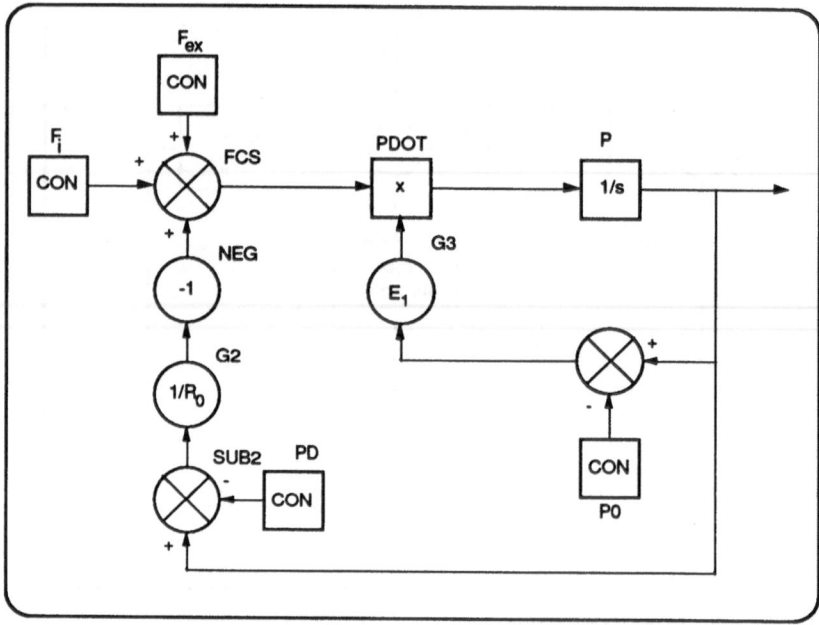

Figure 22.3. Liquor circulation model for constant liquor formation (example LIQUOR1)

This gives

$$P_b = (P_a R_o + P_d R_i) / (R_o + R_i) \tag{22.12}$$

and

$$P_{eq} = R_i R_o F_{ex} / (R_i + R_o) + P_b \tag{22.13}$$

where P_b is the baseline pressure and P_{eq} the new equilibrium ICP.

22.5   Exercises

The CSF circulation model has been developed to gain insight into the variables affecting the circulation of CSF.

Since the volume of CSF is most capable of providing spatial compensation and since the model includes the volume-pressure relationship as well, the model may also be expected to provide knowledge on how and to what extent the ICP is changed when the normal physiological equilibrium is disturbed by intracranial or extracranial processes.

Under physiological steady state conditions the CSF formation rate is balanced by

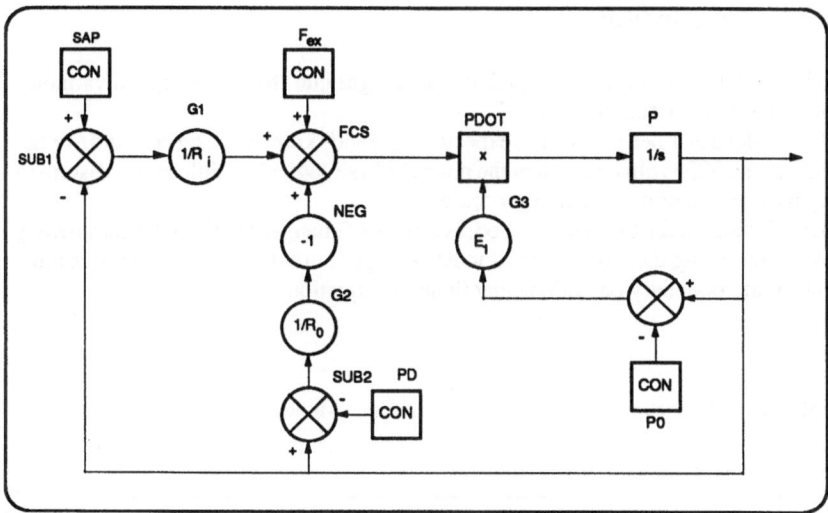

Figure 22.4. Liquor circulation model for pressure dependent liquor formation (example LIQUOR2)

the absorption rate. This implies that the total volume of CSF stored in the craniospinal system does not change, so that the ICP is maintained at a constant baseline or resting level.

22.5.1   The Default Demo

The model describes the pressure course in case the physiological equilibrium conditions are disturbed by external means, i.e. a constant rate infusion. The pressure increases until a new equilibrium pressure is reached at which inflow (physiological CSF production and constant rate infusion) is equal to the outflow (absorption) of CSF.

22.5.2   Hydrocephalus

The model can be used to diagnose cases of hydrocephalus, which is characterized by an increased outflow resistance R_0. The effect of hydrocephalus can be demonstrated by increasing the outflow resistance value of the model (thus decreasing the model variable G2), while the other parameters are kept constant. The equilibrium pressure during constant rate infusion is now reached at a higher ICP level.
In childhood this measured ICP will result in larger ventricles in the brain and therefore stimulate growth in the skull circumference.

22.6   Conclusion

CSF models have been developed to gain insight into the physiological variables governing the circulation of CSF.

The model described here can be expected to provide knowledge on how and to what extent, the ICP is changed when the normal physiological equilibrium is disturbed by intracranial or extracranial processes.

The ICP course can be demonstrated, as a result of changes in the model parameters. When changing the parameters, it must be kept in mind that the parameters must fulfill the boundary conditions mentioned in equations 22.4, 22.8 and 22.11.

Table 22.1. Parameters used in the Liquor Circulation Model

A.   Pressure independent liquor formation

liquor formation rate	F1	0.22	ml/min
liquor infusion rate	Fex	1.46	ml/min
outflow resistance	Ro	12.81	mmHg/(ml/min)
elastance coefficient	E1	0.2343	1/ml
sinus pressure	Pd	0	mmHg
constant pressure term	P0	2.34	mmHg

B.   Pressure dependent liquor formation

liquor infusion rate	Fex	1.46	ml/min
inflow resistance	Ri	358.0	mmHg/(ml/min)
outflow resistance	Ro	12.81	mmHg/(ml/min)
elastance coefficient	E1	0.2343	1/ml
sinus pressure	Pd	0	mmHg
syst. art. pressure	SAP	90	mmHg
constant pressure term	P0	2.34	mmHg

References

Agarwal G.C., Berman B.M. and Stark L.: "A lumped parameter model of the cerebrospinal fluid system". IEEE Trans. Biomed. Eng., 16, 1969, pp. 45-53.

Avezaat C.J.J. and Eijndhoven J.H.M. van: Cerebrospinal fluid pulse pressure and craniospinal dynamics. PhD thesis, Erasmus University Rotterdam, 1984.

Benabid A.L.: Contribution à l' étude de l'hypertension intracranienne modèle mathématique. MD thesis, Grenoble University, 1970.

Bloch R. and Talalla A.: "A mathematical model of cerebro-spinal fluid dynamics". J. Neurol. Sci., 27, 1976, pp. 485-498.

Borgesen S.E., Gjerris F. and Sorensen S.C.: "Intracranial pressure and conductance to outflow of cerebrospinal fluid in normal-pressure hydrocephalus". J. Neurosurg. 50, 1979, pp. 489-493.

Cserr H.F.: "Physiology of the choroid plexus". Physiol. Rev., 51, 1971, pp. 273-311.

Cushing H. and Goetsche E: "Concerning the secretion of infundibular lobe of the pituitary body and its presence in the cerebrospinal fluid". Am. J. Physiol., 27, 1910, pp. 60-86.

Davson H: Physiology of the cerebrospinal fluid. Churchill, London, 1967.

Guinane J.E.: "An equivalent circuit analysis of cerebro-spinal fluid hydrodynamics". Am. J. Physiol., 223, 1972, pp. 425-430.

Löfgren J., Essen C. von and Zwetnow N.N.: "The pressure-volume curve of the cerebrospinal fluid space in dogs". Acta Neurol. Scand., 49, 1973, pp. 557-574.

Löfgren J. and Zwetnow N.N.: "Cranial and spinal components of the cerebrospinal fluid pressure-volume curve". Acta Neurol. Scand., 49, 1973, pp. 575- 585.

Löfgren J.: "Effects of variations in arterial pressure and arterial carbon dioxide tension on the CSF pressure-volume relationships". Acta Neurol. Scand., 49, 1973, pp. 586-598.

Marmarou A.: A theoretical and experimental evaluation of the cerebrospinal fluid system. PhD Thesis, Drexel University, 1973.

Sullivan H.G., Miller J.D., Griffith R.L., Carter W. and Rucker, S.: "Bolus vs. steady-state infusion for determination of CSF outflow resistance". Ann. Neurol., 5, 1979, pp. 228-238.

Welch K: "The principles of physiology of the cerebrospinal fluid in relation to hydrocephalus including normal pressure hydrocephalus:. Adv. Neurology, 13, 1975, pp. 247-332.

23
Blood Glucose Regulation
by the Pancreas and the Kidney

Dirk L. Ypey, Alettus A. Verveen, and Bert van Duijn

23.1   Introduction

Endocrine control systems greatly resemble industrial production processes designed by (bio)chemical engineers. In both kinds of control systems, a certain substance often has to be maintained at a given concentration, despite the fact that the substance is involved in chemical reactions that change its concentration. In many cases it is essential that the concentration of a substance is kept within narrow limits, because the rate of some vital process critically depends on it. In other cases, it is important that a given substance follows a programmed concentration change in order to provide a production plant or an organism with the right potential to respond to applied or spontaneously occurring changes.

There is, however, one significant difference between industrial and "living" control systems. Industrial control systems have been designed by scientists, while the "living systems" have to be discovered by scientists through experimentation, followed by theoretical modeling. This may be much more difficult than designing a control system, because living control systems are usually much more complex.

The purpose of this chapter is to introduce the student of physiology and biomedical technology into the "discovery" of endocrine control systems. The student is provided with examples of simple control systems derived from the complex endocrine system maintaining a constant glucose concentration in the blood. It is our intention to illustrate some basic properties of endocrine control systems based on negative feedback. Such systems serve to maintain constant concentrations (homeostasis) despite disturbances, while the controlled value of such a concentration can easily be programmed by other parts of the body, such as the autonomic nervous system.

The student is encouraged to learn about system behavior from qualitative descriptions by experimentalists who have analyzed the system into its components. For these descriptions we will preferably use standard textbooks instead of specific journal articles, since these textbooks are easier to obtain and understand.

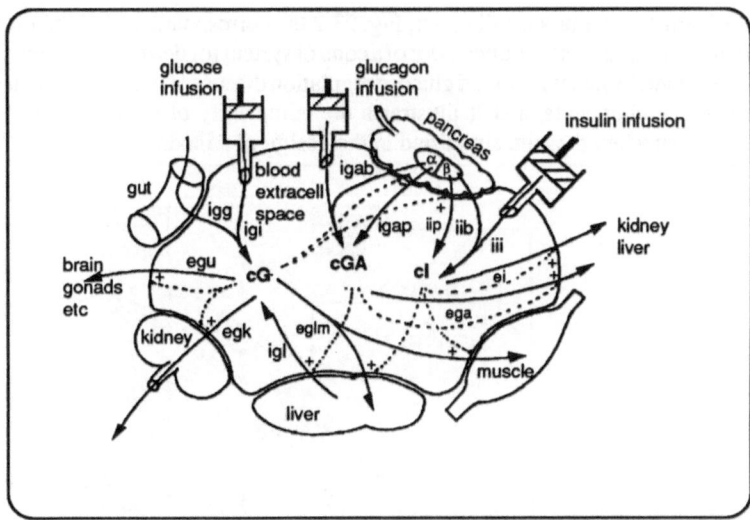

Figure 23.1. Process diagram of blood glucose (G) regulation by the pancreas and the kidney. The various organs involved are positioned along the wall around the simple and single blood reservoir. Infusions are inserted into the blood for experimental manipulation. Two pancreatic hormones are considered in this model, insulin (I), secreted by the beta-cells and glucagon (GA), secreted by the alpha-cells of the islets of Langerhans of the pancreas. Substance flows are indicated by arrows. Inflow abbreviations start with a small i, outflow (efflow) abbreviations start with e. Then follows a small letter(s) indicating the substance flowing. The last letter indicates the origin, the nature or the destination of the flow. For example: igg = inflow of glucose from the gut; igap = inflow of glucagon from the pancreas; iib = basic inflow of insulin. The substance flows are constant or influenced by the concentrations (c) of the three substances involved, cG, cI and cGA. These influences are indicated by dashed-line effect indicators labelled with + (stimulation) or - (inhibition)

The key to translating a qualitative or semi-quantitative system description into an exactly formulated model is to recognize of the various *compartments* involved, the various *flows* of substances into, out of, or between these compartments, and the *system elements* that control the sizes of these flows by relating them to the concentrations of the same or other substances. To practice this translation process, we will construct so-called *process diagrams* (cf. Fig. 23.1), in which these flows, compartments and flow-concentration relationships are identified. The next step will be a *mathematical formulation* of the model, followed by a *block diagram description* (cf. Fig. 23.2; Table 23.1) in terms of our simulation language BIOPSI. This language provides the grammar and idiom for a description of our mathematical system equations in terms of a computer model, which numerically solves these equations to obtain the time behavior of the system.

Fig. 23.1 illustrates a process diagram, Fig. 23.2 the corresponding BIOPSI block diagram and Fig. 23.3 the time behavior of a control system for defined conditions. It is the "complete" system for blood glucose regulation discussed in this section in its three major components, and it illustrates the complexity of endocrine control systems, even when they are simplified in the design of a model.

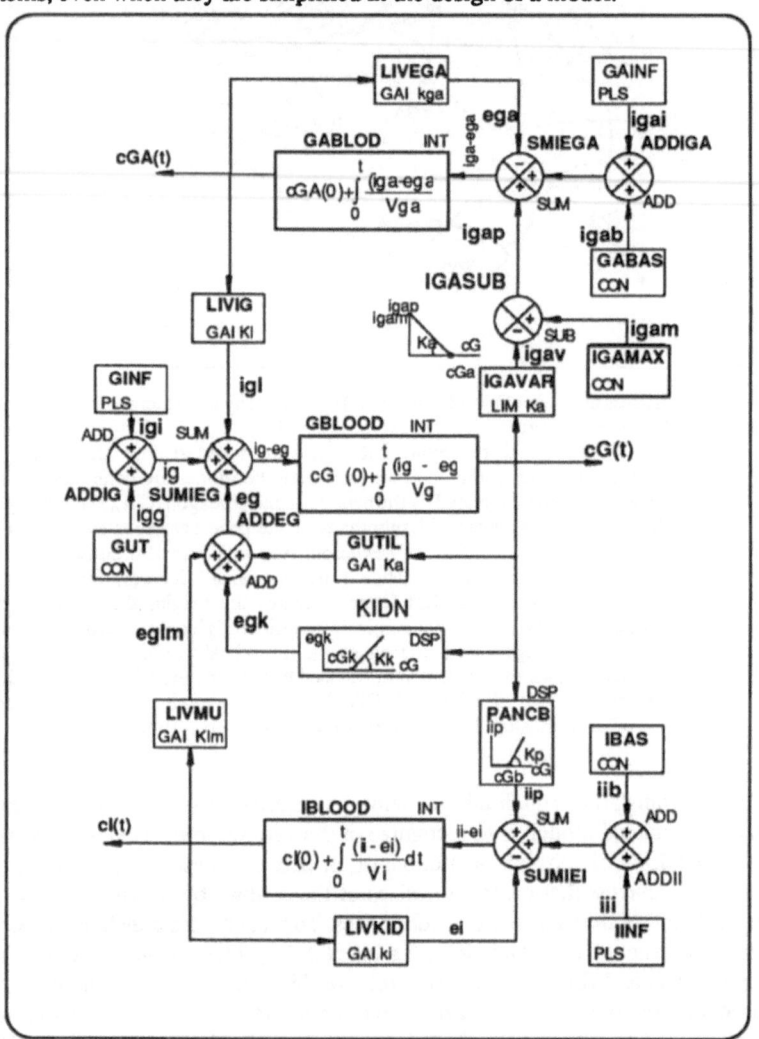

Figure 23.2. Block diagram of the system illustrated in Fig. 23.1. The model is called PAN-KID, because it includes the hormones insulin and glucagon of the PANcreas and the KIDney. The way such a block diagram is constructed from a process diagram is explained after the general introduction, using more simple components derived from this complete system

From a physiological point of view, the endocrine control systems that maintain constant concentrations may be divided into two classes. The first class is "designed" to remove a given substance if it accumulates to above-normal concentrations. We will call this type of regulation *"control by (excess) removal"*; a well known example is the removal of glucose from the blood by the hormone insulin (cf. Fig. 23.1) in case of hyperglycemia (above-normal glucose concentration in the blood). This hormone builds up with the development of *hyper*glycemia and causes the uptake of glucose from the blood into various organs for storage and utilization. The second class of these endocrine control systems maintains constant concentrations by adding a given substance to the blood, if that substance drops to below-normal concentrations. Blood glucose regulation, for example, displays this *"addition-control"* by adding the hormone glucagon (cf. Fig. 23.1) if *hypo*glycemia (below-normal glucose concentration in the blood) develops. Removal and addition control systems have also been called right and left regulation systems, respectively (Verveen, 1979).

In this chapter we will study the behavior of the glucose regulation system. From this system we will derive examples of regulation based on excess removal as well as on addition control for depletion compensation. We will learn that this system (cf. Fig. 23.1) essentially divides into three subsystems. The first may be considered – by approximation – as a first-order excess removal control system that operates through the action of only one organ, the kidney,which serves as both as a sensor and effector organ (section 23.2). The second is at least as complex as a second-order system and is of greater importance. It is also an excess removal system, but derives its properties from a separate sensor organ – the endocrine pancreas – for insulin secretion, and from separate effector organs – liver and muscle tissue – for glucose uptake (section 23.3). The third system is also a second-order control system, but adds glucose to the blood when the hormone glucagon, secreted by the endocrine pancreas, signals glucose depletion (section 23.4).

The demonstration models, and even the more complete model of Fig. 23.1, are ultimate reductions of the real-life system. They illustrate some basic properties of the subsystems for blood glucose regulation but are not intended to reflect the state-of-the-art knowledge and modeling of blood glucose regulation (cf. Fischer et al., 1987). The models will primarily be used to learn about the glucose regulation system, to practice modeling and to explain various diseases resulting from misregulation of blood glucose concentration. Furthermore, the blood glucose regulation system has been chosen for this introduction in endocrinological modeling because it has components with functional structures that are representative of many regulation systems in endocrinology. To complete the treatment of endocrine models, the next chapter provides a state-of-the-art model of gastric acidity regulation.

The design of these models is particularly suited to predict the behavior of the glucose control system under experimental conditions, in which glucose, insulin or glucagon infusions (Fig. 23.1) are given to human subjects or to animals.

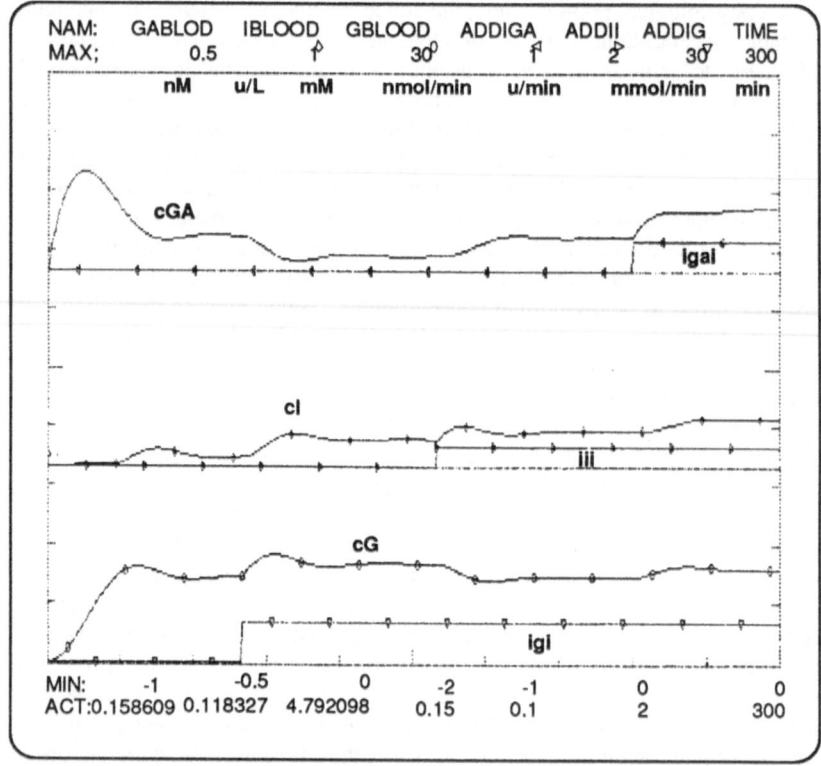

Figure 23.3. Example of a simulation experiment with PANKID. The model calculates cG
(GBLOOD), cI (IBLOOD) and cGA (GABLOD) as a function of time upon the
start of the simulation at t=0 min and upon starting infusions of glucose (igi, as
monitored by ADDIG, at t=80 min), of insulin (iii, as monitored by ADDII, at
t=160 min) and of glucagon (igai, as monitored by ADDIGA, at t=240 min).
Time scale is in minutes. The x-axis of each frame is at y=0. The y-axis of each
quantity extends from the very top to the bottom of the figure. Maximal values
of all these y-axes are given on top of the figure for each block name, with a
specific symbol to label the various curves. Units are also indicated. Minimal
values of all y-axes are given below the figure, together with the actual values
of the quantities at t=300 min. For further explanation, see the various models
discussed below and exercise 23.4.5.5.

The basic knowledge obtained may, finally, be useful in the design of an artificial
pancreas for diabetes patients [cf. Salzsieder et al., 1985; Fischer et al., 1987],
programmed to deliver insulin to the blood upon hyperglycemia as in normal,
healthy people.
But most of all, this chapter serves to let the student experience learning by making
models. This is a very effective way of learning about endocrinology, because a
model not only summarizes knowledge and understanding, but also reveals
insufficient knowledge and understanding.

Table 23.1. Listing of all blocks constituting the block diagram of PANKID. For each block (see first column of block names), block type is specified (2nd column) as well as the interconnection of blocks (3rd to 5th columns). The last three columns specify parameter values of the block diagram structure of Fig. 23.2. For further explanation, see the BIOPSI manual and the following sections of this chapter.

Block	Type	Input1	Input2	Input3	Par1	Par2	Par3
ADDEG	ADD	GUTIL	KIDN	LIVMU			
ADDIG	ADD	GUT	GINF				
ADDIGA	ADD	GABAS	GAINF				
ADDII	ADD	IBAS	IINF				
GABAS	CON				.0000		
GUT	CON				.0000		
IBAS	CON				.0000		
IGAMAX	CON				.5500		
KIDN	DSP	GBLOOD			.0000	10.00	.1000
PANCB	DSP	GBLOOD			.0000	4.000	.1000
GUTIL	GAI	GBLOOD			.1000		
LIVEGA	GAI	GABLOD			1.400		
LIVIG	GAI	GABLOD			13.00		
LIVKID	GAI	IBLOOD			1.500		
LIVMU	GAI	IBLOOD			30.00		
GABLOD	INT	SMIEGA			.0000	7.0000E-02	
GBLOOD	INT	SUMIEG			.0000	7.0000E-02	
IBLOOD	INT	SUMIEI			.0000	7.0000E-02	
IGAVAR	LIM	GBLOOD			.0000	.5500	.1000
GAINF	PLS				240.0	300.0	.1500
GINF	PLS				80.00	500.0	2.000
IINF	PLS				160.0	400.0	.1000
IGASUB	SUB	TGAMAX	IGAVAR				
SMIEGA	SUM	ADDIGA	IGASUB	LIVEGA	1.000	1.000	-1.000
SUMIEG	SUM	ADDIG	ADDEG	LIVIG	1.000	-1.000	1.000
SUMIEI	SUM	ADDII	PANCG	LIVKID	1.000	1.000	-1.000

23.2 Glucose Regulation by the Kidney Alone: GLUKID

23.2.1 Introduction

In this section you will study the design of a simple, first-order model for maintaining a constant concentration of a given substance in the blood by removing excess amounts of the substance. In particular, you will study the design of a model for blood *glu*cose regulation by the *kid*ney (GLUKID) alone. All other system components involved in blood glucose regulation are as yet neglected for didactic reasons. The model is basic to almost any endocrine control system. After studying the design procedure – from qualitative description to quantitative model structure –

you will investigate the properties of the model by running it on your PC at various parameter settings that are representative of physiological experimentation or clinical conditions. The model GLUKID is intended as a first step in the design of more complete models of blood glucose regulation.

23.2.2 Physiological Background

Glucose is an essential fuel substance in our body [Guyton, 1986, Ch. 78; Ganong, 1987, Ch. 19]. Our life would be impossible without glucose circulating through the body to feed the cells of various tissues and to support the metabolic activity of these cells. This is evident from the condition "hypoglycemic shock" [Guyton, 1986, pg. 936], in which diabetic patients develop severe hypoglycemia (a below-normal blood glucose concentration) and lose consciousness (coma). Some organs, such as the brain, depend strongly on glucose, which explains the condition of hypoglycemic coma.

A properly functioning control system for regulating glucose concentration in the blood maintains the right glucose concentration (around 5-6 mM) despite a wide variety of demands on the body.

The complete system for blood glucose control is extremely complex [Ganong, 1987, pg. 292-294]. Even the complex system illustrated in Fig. 23.1.2 is an oversimplification. The system involves a large number of components, subject to many influences from all parts of the body. Therefore, the important thing is to decide which processes are to be considered in designing a model. Fortunately, some components are more important than others. The endocrine pancreas (the islets of Langerhans), for example, and the liver and muscles are the most important organs in the control system. The endocrine pancreas produces the hormones insulin and glucagon. Insulin [Guyton, 1986, pg. 923-930] is secreted under hyperglycemic conditions; it remove excess glucose from the blood, mainly by storing it in the liver and muscles. Glucagon is secreted when hypoglycemia develops and serves to mobilize extra glucose from the liver to compensate for the hypoglycemia [Guyton, 1986, pg. 930-932].

Besides this double endocrine control system, the body is provided with an emergency control system in case the insulin fails to be effective, and hyperglycemia could develop. In that case the kidney takes over excess glucose removal [Guyton, 1986, pg. 408-409]. Glucose excesses are then lost with the urine instead of being stored in the liver or muscles.

In the first model (GLUKID), we only consider the latter example of glucose regulation, since that is the most simple system to begin with. We neglect the existence of all other control processes. Such assumptions are justified here for educational reasons, and they are also essential in designing a well-planned model. It is wise to develop model, complexity and insight into the system hand in hand and to complete the model with accessory components and processes in later stages. In the final model process-neglect assumptions should be justified by the existence of conditions, in which the neglected processes do not play a significant role in the system, as for example in conditions of disease and of experimental control.

23.2.3 Process Diagram

The process diagram (cf. Fig. 23.4) summarizes the processes and system elements selected for consideration in the model. It is a cartoon of the compartments (blood, etc.) and other components (organs, cells, etc.) of the system, including the various flows (arrows) into or out of the compartments as well as the effects of the substance concentrations on these flows.

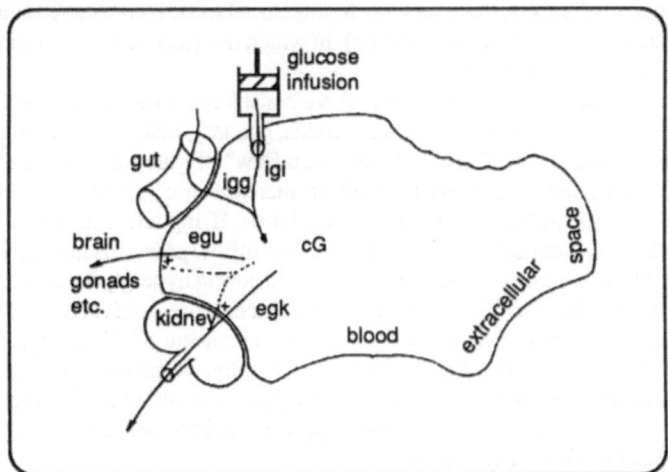

Figure 23.4. Process diagram of GLUKID. Meaning of abbreviations: igg = inflow of Glucose from the Gut; igi = Inflow of Glucose from the Infusion; egu = Efflow of Glucose due to Utilization of glucose by the tissues; egk = Efflow of Glucose through the Kidney; cG = Concentration of Glucose in the blood. Arrows are substance flows, dashed lines are effect indicators, stimulatory (+) in this case.

The nature of these effects, positive (+) or negative (-), is indicated in the diagram by dashed-line "effect indicators" (not by arrows!).

The disease called juvenile diabetes mellitus [type I; Ganong, 1987, pg. 295] justifies the design of a model based on blood glucose regulation by the kidney (GLUKID) only. In this type of diabetes the endocrine pancreas no longer produces insulin; dangerously high blood glucose concentrations may develop, which can only be removed by spilling the glucose away through the kidney into the urine. From a model without the endocrine pancreas we can learn how to balance glucose utilization with glucose intake in diabetes mellitus patients and in animals with experimental diabetes.

When we provide the system with a continuous glucose input from the gut (or the liver) resulting in near-normal blood glucose values, we may also neglect the glucagon controlled mobilization of glucose from the glucose stores. These two principal process-neglect assumptions allow us to arrive at the qualitative model description, summarized in the process diagram of Fig. 23.4. This diagram further defines the processes to be considered and the components and compartments involved. It is one of the three basic blocks of the more complete system illustrated in Fig. 23.1: Glucose (G) enters the blood via the gut with an inflow (igg) (Fig. 23.4). In reality igg does not reach the general circulation unaltered, since it first passes through the liver. This organ already removes part of the inflowing glucose, especially when some insulin is circulating. We neglect this glucose efflow. Thus, igg may also be considered as a constant output from the liver. For simplicity, the plasma space of the blood together with the interstitial space is considered as a single compartment with constant volume (Vg), in which the glucose concentration (cG) builds up with the inflow (igg).

Two glucose outflow processes (eg) are considered. First, glucose is utilized and metabolized in various organs (brain, retina, gonads), which require a sufficient inflow of glucose. We call the "utilization outflow" of glucose (egu). The effect indicator shows that egu increases with an increase in cG. Thus, a growing G promotes its own utilization (negative feedback). If igg and egu are in perfect balance, then cG is constant. The second glucose outflow process is mediated by the kidney. We call it egk. This outflow only occurs during hyperglycemia at cG > 10 mM [Guyton, 1986, pg. 408-409]. Below this glucose threshold (cGk) egk is zero. The process diagram shows that egk grows with increasing cG. Thus, a growing G promotes its own removal (negative feedback). Finally, we provide the system with a glucose infusion delivering a glucose inflow (igi) to the blood. This infusion can be used to model effects of changes in glucose inflow, as may be applied in experimental or clinical conditions.

The process diagram has been drawn with empty places for the system components of the more complete system (compare Fig. 23.4 to Fig. 23.1), which makes it easier to compare this reduced system to the more complete one.

23.2.4 Model Description

23.2.4.1 Foundations and Assumptions

System Component Properties: Flow Definitions

With the help of the process diagram (Fig. 23.4), we further define the model by assuming properties of the various parts of the system. These definitions lead to flow-concentration relationships for the various flows.

The variables igg and igi may be considered as the inputs into the system. In our model, either inflow is kept at a constant value or is stepped from a given value to another chosen value, but the inflows can also be forced to follow another defined time course. Both inflows add up to the total glucose inflow, ig,

$$ig = igg + igi \qquad (23.1)$$

In real life a constant igg is approached if a person regularly eats small amounts of glucose at small intervals of time. Steps in ig can be realized by increasing the regular glucose intake igg or by applying a glucose infusion igi.

The variable cG is considered to be the output of the system. The blood compartment is considered to be a well-stirred, homogeneous volume due to the pumping action of the heart. The volume of the blood for glucose distribution, Vg, is assumed to be constant. In real life this need not to be the case, since drastic changes in cG may cause osmotic water flows between the blood and the tissues.

The outflow of G through metabolic breakdown is assumed to be proportional to cG according to

$$egu = kgu \bullet cG \qquad (23.2)$$

In this equation kgu is the rate constant that determines how strongly egu depends on cG. Since egu becomes larger with cG, glucose facilitates its own removal (negative feedback).

The relationship between glucose utilization (egu) and concentration (cG) may be more complex than equation (23.2), but the mere existence of hypoglycemic coma suggests that egu in the brain at least co-varies with cG in a certain range of cG.

The outflow of G through the kidney is defined as follows:

$$egk = Kk \bullet [cG - Gk] \qquad \text{for } cG > cGk \qquad (23.3a)$$

$$egk = 0 \qquad \text{for } C \leq cG \leq cGk \qquad (23.3b)$$

This relationship has been drawn in block KIDN in Fig. 23.5.
The "kidney constant" Kk determines how sharply egk increases with an increase in cG above the blood glucose concentration threshold for the kidney, cGk. The shape of the function of equation (23.3) is an approximation by Guyton [1986, pg. 408]. As in equation 23.2, glucose promotes its own removal (negative feedback).

The implicit assumption of equations (23.1) and (23.2) are that kgu and Kk are constant and independent of the time scale considered in the simulations.
Both outflows egu and egk add up to the total outflow, eg,

$$eg = egu + egk \qquad (23.4)$$

Equations and the system behavior

Differential equations
The dynamic behavior of the system can be described by the differential equation describing the changes in blood glucose as a function of inflow and outflow. This equation is, in fact, a mass-balance equation, expressing that the change dG/dt in the number of moles of G in the blood is given by the sum of all inflows and outflows:

$$dG/dt = ig - eg \qquad (23.5)$$

Changing to concentration results in

$$\frac{dcG}{dt} = \frac{ig - eg}{Vg} \qquad (23.6)$$

with Vg being the (constant) volume of the extracellular compartment, in which glucose is distributed.
After substitution of the equations (23.4), and then (23.2) and (23.3) in (23.6) we obtain for $cG > cGk$

$$\frac{dcG}{dt} = \frac{ig - kgu \cdot cG - Kk \cdot (cG - cGk)}{Vg} \qquad (23.7a)$$

For $0 \leq cG \leq cGk$ we obtain

$$\frac{dcG}{dt} = \frac{ig - kgu \cdot cG}{Vg} \qquad (23.7b)$$

Analytical Solutions of Steady-State and Transient Behavior

These first-order differential equations can be solved both for the stationary condition ($dcG/dt = 0$) and for the transient behavior.
Equation (23.8a) applies to the stationary glucose concentration cGs ($cG > cGk$)

$$cGs = \frac{Kk \cdot cGk + ig}{Kk + kgu} \qquad (23.8a)$$

Equation (23.8b) describes cGs for $0 \leq cG \leq cGk$:

$$cGs = \frac{ig}{kgu} \qquad (23.8b)$$

Equation (23.9a) applies to the transient behavior for $cG > cGk$:

$$cG(t) = cGs + [cG(0) - cGs] \cdot e^{-t/tk} \qquad (23.9a)$$

with cGs as in equation (23.8a)

and
$$tk = \frac{Vg}{Kk + kgu} \qquad (23.10a)$$

Equation (23.9b) applies to the transient behavior for $0 \leq cG \leq cGk$

$$cG(t) = cGs + [cG(0) - cGs] \cdot e^{-t/tu} \qquad (23.9b)$$

with cGs as in equation (23.8b)

and
$$tu = \frac{Vg}{kgu} \qquad (23.10b)$$

Thus, the addition of control by the kidney makes the system respond faster to changes in ig (tk < tu, compare eq. 23.10a to 23.10b).

We have derived these equations in order to compare them with the behavior of the simulation model in the exercises. In addition, they are useful to estimate system parameters under certain experimental conditions. In many more complex models it is impossible to analytically solve both the stationary and transient behavior.

Integral Equation

Finally, we derive an integral equation, from the differential equation (23.6) and it can be solved numerically by the simulation program BIOPSI:

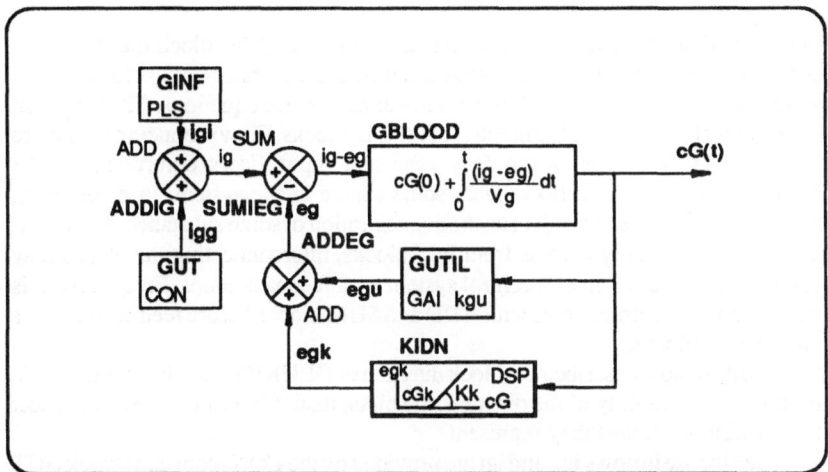

Figure 23.5.   Block diagram of GLUKID. For each block, block name, block type and parameters are indicated. Further explanations are in the text.

Table 23.2.   Blocks, block types, block connections (through inputs) and parameter values
(default setting) of GLUKID

Block	Type	Input1	Input2	Input 3	Par 1	Par 2	Par3
ADDEG	ADD	GUTIL	KIDN				
ADDIG	ADD	GUT	GINF				
GUT	CON				2.000		
KIDN	DSP	GBLOOD			.0000	10.00	.1000
GUTIL	GAI	GBLOOD			.1000		
GBLOOD	INT	SUMIEG			.0000	7.0000E-02	
GINF	PLS				100.0	200.0	2.000
SUMIEG	SUM	ADDIG	ADDEG		1.000	-1.000	

$$cG(t) = cG(0) + \int_0^t \frac{1}{Vg} \cdot (ig - eg) \cdot dt \qquad (23.11)$$

Substitution of equations (23.4), and then (23.2) and (23.3) in (23.11) results in
integral equations equivalent to the differential equations (23.7a) and (23.7b). The
integral equation (23.11) can be found as such in one of the blocks of the GLUKID
block diagram (Fig. 23.5).

23.2.4.2   Block diagram

The translation of the integral equation (23.11) to a BIOPSI block diagram (Fig.
23.5) is straightforward. You may start to define the integrator (INT) block(s). The
number of INT blocks is equal to the number of integral equations. Then you can
define flow blocks feeding the input(s) of the INT blocks. Flow producing blocks are
constants (CON blocks) when they represent input flows controlled by the
experimenter or constant flows from some source in the system. In other cases,
blocks produce flows dependent on the concentration of some substance. Since concentrations are usually outputs from INT blocks, interconnecting the blocks may
result in the construction of a control system with feedback loops. Frequently, it is
useful to add or subtract flows with ADD and SUM blocks before feeding the flows
into the INT blocks.

You have now described the block diagram of GLUKID from the input side. To
improve the readability of the diagram, the block names have been made to reflect
the functions or organs they represent.

The glucose inflows igg and igi are provided by the CON (constant) block GUT
and the PLS (pulse) block GINF (glucose infusion). They are added in the ADD
(adder) block ADDIG (addition of glucose inflows). The output of ADDIG (ig, the
total glucose inflow in this model) feeds the input of the SUM block SUMIEG
(summer of inflow ig and outflow eg of glucose). The output of SUMIEG feeds one
input of the INT (integrator) block called GBLOOD, which lets you choose an

initial condition cG(0) and extracellular volume Vg. The output of GBLOOD has been connected to the input of the GAI (gain) block GUTIL, which produces an output (egu) proportional to cG, the output of GBLOOD. The block GBLOOD has also been connected to the input of the DSP (dead space) block KIDN, which produces an output egk proportional to (cG - cGk) above cGk. The glucose outflow egu and egk are finally added to obtain the total outflow, eg, by feeding the outputs of GUTIL and KIDN to two inputs of the ADD (adder) block ADDEG. Finally, the block diagram is completed by subtracting eg from ig in the SUM block SUMIEG.

Table 23.2 gives a listing of the blocks by name, type, inputs and parameter values. The connections between the blocks in this listing are consistent with the description given above for the block diagram.

Notice that the block diagram of GLUKID is one of the three building blocks of the block diagram of the more complete system in Fig. 23.2.

23.2.4.3 Parameter values

The next parameters (p) and parameter values occur in the model (see Table 23.2):

igg	(p1 of GUT)	= 2 mmol.min^{-1}
igi	(p3 of GINF)	= 2 mmol.min^{-1}
cG(0)	(p1 of GBLOOD)	= 0 mM
Vg	(1/p2 of GBLOOD)	= 14 L
kgu	(p1 of GUTIL)	= 0.1 L.min^{-1}
cGk	(p2 of KIDN)	= 10 mM
Kk	(p3 of KIDN)	= 0.1 L.min^{-1}

These values have been used in the default demo. The parameter kgu was estimated from the tail of the glucose tolerance curve of severe type I diabetics (cf Fig. 78-11, pg. 935 of Guyton, 1986), assuming that the total extracellular volume Vg = 14 L [Ganong, 1987, pg. 9] and that equation 23.9b applies (kgu = Vg/tu = 14/140 = 0.1 L/min). The value of igg = 2 mmol/min has been chosen to obtain severe hyperglycemia to above cGk. Without the kidney the final cG (cGs) would be 20 mM. The size (2 mmol/min) of the glucose infusion pulse and its timing were chosen to illustrate the control action of the kidney just above cGk. Glucose inflows of 1-2 mmol/min have physiological significance, since they would provide the approximate energy to the body for living under resting conditions (sleeping, sitting in a chair [cf Guyton, 1986, pg. 846]. Guyton's Fig. 34-14 [1986, pg. 408] was used to choose cGk = 10 mM and Kk = 0.1 L.min^{-1}. Kk was calculated as the slope of the curve after transforming the X-axis of this figure to cG in mM. The initial condition cG(0) was set to 0 mM in order to demonstrate the transient behavior of cG in the uncontrolled condition. It should, of course, be kept in mind that such a cG(0) = 0 is incompatible with life.

23.2.4.4 Possibilities and limitations

The model GLUKID – glucose regulation by the kidney alone – is only a first step in the construction of a more complete model of the regulation of blood glucose. It is, however, an essential step, since it makes the student familiar with the basic organ structures in which glucose regulation occurs in our body and with the order of magnitude of glucose flows and glucose concentrations relevant for the function of more complete systems. Despite its simplicity, the model already has an instructive value for explaining blood glucose behavior in humans or animals without a functional endocrine pancreas.

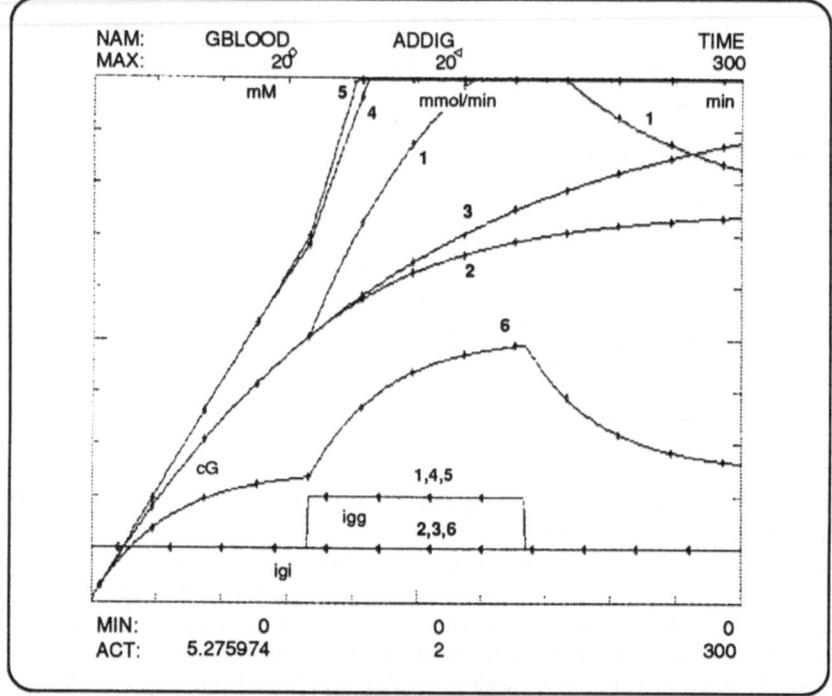

Figure 23.6.   The behavior of GLUKID in the default setting (curve 1) and in various other conditions (curves 2-6, see explanation in the text)

23.2.5 Exercises

23.2.5.1a The default demo

This demo (Fig. 23.6, curve 1) shows how cG changes from cG(0) = 0 mM upon starting igg = 2 mmol/min. The slow increase in cG occurs with a tu = 140 min towards the uncontrolled (as if KIDN was not present) cGs = 20 mM (igg/kgu). Upon approaching cGk = 10 mM of KIDN, an extra 2 mmol/min glucose inflow is given by GINF (at t = 100 min). The cG transient above cGk has a time constant tk. This is smaller than the time constant tu below cGk, consistent with equations (23.10a) and (23.10b). After stepping back to ig = igg = 2 mmol/min (at t = 200 min), cGs stabilizes at a lower value, one that would have been reached without control by the kidney. Verify this value with the analytical solution, equation (23.8a).

23.2.5.1b Compare the default demo with a system without Kidney and without Glucose infusion

Run the model without the glucose pulse, i.e. after making p3 = 0 of GINF (Fig. 23.10, curve 2). Press RX, since this superimposes the curve of this run on that of the demo. Subsequently, run the model (by RX) for the state that occurs after "removal of the kidney" by making p3 = 0 in block KIDN (Fig. 23.10, curve 3).
Comparison of the three curves (Fig. 23.6) again shows that cGs levels above cGk are both lower and reached faster when the kidney is present in the system. Notice that cG control by the kidney is significant, but not very powerful.

23.2.5.2 Effects of changes in kgu and cGk

By making p1 = 0 in GUTIL (kgu = 0) from the default setting, one obtains a condition of metabolic poisoning in which glucose is no longer utilized by the organs. Then the model only consists of an integrator for cG < cGk (Fig. 23.6, curve 4), provided the kidney does not become poisoned in its threshold function. Simulate this condition, followed by the condition that the kidney is no longer working, either because cGk has become zero or Kk has become zero (Fig. 23.6, curve 5). Also study the consequence of increasing metabolic activity by making p1 (kgu) = 0.4 of GUTIL, as for example may occur during exercise or fever. Watch the effect on the time constant (Fig. 23.6, curve 6).
    The kidney may change the threshold concentration cGk for other reasons. Study the consequences of various glucose inflows. By adjusting the threshold in threshold curves like that of the kidney, the body can adjust a control system to certain loads such as in a servo-system. Lowering cGk at high igg would improve the performance of the system.

23.2.5.3 Application to other control systems

Apply the model of GLUKID to other control systems in which a substance concentration is regulated in a similar way by the kidney. For example, the GLUKID structure applies to the blood calcium regulation system if we neglect the actions of parathyroid hormone, calcitonin, vitamin D and the presence of calcium binding proteins in the blood. After adjusting the parameters [see Ganong, 1987, Ch. 21] and considering G as a symbol for calcium, GLUKID applies.

23.2.6 Conclusions

In this chapter you have learned how to design and construct a simple (first order) model for regulating the concentration of a substance in the blood. This model is basic to more complex endocrine control models. You started by studying the physiology of the system, and then selected the components and the processes to be incorporated in the model (process diagram). Subsequently, the components and processes were mathematically defined, based on experimental data or assumptions. The system behavior was defined in terms of a differential equation, which was translated – via an integral equation – into a BIOPSI simulation model. Finally you studied the behavior of the simulation model and compared the results with the mathematical equations. You also studied the effects of parameter changes that represent various diseases.

23.3 Glucose Regulation by Insulin : INSUL

23.3.1 Introduction

In this section you will study the design of a simple second-order endocrine system for maintaining a constant concentration of a substance in the blood. It is a two-substance/one-compartment negative feedback model, characterized by two linear first-order differential equations, one for each substance. From a functional point of view the system is based on excess removal. The substance to be controlled is glucose, and the controlling substance is the hormone insulin, which is secreted by the endocrine pancreas. The model is called INSUL and represents blood glucose regulation by *insul*in alone. The structure of the model is basic for any two-substance/one-compartment excess removing control system, such as the basic structure of the calcium/calcitonin system for blood calcium control [Ganong, 1987, Ch. 21] and the potassium/aldosterone system for blood potassium control [Guyton, 1986, pg. 433-436].

As with the model based on glucose regulation by the kidney (GLUKID), you will follow the design from qualitative description to quantitative structure. Subsequently, you will become familiar with the properties of a control system of this type through simulation experiments under conditions (parameter value settings) that represent normal and pathological function. Various diseases will be simulated, including diabetes mellitus type I and type II and hyperinsulinism [cf. Guyton, 1986, 933-936].

23.3.2 Physiological Background

The hormone insulin (I) is produced and secreted into the blood by the beta-cells in the islets of Langerhans of the pancreas. Hyperglycemia stimulates insulin secretion, and this secretion stops as soon as the glucose concentration changes to hypoglycemic values. Insulin stimulates the uptake of glucose by the liver and muscle tissue. Thus the pancreas/liver-muscle system is a negative feedback regulation system for blood glucose based on excess removal. An increase in the glucose concentration cG increases the secretion of I, giving rise to an increased insulin concentration cI. This rise in cI stimulates glucose uptake by the liver and the muscles, opposing the initial increase in cG. For further details, see Ch. 19 of [Ganong, 1987] and Ch. 78 of [Guyton, 1986]. Details relevant to INSUL are described below in the process diagram and in the quantitative model description.

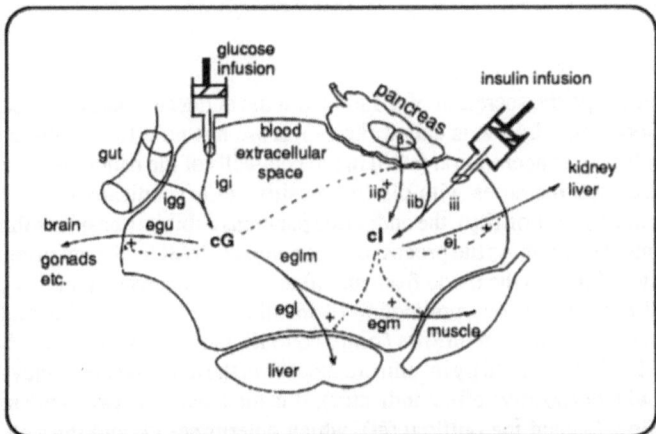

Figure 23.7. Process diagram of INSUL. Meaning of new abbreviations:
ipp =Insulin Inflow into the blood from the Pancreas; iii = Insulin Inflow from the Infusion; iib = Basic Insulin Inflow from pancreas; ei = Efflow of Insulin due to inactivation of the hormone in the kidney or liver; eglm = Efflow of Glucose from the blood into the Liver or the Muscles; cI = Concentration of Insulin in the blood. The other abbreviations have been explained in the legend of Fig. 23.4

23.3.3   Process Diagram of INSUL

The process diagram in Fig. 23.7 summarizes all the processes considered in INSUL that play a role in blood glucose regulation by the hormone insulin. It represents the organs involved, including the blood and interstitial space for glucose and insulin accumulation, the substance flows (arrows) and the ways they are affected by substance concentrations (dashed-line effect indicators). It is one of the three components of the more complete system, depicted in Fig. 23.1.

Glucose flows

Glucose enters the blood from the gut by an inflow (igg). Again, we neglect – for didactic reasons – that igg is reduced during passage through the liver before it enters the blood. Therefore, calculated values of cG in the model may be higher than in reality. We also consider a glucose inflow (igi), which can be applied through glucose infusion into the blood. Glucose also disappears from the blood by an outflow (egu), representing the necessary uptake of glucose by various organs (brains, gonads, etc.). Finally, excess glucose is removed from the blood by uptake into the liver and into the muscles for storage. The total outflow is called eglm. Both separate outflows, (egl) as well as (egm), are controlled by the concentration of insulin (cI), and both outflows increase with an increase in cI (see + effect indicators). However, egm is of much greater importance than egl.

Insulin flows

Glucose promotes the secretion of insulin from the endocrine part of the pancreas into the blood. This inflow is called (iip). Glucose has no effect below a certain threshold glucose concentration (cGb) in the beta-cells of the pancreas. Above that concentration ipp increases with cG (see positive effect indicator). Besides this glucose evoked secretion (iip), the endocrine pancreas exhibits a basic insulin secretion (iib) into the blood. As the process diagram shows, INSUL also incorporates an insulin inflow (iii) into the blood from an infusion, so that experimental or clinical manipulation of (cI) or overproduction of insulin by a tumor can be studied. In addition to the three insulin inflows (summed to ii), there is one outflow of insulin out of the blood (ei) caused by insulin breakdown in the liver and the kidney. This ei increases with cI (positive effect indicator). It is the balance between the summed insulin inflow (ii) and the outflow (ei), which determines cI, and through cI, the storage of G into the liver (egl) and into the muscles (egm). Thus, an increase in cG causes, via the increased insulin secretion, an increased removal of G, counteracting the increase in cG (negative feedback).

Influences of various other hormones [Guyton, 1986, pg. 930], for example, and of the autonomic nervous system [Ganong, 1987, pg. 288] on insulin secretion are not included in INSUL. In addition, the nature of the circulatory system is neglected: the blood compartment is just a simple, but well-stirred reservoir. Below, we will even extend it to represent the total extracellular space. This implies that the relationship between cI and iip may be expected to be different from reality, since the insulin secreted first passes the liver, where it is partially degraded, before entering the circulatory system. However, all these missing influences can more easily be incorporated in later stages of model design, once the basic model structure is established.

23.3.4   Model Description

23.3.4.1   Foundations and assumptions

First we define and justify the various flows in the system. We then develop the differential equations describing the behavior of the system, based on the definitions of the flows. Finally, we arrive at integral equations, which can be transformed into a BIOPSI block diagram.

Glucose flows

The equations for the glucose inflows ig and outflow egu of GLUKID also apply to INSUL. Thus,

$$ig = igg + igi \qquad\qquad (23.12)$$

and

$$egu = kgu \bullet cG \qquad\qquad (23.13)$$

The inflows igg and igi in equation (23.12) are constant and adjustable.
The other glucose outflow, eglm, for glucose storage into the liver and muscles (see process diagram), is assumed to be proportional to cI:

$$eglm = Klm \bullet cI \qquad\qquad (23.14)$$

This equation does not express the different glucose uptake mechanisms of the liver and muscles [Guyton, 1986, pg. 924-925]. It also assumes that eglm does not depend on cG, i.e. that the glucose concentrations are usually in the submaximal range for the kinetics of the uptake process.
The total glucose outflow, eg, is thus

$$eg = egu + eglm \qquad\qquad (23.15)$$

Insulin flows

The way glucose determines insulin secretion by the beta cells of the pancreas (see process diagram) is defined as follows:

$$iip = Kb \bullet (cG - cGb) \quad cG > cGb \qquad\qquad (23.16a)$$

$$iip = 0 \qquad\qquad 0 \leq cG \leq cGb \qquad\qquad (23.16b)$$

This relationship (drawn in block PANCB, Fig. 23.8) is an approximation of Fig. 19-13 of Ganong (1987, pg. 288), with $cGb = 100$ mg/dL ($= 5$ mM). The effect of cG on iip is assumed to be instantaneous. Thus, time dependent effects after the onset of glucose induced insulin secretion [Ganong, 1987, pg. 289] have not been included in the model. However, the almost immediate halt of insulin release upon cessation of glucose stimulation [Ganong, 1987, pg. 289] is consistent with the assumed instantaneous glucose effects on iip. Furthermore, equation 23.16 does not show saturation, as in Fig. 78-8 of [Guyton, 1986 (pg. 929)]. This saturation can easily be included in a later stage of the design.

Besides iip, the model includes a basal insulin secretion (iib) from the pancreas and an insulin inflow (iii) from an infusion. The total insulin inflow (ii), then is

$$ii = iip + iib + iii \qquad (23.17)$$

Insulin breakdown, which mainly occurs in the liver and the kidney, has been assumed to be a first-order process proportional to cI and defined by

$$ei = ki \bullet cI \qquad (23.18)$$

We have chosen a simple first-order insulin breakdown process, since in the literature it is often characterized by a half-decay time [Guyton, 1986, pg. 924; Ganong, 1987, pg. 279]. Furthermore, deviations from first-order kinetics are not expected to strongly influence the qualitative behavior of the system as long as ei remains an increasing function of cI (see exercises).

Mass-balance equations

The behavior of the system, as it is defined, is exactly described by two mass balance equations. One is the differential equation for dcG/dt as a function of the glucose inflows and outflows. The other is the differential equation for dcI/dt as a function of the insulin inflows and outflows.

The first equation is derived from the expression for the change of the number of moles of G

$$\frac{dG}{dt} = ig - eg \qquad (23.19)$$

After dividing the left and the right term by the volume for G distribution, Vg, we obtain a differential equation for cG:

$$\frac{dcG}{dt} = \frac{ig - eg}{Vg} \qquad (23.20)$$

Vg is assumed to be constant. It represents both the blood plasma and the interstitial space, since G rapidly distributes between these two spaces.

The complete expression of (23.20) is obtained, if equations (23.12), and (23.15) and then (23.13) and (23.14) are substituted:

$$Vg \cdot \frac{dcG}{dt} = igg + igi - kgu \cdot cG - Klm \cdot cI \qquad (23.21)$$

The second differential equation derives from the mass balance of the change in I,

$$\frac{dcI}{dt} = ii - ei \qquad (23.22)$$

Dividing (left and right) by Vi, the volume for I distribution, we obtain the differential equation for the change in cI:

$$\frac{dcI}{dt} = \frac{ii - ei}{Vi} \qquad (23.23)$$

Vi is assumed to be constant. It again represents both the blood plasma and the interstitial space, since I rapidly distributes between the two.
The complete expression of (23.23) is obtained, if equations (23.17) and then (23.16) and (23.18) are substituted. Equation (23.24a) results for the condition cG > cGb:

$$Vi \frac{dcI}{dt} = Kb \cdot (cG - cGb) + iib + iii - ki \cdot cI \qquad (23.24a)$$

Equation (23.24b) results for the condition $0 \leq cG \leq cGb$:

$$Vi \cdot \frac{dcI}{dt} = iib + iii - ki \cdot cI \qquad (23.24b)$$

Since dcG/dt depends on cI (23.21) and dcI/dt on cG (23.24), both equations are coupled.

Steady-state control

We derive the solution of the simultaneous differential equations for the steady-state concentrations of G (cGs) and I (cIs) so we can verify in our exercises the model behavior for stationary conditions.
In those conditions dcG/dt = dcI/dt = 0. Therefore we can write for equation (23.21):

$$cGs = \frac{igg + igi - Klm \cdot cIs}{kgu} \qquad (23.25)$$

for equation (23.24a) (cG > cGb)

$$cIs = \frac{Kb \cdot (cGs - cGb) + iib + ii}{ki} \qquad (23.26a)$$

and for equation (23.24b) $(0 \leq cG \leq cGb)$

$$cIs = \frac{iib + iii}{ki} \qquad (23.26b)$$

By substitution of (23.26) into (23.25) we obtain cGs as a function of the adjustable glucose and insulin inflows:
For $cG > cGb$:

$$cGs = \frac{Klm \cdot Kb \cdot cGb + ki \cdot (igg + igi) - Klm \cdot (iib + iii)}{Klm \cdot Kb + kgu \cdot ki} \qquad (23.27a)$$

For $0 \leq cG \leq cGb$:

$$cGs = \frac{ki \cdot (igg + igi) - Klm \cdot (iib + iii)}{kgu \cdot ki} \qquad (23.27b)$$

Equation (23.27a) implies that cGs only remains above cGb, if

$$ki \cdot (igg + igi) - Klm \cdot (iib + iii) > kgu \cdot ki \cdot cGt \qquad (23.28a)$$

Equation (23.27b) implies that cGs only remains positive, if

$$ki \cdot (igg + igi) > Klm \cdot (iib + iii) \qquad (23.28b)$$

In both cases, the system requires higher values of ig to maintain physiological values of cGs when some base line insulin inflow exists, apart from the ipp.

We may also obtain cIs as a function of the glucose and insulin inflows if we substitute equation (23.25) into equation (23.26a). Then we find for $cGs > cGb$

$$cIs = \frac{Kb \cdot ((igg + igi) - kgu \cdot cGb) + kgu \cdot (iib + iii)}{Kb \cdot Klm + ki \cdot kgu} \qquad (23.29a)$$

For $0 \leq cGs \leq cGb$, equation (23.26b) applies

$$cIs = \frac{iib + iii}{ki} \qquad (23.29b)$$

Thus, for $cG > cGb$, cIs linearly increases with (igg + igi) with the proportionality constant $Kb/(Kb \cdot Klm + ki \cdot kgu)$. Below cGb, cIs is zero or constant, depending on the presence of a basal insulin inflow.

Integral equations

Finally, we derive integral equations from the differential equations (23.20) and (23.23), in order to make the step from mass-balance equations to block diagram. This is required for defining the INSUL model in terms of a BIOPSI program for numerically solving the differential equations.

The resulting equations are (23.30) and (23.31), respectively.

$$cG(t) = c\,G(0) + \int_0^t \frac{1}{Vg} \bullet (ig - eg) \bullet dt \qquad (23.30)$$

$$cI(t) = cI(0) + \int_0^t \frac{1}{Vi} \bullet (ii - ei) \bullet dt \qquad (23.31)$$

Substitution of the equations for the various flows (ig, eg, ii, ei) yields the complete expressions for cG(t) and cI(t).

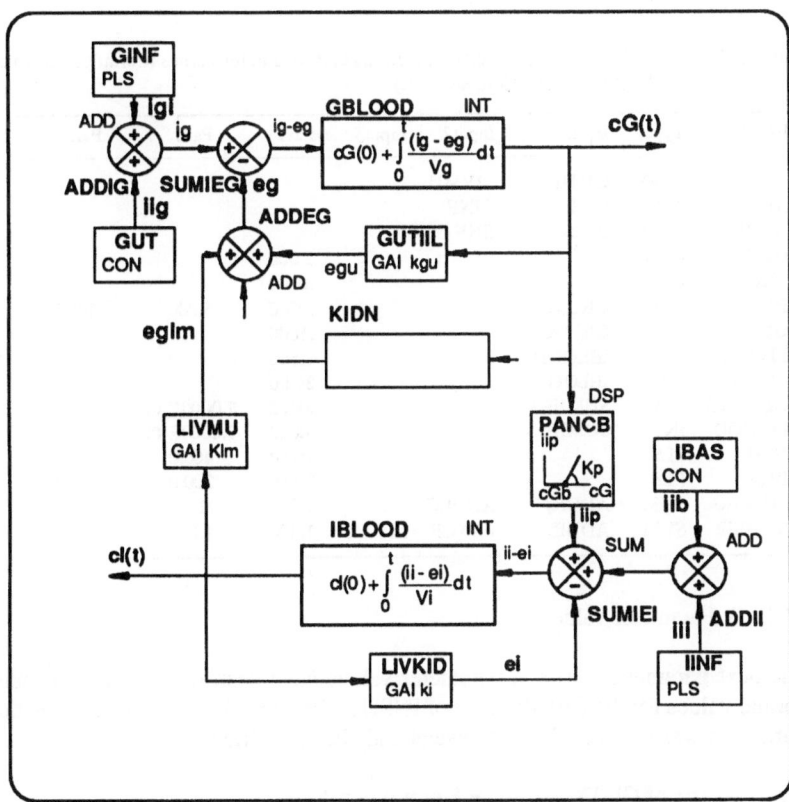

Figure 23.8.   Block diagram of INSUL (for explanation, see text)

23.3.4.2 Block diagram

The block names and types for INSUL are listed in Table 23.3, which also gives the connections between the blocks and the parameter values (see below). A diagram of the INSUL block structure, consistent with Table 23.3, is drawn in Fig. 23.8. The key difference in INSUL is that it contains two integrators (the INT blocks GBLOOD and IBLOOD for the total extracellular volumes of glucose and insulin, respectively) instead of one, as in GLUKID. This follows directly from the two differential equations (23.20) and (23.23), describing the system behavior. Since there are no new block types in the diagram, its structure can be derived from the equations and the process diagram of INSUL, as explained in GLUKID.
Notice that INSUL is one of the three building blocks of the more complete system illustrated in Fig. 23.2.

Table 23.3. Blocks, block types, block connections and parameter values (default setting) of INSUL (for explanation, see text)

Block	Type	Input1	Input2	Input3	Par1	Par2	Par3
ADDEG	ADD	GUTIL	LIVMU				
ADDIG	ADD	GUT	GINF				
ADDII	ADD	IBAS	IINF				
GUT	CON				2.000		
IBAS	CON				.0000		
PANCB	DSP	GBLOOD			.0000	5.000	.1000
GUTIL	GAI	GBLOOD			.1000		
LIVKID	GAI	IBLOOD			1.500		
LIVMU	GAI	IBLOOD			30.00		
GBLOOD	INT	SUMIEG			.0000	7.0000E-02	
IBLOOD	INT	SUMIEI			.0000	7.0000E-02	
GINF	PLS				100.0	400.0	2.000
IINF	PLS				200.0	300.0	.1500
SUMIEG	SUM	ADDIG	ADDEG		1.000	-1.000	
SUMIEI	SUM	ADDII	PANCB	LIVKID	1.000	1.000	-1.000

23.3.4.3 Parameter values

The next parameters (p) and parameter values have been used as approximate normal values for the default demo of INSUL. The first five parameters have the same values as in GLUKID and have already been justified.

igg   (p1 of GUT) $= 2\ \mathrm{mmol.min^{-1}}$
igi   (p3 of GINF) $= 2\ \mathrm{mmol.min^{-1}}$
cG(0)   (p1 of GBLOOD) $= 0\ \mathrm{mMol}$
Vg   (1/p2 of GBLOOD) $= 14\ \mathrm{L}$
kgu   (p1 of GUTIL) $= 0.1\ \mathrm{L.min^{-1}}$

Kb	(p3 of PANCB)	= 0.1 L.min^{-1}
cGb	(p2 of PANCB)	= 5 mM
iib	(p1 of IBAS)	= 0 unit.min^{-1}
iii	(p3 of IINF)	= 0.15 unit.min^{-1}
Vi	(1/p2 of IBLOOD)	= 14 L
ki	(p1 of LIVKID)	= 1.5 L.min^{-1}
Klm	(p1 of LIVMU)	= 30 mmol G .min-1 per unit I .L^{-1}
cI(0)		= 0 unit.L^{-1}

Kb and cGb are approximate values, based on Fig 19-13 of [Ganong, 1987, pg. 288] and on [Guyton ,1986, pg. 929]. Basal insulin secretion (iib) is < 0.1 times maximal secretion cf [Guyton, 1986, pg. 929], but is assumed to be zero in the default. The insulin infusion (iii) has been arbitrarily chosen within the physiological range at iii = 0.15 unit/min.

Insulin breakdown occurs rapidly, with a time constant (ti) of about 9 min [Ganong, 1987, pg. 279]. For an extracellular volume Vi = 14 L [Ganong, 1987, pg. 9] for insulin distribution, ki follows from ki = Vi/ti = 1.5 L/min. The rate constant Klm for insulin-induced glucose uptake by the liver and muscles has been estimated from data on stationary glucose levels during regular glucose uptake by mouth in healthy students (unpublished observations) and from [Nijs et al. , 1988].

23.3.4.4 Possibilities and limitations

INSUL provides the basic structure of endocrine functions in the blood glucose control system based on excess removal. The model produces many of the phenomena characteristic of the normal behavior and of various diseases. However, INSUL has limitations, some of which are of a quantitative nature, since the assumptions have greatly simplified the system. Other limitations result from process-neglect assumptions, such as the absence of influences of other hormones (glucagon, cortisol, gastro-intestinal hormones, growth hormone; cf [Guyton, 1986, pg. 930] and of the autonomic system [Ganong, 1987, pg. 288-289]. But once the basic structure of the control system is given, the process definitions can easily be improved and the neglected processes and influences can be introduced. So far, we have only used a simplified insulin system to illustrate the properties of a simple second-order control system.

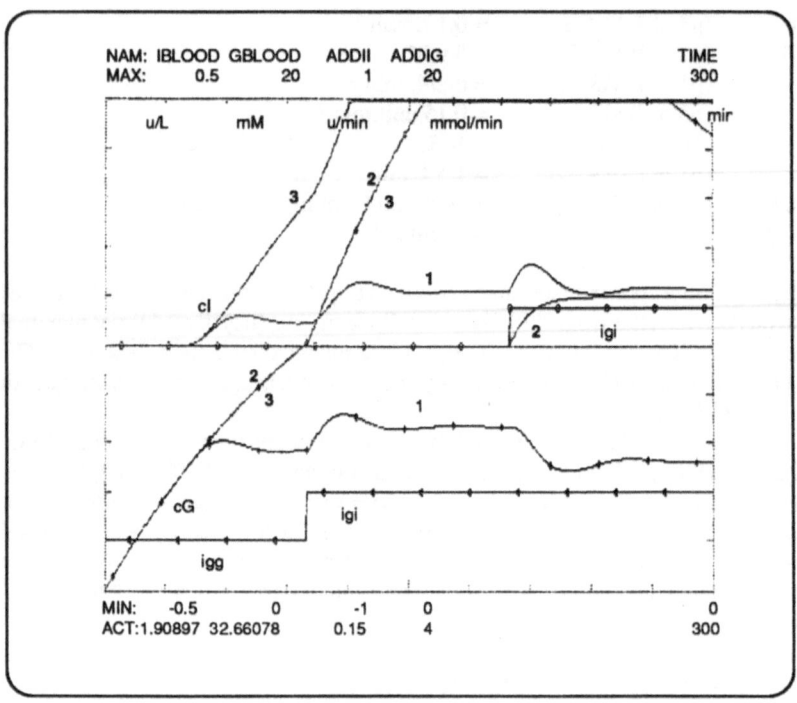

Figure 23.9. The behavior of INSUL in the default setting (curves 1), in the condition of the absence of functional beta-cells (Kb = 0, curves 2) and in the absence of insulin induced glucose uptake (Klm = 0; curves 3)

23.3.5 Exercises

23.3.5.1 The default demo

Fig. 23.9 (curve 1) shows the behavior of INSUL for the default setting of the parameters. The simulation is a 3-phase "experiment". In the first 100 min. the model is allowed to establish steady-state conditions from cG=0 mM and cI=0 u/L for a standard glucose inflow from the gut of 2 mmol/min starting at t=0. In the second phase (100-200 min.) you can see the model response to an extra glucose inflow (2 mmol/min) from a glucose infusion into the blood. The third phase (200-300 min.) is the model response to an insulin infusion of 0.15 units/min.

In the beginning, cG rises rapidly (only opposed by glucose utilization), until cGb=5 mM. Then the pancreas starts to secrete insulin, so that cI also starts to rise. The increase in cI causes an extra outflow of glucose from the blood, the uptake by the liver and muscles, and opposes a further increase in cG. Both cG and cI then settle at a constant, stationary value. The step increase in glucose inflow at t = 100 min again causes a rapid increase in cG, but now cI also rapidly increases, since cG > cGb. Again, the increased cI opposes a further increase of cG because of an increased uptake of G into the liver and muscles; steady-state conditions are established, but now at a higher cG and cI. Thus, the changes in cG and cI upon glucose infusion are in the same direction, while cI follows the changes in cG. Furthermore, stationary cG and cI are of the right order of magnitude for humans [Cahill and Soeldner, 1969, Fig. 27 on pg. 114 derived from Williams et al., Cli. Res. 14, 356, 1966]. Dogs show a transient behavior similar to that in Fig. 23.9 [Salzsieder et al., 1985].

The response to the step increase in insulin inflow at t = 200 min. is characteristically different from that to step increases in glucose inflow. The increase in insulin inflow causes a rapid increase in cI, which causes an increased storage of glucose in the liver and muscles so that cG soon decreases. This cG decrease, however, reduces the drive of the pancreas to secrete insulin, causing a decrease of cI after the initial increase. Finally, both cG and cI settle at stationary values, cG below and cI above the preceding stationary values. Thus, the cI and cG changes with insulin infusion are in opposite directions.

The off-responses of the system to the cessation of infusions can be seen when the simulation is continued by pressing C or CX.

23.3.5.2 Effects of changes in Kb

In juvenile diabetes mellitus (type I diabetes) the beta-cells of the endocrine pancreas gradually deteriorate and become unable to secrete insulin upon glucose stimulation. This means, in terms of our model, that Kb (p3 of PANCB) becomes zero. This final stage is illustrated in curves 2 of Fig. 23.9: cG grows to very large stationary values (20 mM in period 1 and 40 mM in period 2 (out of screen) because cI remains zero; the feedback control is completely absent. However, an insulin infusion pulls cG down again (final part of curves 2).
Study the effect of gradually decreasing Kb and find out the required insulin infusion to keep cG within 5-8 mM for a few values of Kb at a given glucose input. Also study the effect of increasing Kb on the responses of cG and cI to the glucose and insulin infusions.

23.3.5.3 Effects of changes in Klm

In liver disease, the uptake of glucose under the influence of insulin may become decreased. An extreme case (Klm = p1 of LIVMU has been made zero from the default setting) is illustrated in Fig. 23.9 (curve 3). It is the same as the case Kb = 0 in that the same stationary cG is reached, but here cI rises sharply upon starting the simulation at t=0 because there is no feedback effect of insulin on cG through the liver. Therefore, cG as well as cI remain high.

A reduced insulin-induced uptake of glucose by fat cells is characteristic for maturity-onset diabetes (type II diabetes; see [Ganong, 1987, pg. 295]). When we include glucose uptake by fat cells in the parameter Klm, we can simulate the increased blood glucose levels in these patients by decreasing Klm.

Study the effect of gradually decreasing and increasing Klm on the time dependent and stationary behavior of the system.

23.3.5.4 Hyperinsulinism

Some types of tumors (insulinomas) may spontaneously secrete insulin, regardless of the glucose concentration [Guyton, 1986, pg. 936]. Simulate this condition by varying the insulin inflow iib at a given glucose inflow igg. For each iib find the required extra glucose intake, for example by glucose infusion (igi), to maintain cG within 5-8 mM. You may find that cG becomes negative at the higher iib's. This shortcoming of the model is due to the assumptions underlying equation (23.14). How could this problem be solved?

23.3.5.5 Effects of changes in cGb

Many factors may influence blood glucose control by changing the secretion properties of the beta-cells of the pancreas [Ganong, 1987, pg. 288], including autonomic nervous system activity. These influences may concern isolated changes of iib, Kb and/or cGb. In particular study the changes in cGb by changing its value during a 300-min simulation without applying glucose and insulin infusions (make p3=0 of GINF and of IINF).

23.3.5.6 Insulin turnover and glucose utilization

Change the default parameter setting as follows: p3=0 of PANCB (removal of beta-cells); p1=0 of LIVMU (liver and muscles unresponsive to insulin); p1=0 of GUT (no glucose uptake from food in the intestines); p1=5 of GBLOOD (cG value at t=0); p1=1 of IBLOOD (cI value at t=0). Run the model to observe spontaneous clearance of insulin and glucose from the blood and the establishment of steady-state concentrations upon starting infusions of glucose and insulin.

23.3.5.7 Combination of INSUL and GLUKID

A model INSKID is available, which combines INSUL and GLUKID. Run it in its default setting and then lower Kb or cGk to observe that the kidney contributes to blood glucose regulation in such conditions. This model is very similar to that of

[Stolwijk and Hardy, 1974], and it was used by [Randall, 1980] as a BASIC programmed simulation model in teaching physiology.

23.3.6 Conclusion

In this section you learned to design and construct a second-order model for regulating the concentration of a substance in the blood based on excess removal. The model was a two-substances/one-compartment model, derived from the glucose-insulin/blood system maintaining a constant glucose concentration in the blood. You started with a qualitative description of the system, selected the processes to consider in a process diagram, defined the processes quantitatively, and constructed a BIOPSI model of the system in terms of a block diagram. Finally, you learned about the behavior of the system by studying the effect of changing the various parameters of the model, as may occur during disease.

23.4   Glucose Regulation by Glucagon: GLUCAG

23.4.1 Introduction

In this section you will again study the design and behavior of a simple second-order (two-substance/one-compartment) model of an endocrine system maintaining a constant concentration of a substance in the blood. However, this time the model is based on controlled *addition* of the substance, in case the substance becomes depleted. The substance controlled is, again, glucose (G) and the hormone involved is glucagon (GA). Like insulin, this hormone is also produced by the endocrine pancreas. The combined action of removal control by insulin and addition control by glucagon provides the body with an efficient system for the control of the blood glucose concentration.

The model to be discussed is called GLUCAG, representing blood glucose regulation by the hormone GLUCAGon. Its structure is basic to many endocrine models, such as the system for blood calcium control by PTH (parathyroid hormone). Other examples of such endocrine systems are the regulation of the blood concentrations of cortisol and thyroid hormone. All these systems are based on addition-control and have at least the complexity of a second-order system.

The second-order glucagon system is not supported by a first-order system based on controlled addition, as was the case for the insulin system, which was supported by the kidney. We are not aware of any first-order addition-control system in endocrinology. Therefore, we do not discuss this most simple type of control system as a separate model, but refer you to GLUKID and to exercise 23.4.5.4.

Unlike INSUL, GLUCAG cannot be used to illustrate mechanisms of many diseases, since diseases based on isolated misregulation of the glucagon system seem to be rare. However, GLUCAG is still useful to explain consequences of experimental manipulations of the system, and these are representative of diseases in other comparable control systems.

After having studied the models GLUKID, INSUL and GLUCAG, you have the opportunity to do exercises with a more complete model (PANKID), which includes these three models as units.

23.4.2 Physiological Background

The hormone glucagon (GA) is produced and secreted by the alpha-cells of the islets of Langerhans of the pancreas. Hypoglycemia stimulates and hyperglycemia suppresses GA secretion. Thus, the way GA secretion responds to changes in cG is the opposite of the response to insulin secretion. The effects of GA in the body are also opposite to those of insulin. GA mobilizes glucose from the liver into the blood, while insulin stimulates glucose storage from the blood into the liver and the muscles. These properties of GA make the pancreas-liver system a negative feedback regulation system for blood glucose, based on addition-control: lowering of cG causes an increased GA secretion, which results in an increased glucose inflow into the blood from the liver, opposing the initial decrease of cG. For further details, see Ch. 19 of [Ganong, 1987] and Ch. 78 of [Guyton , 1986]. Details relevant to GLUCAG are described below in the process diagram and in the quantitative model description.

23.4.3 Process Diagram

The process diagram (Fig. 23.10) summarizes the processes considered in GLUCAG. Compare it with Fig. 23.1 to see its relationship to the more complete system.

Glucose flows

As in GLUKID and INSUL, glucose may enter the blood through inflows from the gut (igg) and from a glucose infusion (igi). However, in GLUCAG there is one additional glucose inflow, (igl), which is provided by the liver and which serves to supply the blood with glucose when the glucose concentration (cG) becomes too low. This igl is stimulated by an increase in the concentration of glucagon (cGA) (see + effect indicator; [Guyton, 1986, pg. 931]). Besides the three inflows mentioned we consider one (pooled) outflow (egu) of glucose, resulting from glucose utilization (metabolism) in various tissues. This outflow egu is assumed to increase with increasing cG (see + effect indicator; negative feedback), as in the previous models.

Glucagon flows

The endocrine pancreas (the alpha-cells of the islets of Langerhans) spontaneously secrete glucagon (GA) in the absence of glucose in the blood [Ganong, 1987, pg. 288]. However, when cG rises, this Inflow of GlucAgon from the Pancreas (igap) is suppressed see - effect indicator; [Guyton, 1986, pg. 931]. In addition to this igap, GA enters the blood by a basal GA inflow (igab) from the pancreas and/or from a GA infusion (igai). These three GA inflows contribute to building up the concentration of GA in the blood, cGA. At the same time GA disappears from the blood through an outflow (ega), caused by breakdown of GA, mainly in the liver [Ganong, 1987, pg. 291].

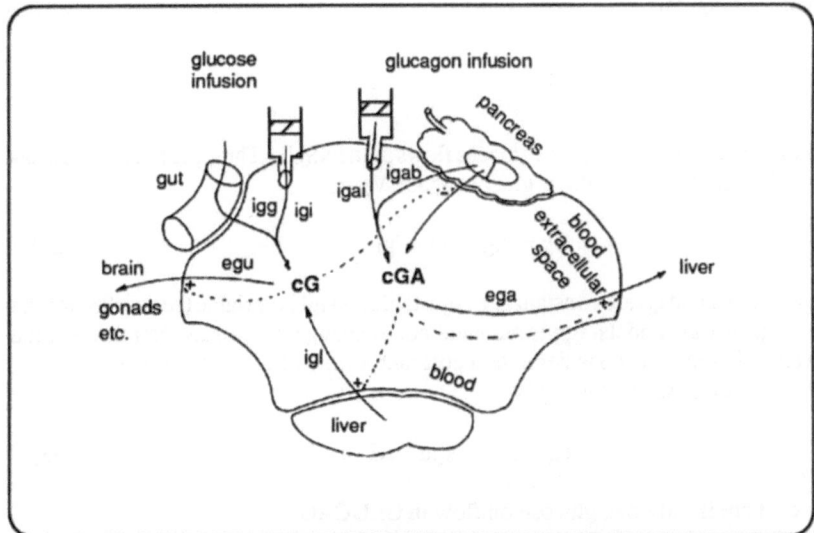

Figure 23.10. Process diagram of GLUCAG. Meaning of new abbreviations: igap = Inflow of GlucAgon into the blood from the Pancreas; igai = Inflow of GlucAgon from the infusion; igab = Basic Inflow of glucAgon; ega = Efflow of GlucAgon from the blood by breakdown in the liver; igl = Inflow of Glucose from the Liver, which is stimulated (+ effect indicator) by an increase in the glucagon concentration cGA. Other abbreviations have been explained in the legends of Fig. 23.4 and Fig. 23.7

Finally, GA stimulates (see + effect indicator) glucose mobilization from the liver igl [Ganong, 1987, pg. 290]. Since G suppresses igap, increases in cG suppress igl through the cGA decrease. Thus, an increase of cG suppresses a further accumulation of G (negative feedback).

This process diagram does not include the many factors affecting glucagon secretion, such as the activity of the autonomic nervous system, the presence of the glucose controlling hormone insulin [Ganong, 1987, pg. 291] and the direct influence of insulin on the glucagon system (and vice versa). These factors can be incorporated in a later stage of model design.

23.4.4 Model Description

23.4.4.1 Foundations and assumptions

As in GLUKID and INSUL, we follow the same procedure from process definition to block diagram.

Glucose flows

Similar equations apply to the glucose flows as in INSUL. Thus, the total glucose inflow ig into the extracellular glucose space Vg is

$$ig = igg + igi + igl \qquad (23.32)$$

in which igg and igi are constant and adjustable and igl is variable (see eq. 23.34). As in the previous models, igg is assumed not to change by passing first through the liver, or igg may be considered as a constant inflow of G from the liver.
The total glucose outflow eg is

$$eg = egu = kgu \cdot cG \qquad (23.33)$$

since there is only one glucose outflow in GLUCAG.
The inflow of glucose from the liver igl is defined by

$$igl = Kl \cdot cGA \qquad (23.34)$$

assuming instantaneous cGA effects, which is a fair approximation based on [Wilson and Foster, 1985, Fig. 26-11, pg. 1031].

Glucagon flows

The effect of cG on igap is defined by

$$igap = Ka \bullet (cGa - cG) \qquad 0 \leq cG \leq cGa \qquad (23.35a)$$
$$igap = 0 \qquad\qquad\qquad cG > cGa \qquad\qquad (23.35b)$$

This relationship has been drawn, in block IGASUB of Fig. 23.11, producing the glucagon flow igap from the alpha-cells of the islets of Langerhans. The proportionality constant Kpa derives its value from the properties of these alpha cells, while cGa is the threshold concentration of cG, below which the alpha-cells secrete glucagon.

The total glucagon inflow iga into the blood is

$$iga = igap + igab + igai \qquad (23.36)$$

The basal glucagon inflow igab and the inflow igai from the infusion are constant and adjustable (see process diagram). Glucagon breakdown, mainly by the liver, is assumed to depend on its concentration and to follow first-order kinetics, consistent with half-time values reported [Ganong, 1987, pg. 291]. The resulting efflow ega is then given by

$$ega = kga \bullet cGA \qquad (23.37)$$

In these process definitions, we have neglected the nature of the circulatory system, in which the organ topography is such that GA affects the liver cells and is partially broken down by them, before it enters the general circulation [Ganong, 1987, pg. 291]. Instead, we consider the glucagon space as a well-stirred constant volume (Vga), including the blood plasma and the interstitial space, since it is our first goal to design a simple model with the right qualitative behavior. From the model for gastric acid secretion (Chapter 24) you will learn how to combine fluid flow equations with substance break-down equations, as should be done in the design of a more realistic INSUL and GLUCAG model.

Mass-balance equations

The change in cG is given by

$$\frac{dcG}{dt} = \frac{ig - eg}{Vg} \qquad (23.28)$$

The complete expression of (23.38) is obtained if equation (23.32) and then (23.33) and (23.34) are substituted:

$$Vg \bullet \frac{dcG}{dt} = igg + igi + Kl \bullet cGa - kgu \bullet cG \qquad (23.39)$$

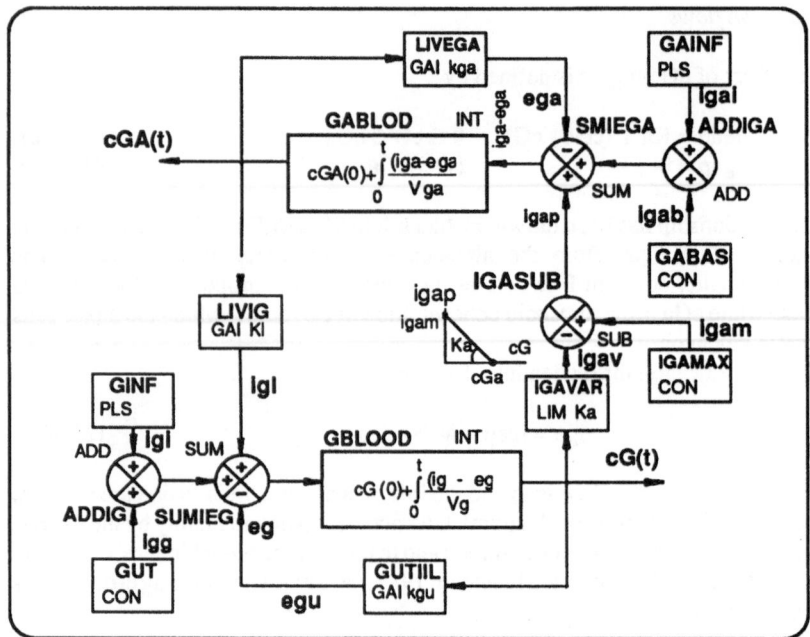

Figure 23.11. Block diagram of GLUCAG (for explanation see text)

The change in cGA is given by

$$\frac{dcGA}{dt} = \frac{iga - ega}{Vga} \qquad (23.40)$$

The complete expression of (23.40) is obtained if equations (23.36) and (23.37), and then (23.35) are substituted. For $0 \le cG \le cGa$ we find

$$Vga \cdot \frac{dcGA}{dt} = igab + igai + ka \cdot (cGa - cG) - kga \cdot cGA \qquad (23.41a)$$

For $cG > cGa$ we obtain

$$Vga \cdot \frac{dcGA}{dt} = igab + igai - kga \cdot cGA \qquad (23.41b)$$

Steady-state control

Equations (23.39) and (23.41) can be solved for the stationary condition dcG/dt = dcGA/dt = 0, as shown with INSUL.

Integral equations

From the differential equations (23.38) and (23.40) we derive the integral equations in order to define GLUCAG in terms of a BIOPSI simulation program for numerically solving equations (23.38) and (23.40). The resulting equations are (23.42) and (23.43), respectively:

$$cG(t) = cG(0) + \int_0^t \frac{1}{Vg} \cdot (ig - eg) \cdot dt \qquad (23.42)$$

$$cGA(t) = cGA(0) + \int_0^t \frac{1}{Vga} \cdot (iga - ega) \cdot dt \qquad (23.43)$$

Substitution of the equations for the various flows yields the complete expressions for cG(t) and cGA(t).

23.4.4.2 Block diagram

The block names and types are listed in Table 23.4, which also gives the connections between the blocks and the standard parameter values (see below). A diagram of the GLUCAG block structure, consistent with Table 23.4, is drawn in Fig. 23.11. This diagram has the same basic structure as INSUL and is, again, centered around the presence of two integrators, GBLOOD for integrating the glucose flows and GABLOD for integrating the glucagon flows. However, the alpha-cells of the pancreas require new types of BIOPSI blocks to realize equation 23.35: a SUB block IGASUB, which subtracts a variable GA inflow (igav, depending on cG) determined by a LIM block IGAVAR from a maximal GA inflow, determined by the CON block IGAMAX. In other respects, the block diagram has the same structure as INSUL. You may also compare GLUCAG with Fig. 23.2 to see the relationship to the more complete system.

23.4.4.3 Parameter values

The next parameters (p) and parameter values have been used as approximate normal values for the default demo of GLUCAG. The first five parameters were also present in GLUKID. Three of these have the same values, and have already been justified.

Table 23.4. Blocks, block types, block connections and parameter values (default setting) of GLUCAG

Block	Type	Input1	Input2	Input3	Par1	Par2	Par3
ADDIG	ADD	GUT	GINF				
ADDIGA	ADD	GABAS	GAINF				
GABAS	CON				.0000		
GUT	CON				.0000		
IGAMAX	CON				.5000		
GUTIL	GAI	GBLOOD			.1000		
LIVEGA	GAI	GABLOD			1.400		
LIVIG	GAI	GABLOD			13.00		
GABLOD	INT	SMIEGA			.0000	7.0000E-02	
GBLOOD	INT	SUMIEG			.0000	7.0000E-02	
IGAVAR	LIM	GBLOOD			.0000	.5000	.1000
GAINF	PLS				200.0	300.0	.1000
GINF	PLS				100.0	400.0	.5000
IGASUB	SUB	IGAMAX	IGAVAR				
SMIEGA	SUM	ADDIGA	IGASUB	LIVEGA	1.000	1.000	-1.000
SUMIEG	SUM	LIVIG	ADDIG	GUTIL	1.000	1.000	-1.000

Parameters:

igg	(p1 of GUT)	$= 0$ mmol.min^{-1}
igi	(p3 of GINF)	$= 0.5$ mmol.min^{-1}
cG(0)	(p1 of GBLOOD)	$= 0$ mM
Vg	(1/p2 of GBLOOD)	$= 14$ L
kgu	(p1 of GUTIL)	$= 0.1$ L.min^{-1}
Ka	(p3 of IGAVAR)	$= 0.1$ nmol glucagon.min^{-1} per mM glucose
cGa	(p2/p3 of IGAVAR)	$= 5$ mM
igab	(p1 of GABAS)	$= 0$ nmol.min^{-1}
igai	(p1 of GAINF)	$= 0.1$ nmol.min^{-1}
Vga	(1/p2 of GABLOD)	$= 14$ L
kga	(p1 of LIVEGA)	$= 1.4$ L.min^{-1}
Kl	(p1 of LIVIG)	$= 13$ mmol glucose.min^{-1} per nM glucagon
cI(0)	(p1 of GABLOD)	$= 0$ mM

Glucose inflow from the gut (igg) has been put at zero in order to demonstrate GA-induced glucose mobilization from the liver. A glucose infusion of igi=5 mmol/min has been chosen in order to demonstrate regulation in the cG range of GA action. Ka and cGa are approximate values taken from Fig. 19-13 of [Ganong, 1987; pg. 288]. Vga has been made equal to Vg, but may very well be smaller; kga has been calculated from the time constant tga=10 min (consistent with [Ganong, 1987, pg. 291]) and Vga according to tga=kga • Vga. Kl has been given a value such that both the transient and the stationary behavior occurred in the physiological range of cGA values.

Furthermore, the block structure of the glucagon secreting cells is such that p1 of IGAMAX (maximal value of igap) is equal to p2 of IGAVAR.

23.4.4.4 Possibilities and limitations

GLUCAG provides the basic structure of that part of the blood glucose control system that is based on substance addition proportional to depletion. It has, therefore, the potential to produce phenomena characteristic of system parameter variations under clinical and experimental conditions in similar control systems with second-order two-substances/one compartment models based on proportional depletion compensation. However, in order to obtain an exact predictive value, the process definitions should be improved and more processes should be considered in the model.

23.4.5 Exercises

23.4.5.1 The default demo

Fig. 23.12 (curve 1) shows the behavior of GLUCAG for the default setting of the parameters. As in INSUL, the simulation is a 3-phase "experiment". In the first 100 min. the model is allowed to establish steady-state conditions from the initial conditions cG = 0 mM and cGA = 0 nM. No glucose or glucagon infusions are given in this first period, and no glucose inflow from the gut or glucagon basal secretion is occurring. In the second phase (100-200 min) 0.5 mmol glucose per minute is infused into the blood. In the third phase, from t = 200 min., 0.1 nmol glucagon per minute is infused into the blood in addition to the continued glucose infusion.

Just after the start of the experiment, igap is maximal since cG = 0 at t = 0. Therefore, cGA rises rapidly. As a result, the liver is stimulated to release glucose, so that a cG increase along with the cGA increase. While cG is rising, igap declines (23.35), explaining a reduced rate of cGA increase. Finally, after a damped oscillation in cG and cGA, the system reaches a steady state in cG and cGA.

Upon starting the glucose infusion at t = 100 min., cG further rises and just reaches cGa, so that igap shuts off completely and cGA becomes zero. Upon starting the glucagon infusion, cGA rises again, but now with first-order kinetics, since the alpha-cells of the pancreas are no longer active. Furthermore, the rise in cGA causes glucose release from the liver, resulting in a rise in cG with second-order kinetics. These responses illustrate the basic properties of the glucagon system for blood glucose control, as explained in standard textbooks of physiology and endocrinology.

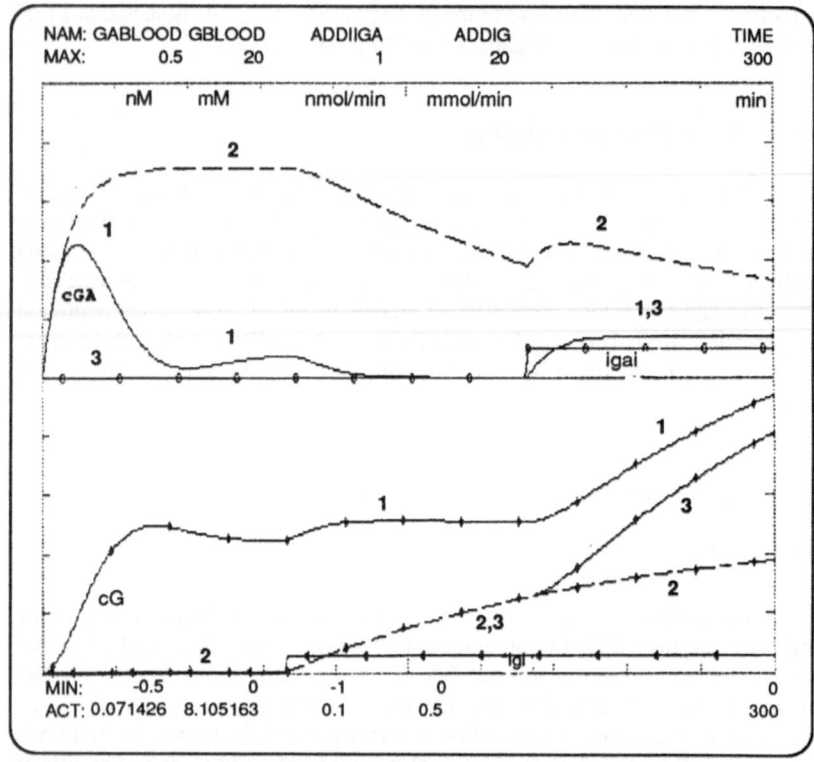

Figure 23.12. The behavior of GLUCAG upon glucose and glucagon infusions in the
              default setting (curve 1), at Kl = 0 (non-functional liver, curve 2) and at
              Ka = 0 (non-functional alpha-cells, curve 3)

23.4.5.2 Effects of changes in Kl

Although diseases due to misregulation of the glucagon control system seem to be
rare, one may, in principle, expect the existence of diseases in which the liver does
not or only weakly responds to glucagon in providing the blood with glucose. An
extreme example (Kl=0) of this condition is illustrated in Fig. 23.12 (curve 2).
Perform this experiment after changing p1 of LIVIG to zero.

23.4.5.3 Effects of changes in Ka

Another possible type of misregulation of the glucagon system is the partial or
complete loss of the ability of the alpha-cells to secrete glucagon. This is illustrated
in Fig. 23.12 (curve 3) for the extreme case that Ka=0. Run this simulation by

making (from the default setting) p1=0 of IGAMAX and by making p1=0, p2=0 and p3=0.001 of IGAVAR. Obviously, only after starting the glucose infusion at t=100 min does cG start to rise with a single-exponential time course. An extra increase in cG follows the single-exponential rise in cGA upon infusion of glucagon at t=200 min. Repeat this type of simulation for intermediate values of Ka. Also study the effects of increased values of Ka on the transient and stationary behavior of the glucagon system.

23.4.5.4 First-order addition control as a reduction of GLUCAG

GLUCAG is a second-order addition control system. By removing LIVIG from the model and by considering igap as a glucose inflow to be connected to SUMIEG instead of igl, one obtains a first order addition control system. Such first-order systems are rare, if they exist at all, but they are instructive nonetheless. In this hypothetical first-order addition control model, the alpha-cells would both serve as a sensor and as an effector, as does the kidney in the first-order glucose excess removal system GLUKID. For reasons of clarity some additional blocks may be removed from GLUCAG: ADDIGA, GABAS, LIVEGA and GABLOD. Furthermore, change the parameters of GINF into p1=100, p2=200 and p3=1. Run the changed model and compare its behavior to that of GLUCAG. The changed model is also available as GLUAD1.

23.4.5.5 Behavior of the complete glucose regulation system

The model PANKID includes GLUKID, INSUL and GLUCAG. Its process diagram is given in Fig. 23.1, its block diagram in Fig. 23.2 and its parameter values in Table 23.1. Run PANKID in its default setting and try to understand its behavior for the default parameter setting from what you learned in the three separate models. Of particular interest is the behavior of cGA upon insulin infusion and of cI upon glucagon infusion: Insulin infusion causes extra glucagon secretion and glucagon infusion causes extra insulin secretion when cGa > cGb is chosen (Fig. 23.3), which is a normal physiological condition [Ganong, 1987, pg. 291-292]. In that case there is a range of cG in which INSUL and GLUCAG are working simultaneously. The insulin infusion may be thought to represent the activity of an insulinoma [Ganong, 1987, pg. 294] and the glucagon infusion a glucagonoma, spontaneously secreting glucagon (a very rare disease, see Wilson and Foster, 1985, pg. 1302). The simulation then shows how the glucagon system counteracts the insulin system and vice versa. This mutual counteraction results in a more precise stationary control of cG for cGb<cG<cGa than would have occurred with only one of the two control systems.

23.4.6   Conclusion

In this chapter you learned to design and construct a second – order, two-substances/one-compartment model for the regulation of the concentration of a substance in the blood based on substance addition proportional to substance depletion. In addition, you investigated its behavior under various conditions in order to become familiar with the properties of this type of model and with the characteristic behavior of the particular system considered, i.e. the glucagon system maintaining a constant glucose concentration in the blood. Finally, you combined all three glucose control models (GLUKID, INSUL and GLUCAG) into a single, complete model PANKID.

23.5   General Discussion

So far, you have been introduced to endocrine modeling, have practiced designing models and have learned about a specific system: blood glucose regulation. You have also experienced the instructive value of making models, since they confront you with insufficiencies in your knowledge and understanding. Now you may want to proceed in improving the glucose regulation models, to apply your skills to other systems or to learn about another endocrine model such as the gastric acidity regulation model of the next chapter. For advanced glucose regulation models we refer to [Salzsieder et al., 1985], which also provides relevant references to other papers. The general paper on modeling [Smith, 1983] may also be appreciated because the approach (making mass-balance equations) of this section will be recognized and because the paper goes further into its analysis. Finally, you may want to improve your theoretical insight in endocrine modeling by studying, for example, [Carson et al., 1983].

The availability of simulation tools such as PSI are of great value in the teaching of and research in physiology. A PSI model is much easier to implement than a model to be written in one of the standard computer languages cf. [Randall, 1980], although writing such a program may also be instructive.

You have discovered that many diseases may be seen as a misregulation of a control system due to altered parameter values. For example, type I diabetes (juvenile diabetes) may be understood as a consequence of a lowered Kb in the model INSUL. Once you know the parameters of a system, you may even predict the existence of certain types of diseases from assumed parameter changes. However, only clinicians can tell whether the predicted diseases exist and how frequently they occur. Furthermore, a model never explains why a certain misregulation causes illness, for example why hypoglycemia may cause unconsciousness, nor can it explain how a parameter change is caused by a disease. To understand this, one needs more general knowledge of body functions.

If diseases may be seen as the result of altered parameters, it may also be expected that a system is more vulnerable to changes in certain parameters than to changes in other ones. How sensitive a system is in its regulation properties can be investigated with a technique called "sensitivity analysis". This technique, which is a standard tool in control engineering, will be explained and applied in Chapter 24.

Finally, a few shortcomings of the models used in this section (but also in published models) should be mentioned: All our models lack time dependent effects of the secretion-concentration relationships, both short-term (minutes-hours) and long-term (hours-days).

Examples of long-term effects are the so-called "exhaustion" of beta-cells by chronic overstimulation [Ganong, 1987, pg. 295] and the down regulation of the sensitivity of target tissues (fat cells) by chronic hyperinsulinaemia, as in type II diabetes [Ganong, 1987, pg. 295].

Incorporation of time dependent effects would, therefore, be a useful improvement of our glucose regulation models.

Another shortcoming of our models is that we have not included saturation effects in the secretion-concentration relationships (eq. 23.14, 23.16, 23.34, 23.35). This can easily be done by using LIM blocks.

A final shortcoming concerns the number of compartments for substance distribution. If experimental data would justify more compartments for glucose, insulin and glucagon distribution, then these compartments can easily be added to the models by the addition of extra differential equations and the corresponding integrator blocks.

However, the shortcomings mentioned for the various model descriptions and those mentioned here are not expected to affect the qualitative behavior of the glucose regulation system. If so, it is simply a challenge to improve the model.

Acknowledgements

We thank Drs. J.K. Radder and Tj. Wieringa (Endocrinology, Academic Hospital, Leiden University) for advice and Dr. J. de Goede (Dept. of Physiology and Physiological Physics, Leiden University) for the many instructive discussions that stimulated this work.

References

Cahill, G.F. and Soeldner, J.S.: "Glucose homeostasis: A brief review". In: *Hormonal Control Systems, Supplement 1, Mathematical Biosciences*, Eds.: B. Stear and A.H. Kadish, American Elsevier Publishing Company, Inc., New York, 1969, pp. 83-114.

Carson, E.R., Cobelli and C., Finkelstein, L.: *The mathematical modelling of metabolic and endocrine systems: model formulation, identification and validation*. J. Wiley, New York, 1983.
Fischer, U., Schenk, W., Salzieder, E., Albrecht, G., Abel, P., and Freyse, E.J.: "Does physiological blood glucose control require an adaptive control stategy?", *IEEE Trans. Biomed. Eng. BME-34*, 8, 1987, pp. 575-582.

Ganong, W.F.: *Review of Medical Physiology, 13th edition*. Appleton & Lange, Norwalk, Connecticut/San Mateo, California, 1987.

Guyton, A.C.: *Textbook of Medical Physiology, 7th edition*. W.B. Saunders Company, Philadelphia, London, 1986.

Nijs, H.G.T., Radder, J.K., Frolich, M. and Krans, H.M.J.: "Insulin action is normalized in newly diagnosed type I diabetic patients after three months of insulin treatment". *Metabolism*, 37, no 5, 1988, pp. 473-478.

Randall, J.E.: *Microcomputers and physiological simulation*. AddisonWesley Publishing Company Inc., London, Amsterdam, 1980.

Salzsieder, E., Albrecht, G., Fischer, U. and Freyse, E.J.: "Kinetic modeling of the glucoregulatory system to improve insulin therapy". *IEEE Trans. Biomed. Eng. BME-32*, no 10, 1985, pp. 846-855.

Smith, R.S.: "Qualitative mathematical models of endocrine systems". *Am. J. Physiol*. 245, 1983, R473-R477.

Verveen, A.A.: "Left- and right-regulating systems". *Biol. Cybern*. 35, 1979, pp. 131-136.

Wilson, J.D. and Foster, D.W.: *Williams Textbook of Endocrinology, 7th edition*, W.B. Saunders Comp., Philadelphia, London, 1985.

24
Regulation of Gastric Acidity

Bert van Duijn, Dirk L. Ypey, and Jacob de Goede

24.1 Introduction

In this section we will practice the modeling method for endocrine systems explained in Chapter 23. A model for the regulation of gastric acidity in humans will be designed, starting with a process diagram and resulting in a block diagram. The process diagram is gathered from data reported in the literature as are the parameter values used in the model.

It will be shown that a simulation language such as BIOPSI can be very useful when the designed model is difficult to analyze analytically. The model described in this section is a three-compartment model that is different from the models described in Chapter 23 in that the volume of one compartment is not constant. Due to this variability, non-linearity is introduced in the model.

In the exercises we will analyze the designed model, and we will examine the influence of different parts of the system on the regulation under various conditions. Parameter values measured in humans under pathological conditions will be used to simulate several diseases known to be caused by de-regulation of gastric acid secretion.

From these exercises it will become clear that even with a relatively simple model of a very complex system many aspects of the regulation can be analyzed and understood.

24.2 Theory

The function of the alimentary tract is to provide the body with a continual supply of water, electrolytes and nutrients. Most of the food we eat must be digested before it can be used directly by the cells of the body. Digestion is the conversion of food substances into simple components that can pass through the intestinal wall (absorption) into the blood to be assimilated by cells.

The digestive and absorptive functions of the gastrointestinal system are regulated both by gastrointestinal hormones and by the nervous system. After swallowing, food enters the stomach via the oesophagus. The stomach has various functions in the gastrointestinal system. First, it serves as a reservoir for the food ingested while regulating the rate of food entering the small intestine. Second, the stomach secretes several substances that begin the process of digestion.

In this section we design a model of the regulation of the acid secretion in the stomach. Gastric acid is of use in various functions in the stomach. It kills most of the bacteria that enter along with the food. In addition, gastric acid activates the enzyme pepsin, which provides for protein digestion. Pepsin is secreted in an inactive form (pepsinogen), which is converted to pepsin. The hydrochloric acid in the stomach initiates this process, and the formed pepsin is only active in an environment with a high acidity (optimum pH of 2). Consequently, effective protein digestion in the stomach is only possible with a well regulated H^+ secretion.

In the control of gastric acid secretion upon food intake, two types of gastric glands play an important role. Oxyntic cells in the oxyntic (or gastric) glands secrete hydrochloric acid, while the peptic cells in these glands secrete pepsinogen. The oxyntic cells act as the effector in the acid secretion control. G-cells in the pyloric glands, which secrete the hormone gastrin (several types of this hormone are known, e.g. G17 and G34) into the blood, act as the sensor for pH, stomach volume and food constituents in the control system. The hormone gastrin stimulates the oxyntic glands to secrete hydrochloric acid. For a general overview of gastric physiology see [Ganong, 1987; Guyton, 1986; Johnson, 1987; Vander et al., 1986].

Three phases can be distinguished in the regulation of the H^+ secretion :

The Cephalic Phase
The sight, smell and thought of food as well as the presence of food in the mouth in-creases gastric acid secretion by the oxyntic cells. Only neural mechanisms are involved in this phase of the regulation, independent of the presence of food and fluid in the stomach and duodenum. [see e.g. Johnson, 1987; Tache, 1988].

The Gastric Phase
The presence of food in the stomach causes a huge increase in gastric acid secretion. Stretching of the stomach wall (i.e. increasing the stomach volume) and chemical stimuli (amino acids and other products of digestion) stimulate the G-cells in gastrin production. On the other hand, increasing acidity in the stomach inhibits the G-cell gastrin production. Gastrin reaches the oxyntic cells via the blood circulation and stimulates acid secretion. In this gastric phase of gastric acid regulation, which only depends on the presence of food in the stomach, the gastrointestinal hormone gastrin and local neural networks are mainly involved [see e.g. Johnson, 1987; Nicholl et al., 1985; Sanders and Soll, 1986; Walsh, 1988].

The Intestinal Phase

The presence of food in the small intestine inhibits both gastric secretion and gastric motility. This inhibition of gastric acid secretion is probably due to the release of the hormones secretin and GIP (gastric inhibitory polypeptide). The regulation of gastric acid secretion in this phase, which is only dependent on the presence of food in the small intestine, is mainly hormonal [see e.g. Johnson, 1987; Nicholl et al., 1985; Walsh, 1988].

The three phases do overlap, and besides psychical factors (such as stress) and circadian influences, other stimulants like alcohol and caffeine influence gastric acid secretion. In a 24-hour cycle the gastric phase accounts for more than two-thirds of the total gastric acid secretion [see e.g. Johnson, 1987].

24.3   Aspects to Be Considered

When modeling complex systems such as the regulation of gastric acid secretion, in which neural, hormonal and mechanical signals cooperate, it is important to make reductions and simplifications. One difficulty is to find experimental data (from the literature) that is suited for use in the model. To begin with, parts of the control system with unknown parameters may be omitted in the model. In addition, components that seem of minor importance in the regulation can also be omitted in a first attempt to model a system. Later on, more complex and less understood components can be incorporated.

Our model of the regulation of the H^+ secretion in the human stomach only considers the processes activated by the presence of food and fluid in the stomach (i.e. the gastric phase). This phase includes the acid dependent and volume dependent gastrin secretion and the gastrin stimulated acid secretion. Furthermore, gastrin inactivation and inflow and outflow of food and fluid are included in the model. In summary, the following processes known to play a role in gastric acidity regulation are *not* involved in the model:

- Processes activated in the cephalic and intestinal phase.
- The gastrin secretion stimulating effect of amino acids and other products of digestion, which play an important role in the digestion of protein-rich meals.
- The buffer capacity of protein-rich food that has entered the stomach and may change the kinetics of the regulation.
- The circadian influences on the gastric acid secretion. This effect becomes visible most clearly during the evening and does not depend on the blood gastrin concentration.
- The direct effect of H^+ ions on the oxyntic cells, which is not clear as yet. In both humans and dogs it has been reported that no direct effect of H^+ ions on the H^+ secreting cells is present. Other reports describe an inhibitory effect of H^+ ions on the oxyntic cells in dogs.

- The existence of different types of the hormone gastrin with different half-life times and different acid secretion stimulating potency.
- The acid secretion stimulating effect of caffeine, alcohol and calcium ions which may enter the stomach with food.

Although we have excluded many processes that play a role in the regulation of the gastric acid secretion, we can design a proper model for illustrating the relative importance of stomach volume controlled acid secretion and acid feedback control. Furthermore, de-regulation in various diseases can be demonstrated.

However, when drawing conclusions from this model the omissions and simplifications should always be kept in mind. Our model simulates an experiment in which nonprotein food or fluid is put in the stomach directly and only the hormone gastrin is present as oxyntic cell stimulator. A more detailed description and analysis of this model can be found in [Van Duijn et al., 1989].

24.4   Model Description

24.4.1   Physical Foundations and Assumptions

Experiments have been reported in which stomach pH, stomach volume and blood gastrin concentration are measured and manipulated in humans. In our model we use data from these publications.

When making a model it is of great use to start with a process diagram as described in Chapter 23. Figure 24.1 shows the process diagram of the regulation of the gastric acid secretion as used in the model. In the diagram the various inflows and outflows of water, food, gastrin and H^+ ions are shown. At the same time the effects of stomach volume, gastrin, and H^+ ions on the different target cells are illustrated.

In addition, the anatomical organization of the stomach in relation to the blood becomes clear in the diagram. The details, physical foundations, assumptions and mathematical description of the processes summarized in the process diagram (Figure 24.1) are given below.

24.4.1.1   Gastrin Flows

• **G17** seems to be the major form of gastrin for biological activity. We therefore assume that it is only this type which plays a role in our control system. Gastrin (G17) is inactivated in the kidney and small intestine. The rate of inactivation –which can be regarded as a gastrin outflow from the blood compartment, **eg** $(ng.s^{-1})$ – is proportional to the blood gastrin concentration, **[G]** $(ng.L^{-1})$.

Thus we can write for the gastrin outflow from the blood:

$$eg = k[G] \qquad (24.1)$$

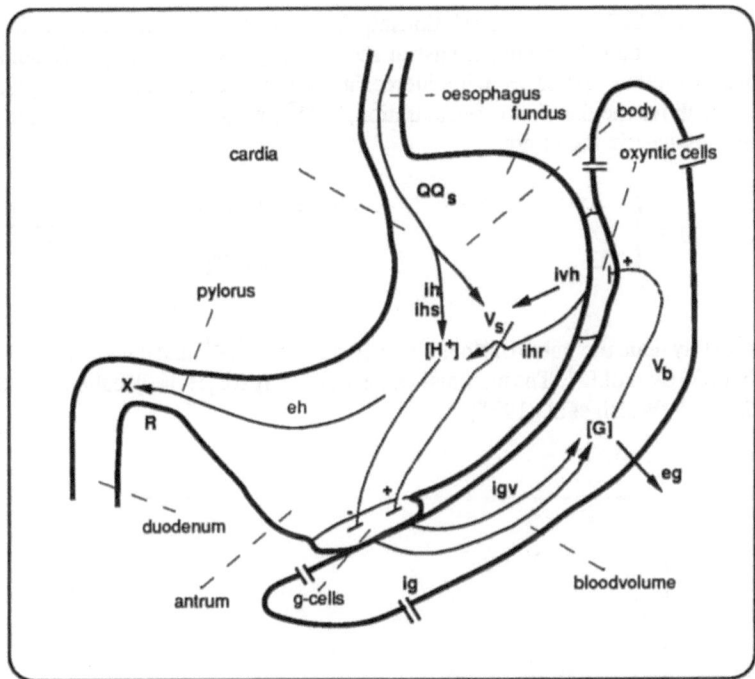

Figure 24.1. Process diagram of the gastric acid regulation system. The arrows represent flows of water, food, gastrin and H^+ ions. The names of the different flows, as used in the model, are placed next to the arrows. Furthermore, the effects of stomach volume, gastrin and H^+ ions on the G-cells and oxyntic cells are indicated by dashed-line effect indicators (+ = stimulation, - = inhibition). The names of the different parts of the stomach are also indicated (dashed lines)

Assuming a blood volume, **Vb**, of 5 L, the calculated value of the constant **k** is 10^{-2} $L.s^{-1}$ (starting from the half-life time of gastrin of 360 seconds). See Walsh et al., 1976.

• The G-cells in the pyloric glands secrete gastrin into the blood. In our model the gastrin secretion depends on the H^+ concentration in the stomach, **[H⁺]** $(mol.l^{-1})$, and on the stomach volume, **Vs** (L).

The stomach volume dependent gastrin secretion, **Igv** $(ng.s^{-1})$, is independent of the H^+ concentration in the stomach and starts above a certain threshold volume, V_0, of about 0.5 L. Figure 24.2A shows the relationship between the stomach volume, Vs, and the gastrin secretion igv. Therefore we can write:

$$ igv = \begin{cases} Kd(Vs - V_0) & \text{for } Vs \geq V_0 \\ 0 & \text{for } Vs < V_0 \end{cases} \qquad (24.2) $$

The value of the constant **Kd** is about 1.8 $ng.L^{-1}.s^{-1}$, [Schiller et al., 1980].

Figure 24.2B shows the relationship between the H^+ concentration in the stomach and the pH dependent gastrin secretion, ig (ng.s^{-1}). The pH dependent gastrin secretion decreases with increasing gastric acidity and is completely blocked above the threshold concentration, $[H^+]^*$ (mol.L^{-1}). We write for the pH dependent gastrin secretion:

$$ig = \begin{cases} Kp([H^+]^* - [H^+]) & \text{for } 0 \leq [H^+] \leq [H^+]^* \\ 0 & \text{for } [H^+] < [H^+]^* \end{cases} \qquad (24.3)$$

In a healthy man, the value of Kp is about 70 ng.L.mol^{-1}.s^{-1}, and the value of $[H^+]^*$ is about 6.10^{-3} mol.L^{-1}. The maximal value of ig (i.e. ig for $[H^+]=0$), called $igmax$, is 0.42 ng.s^{-1} [Walsh et al., 1975].

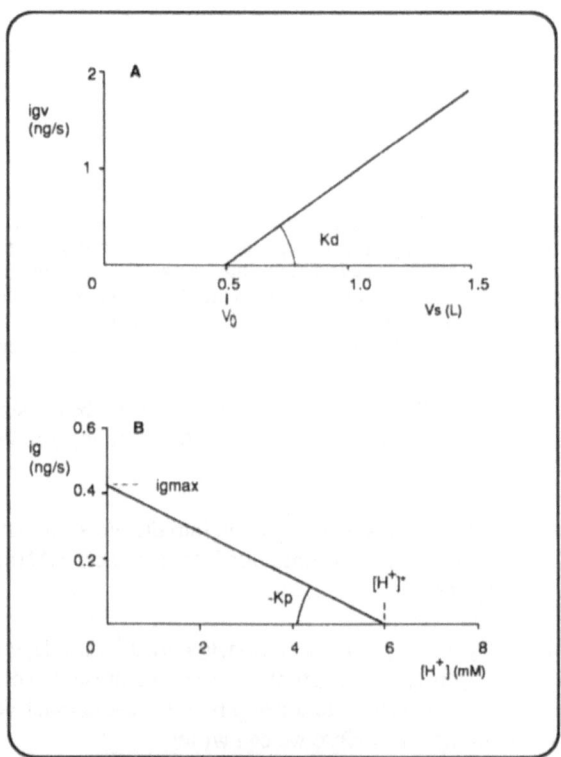

Figure 24.2.   (A) Relationship between the gastrin flow from the G-cells, igv, into the blood and the stomach volume. (B) Relationship between the gastrin flow from the G-cells, ig, into the blood and the gastric acidity, [H+].

24.4.1.2 H⁺ Flows

• The oxyntic cells in the oxyntic glands secrete hydrochloric acid, and these cells are stimulated by the hormone gastrin. Figure 24.3 shows the relationship between the regulated acid secretion, ihr (mol.s^{-1}), and the blood gastrin concentration, $[G]$. The acid production is proportional to $[G]$ up to a certain limit, $[G]^*$. Hence, we can write:

$$ihr = \begin{cases} Kg([G] & \text{for } 0 \le [G] \le [G]^* \\ Kg[G]^* & \text{for } [G] < [G]^* \end{cases} \tag{24.4}$$

The maximal hydrochloric acid production, called $ihrmax$, is about 1.1×10^{-2} mmol.s^{-1} at a blood gastrin concentration $[G]^*$ of about 74 ng.L^{-1}. The value of Kg is about 1.5×10^{-7} mol.L.ng^{-1}.s^{-1} [Lam et al., 1980].

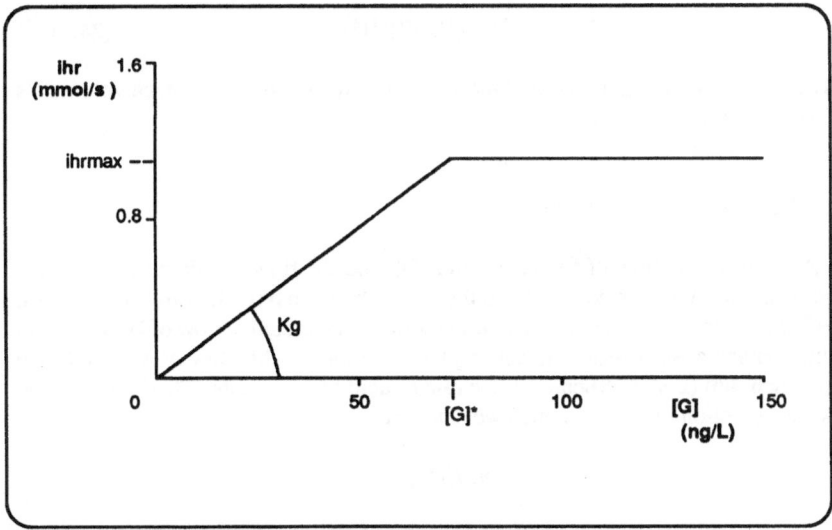

Figure 24.3. Relationship between the acid flow from the oxyntic cells, ihr, into the stomach and the blood gastrin concentration, [G]

• H⁺ ions enter the stomach together with food and fluid. For this inflow, ih (mol.s^{-1}), of H⁺ ions we can write:

$$ih = Q[H^+]_f \tag{24.5}$$

In this equation Q (L.s^{-1}) is the inflow of food and fluid into the stomach, and $[H+]_f$ (mol.L^{-1}) is the H⁺ concentration in this food and fluid. In the first instance we consider $[H^+]_f$ to be 10^{-7} mol.L^{-1} (i.e. a food pH of 7). Q equals 0 when there is no food intake.

H^+ ions also enter the stomach together with saliva. For this inflow, **ihs** (mol.s^{-1}), of H^+ ions we can write:

$$ihs = Qs[H^+]_S \tag{24.6}$$

In this equation **Qs** (L.s^{-1}) is the flow of saliva into the stomach. In our model we assume a constant inflow of saliva of 10^{-5} L.s^{-1}. The pH of saliva is about 7. Therefore, the H^+ concentration in the saliva, $[H^+]_s$, is about 10^{-7} mol.L^{-1}.

• Food and fluid present in the stomach is transported through the pylorus (the stomach exit port) into the duodenum. Together with this outflow of food and fluid, **X** (L.s^{-1}), H^+ ions leave the stomach. For this outflow of H^+ ions, **eh** (mol.s^{-1}) (also called pyloric outflow), we can write:

$$eh = X[H^+] = (V_S / R)[H^+] \tag{24.7}$$

The flow of food and fluid out of the stomach is related to the stomach volume Vs (see section 24.4.1.3 for details).

24.4.1.3 Water and Food Flows

• Apart from the flow of food and fluid (Q), and the flow of saliva (Qs), into the stomach, water is also secreted into the stomach by the oxyntic cells together with H^+ ions. For every 150 mmol of hydrochloric acid secreted, a flow of 1L water into the stomach arises (assuming isotonic HCl secretion). This flow of water into the stomach, **ivh** (L.s^{-1}), depends on the rate of acid secretion, ihr. Therefore, for the water secretion into the stomach we can write:

$$ivh = ihr/[H^+]_0 \tag{24.8}$$

In this equation $[H^+]_0$ is the H^+ concentration in the fluid secreted by the oxyntic cells (i.e. 150.10^{-3} mol. L^{-1}). See section 24.4.1.2 for details about ihr.

• Food and fluid in the stomach is transported through the pylorus to the duodenum. In our model the outflow of food and fluid, called X (L.s^{-1}), is assumed to be proportional to the stomach volume, Vs.
Hence, we can write:

$$X = Vs/R \tag{24.9}$$

By doing this we simulate the single exponential decay of the stomach volume usually observed after taking a meal. In this equation the constant **R** (s) represents the "pyloric time constant". R varies between 500 and 7000 seconds and depends on the fluidity of the food. In our simulations we choose for R a value of 3000 s.

24.4.1.4 Differential Equations

The equations (24.1) through (24.9) can be combined into three differential equations that describe the changes in time of Vs, [G] and [H$^+$].
• The summed food and fluid flows – Q, Qs and the equations (24.8) and (24.9) – give the change in the stomach volume in time. Hence,

$$dVs/dt = Q-X+Qs+iv \qquad (24.10)$$

Note that X has a negative sign, because X is a flow out of the stomach.

• The change in the mass of gastrin molecules (G) in the blood in time consists of the summed gastrin flows (equations 24.1, 24.2 and 24.3):

$$dG/dt = ig+igv-eg \qquad (24.11)$$

Note the negative sign before eg, because eg is a flow out of the blood.
To obtain the change in the gastrin concentration [G] in time, we have to divide G by the blood volume Vb. Therefore, the change in G in time is:

$$Vbd[G]/dt = ig+igv-eg \qquad (24.12)$$

This procedure is only legal when Vb is assumed to be constant.

• The summed H$^+$ ion flows (equations 24.4, 24.5, 24.6 and 24.7) represent the change in the amount of H$^+$ ions in the stomach in time. Hence,

$$dH^+/dt = ihs + ihr + ih - eh \qquad (24.13)$$

From this equation we can derive the change in the H$^+$ concentration in the stomach in time. First we write:

$$dH^+/dt = d(Vs[H^+])/dt = Vsd[H^+]/dt + [H^+]dVs/dt \qquad (24.14)$$

We have to use this procedure because the stomach volume, Vs, is not constant as is the blood volume in equation (24.12). Combining of the equations (24.13) and (24.14) leads to:

$$Vsd[H^+]/dt = ihs + ihr + ih - eh - [H^+]dVs/dt \qquad (24.15)$$

The equations (24.1) through (24.9) can be substituted into the equations (24.10), (24.12) and (24.15). However, this is not necessary for building the model in BIO-PSI. One of the advantages of simulation is that we do not have to solve the differential equations analytically. The system of three first-order differential equations (24.10, 24.12 and 24.15) cannot easily be solved analytically. This is due to the non-linearity of the system.

Q (eating) is considered to be the input of the model. The outputs are $[H^+]$, $[G]$ and Vs. In our simulations we look at the behavior of the outputs in time for different eating behavior (i.e. different Q).

24.4.2   Structure and Parameters

To develop the structure of our model in BIOPSI we distinguish four parts of the control system. We can set apart the stomach volume control, the gastrin control, the H^+ control and the meal taking behavior (Q). First, we design these four parts, which can then be put together easily to become the complete system.

• *Volume Control*

To design the block diagram of the volume control we need the equations (24.8), (24.9) and (24.10). Figure 24.4A shows the structure diagram of the volume control. The block configurations and parameters for the volume control are:

Blockname	Type	Input1	Input2	Input3	par1	par2	par3
Vs	INT	VSUM			0.0312	1	
VSUM	SUM	X	QSUM	ivh	-1	1	1
X	DIV	Vs	R				
QSUM	SUM	Qs	Q		1	1	
Qs	CON				10^{-5}		
R	CON				3000		
ivh	DIV	ihr	[H+]o				
[H+]o	CON				15.10^{-2}		

Note that par1 of VSUM is equal to -1, because input1 is X, which is an *outflow*. The input Q in the SUM block QSUM originates from the block diagram of meal taking behavior. The input ihr in the DIV block is from the acidity regulation block diagram.

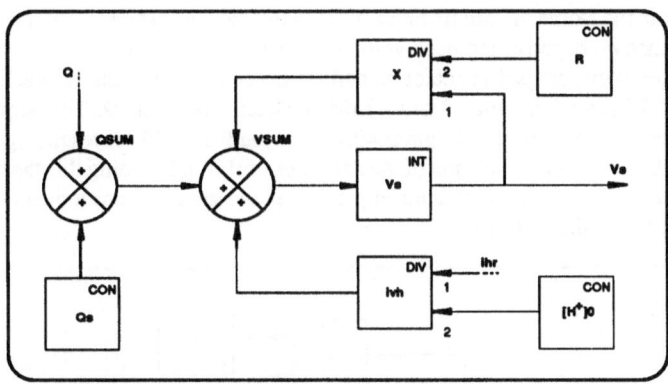

Figure 24.4A. BIOPSI structure diagram of the stomach volume control. Block type is indicated in the upper right corner of each block; the block name is indicated in the middle of each block. Circles indicate a SUM block; names of the sum blocks are next to it.

• *Gastrin Control*

To design the block diagram of the gastrin control we need the equations (24.1), (24.2), (24.3) and (24.12). Figure 24.4B shows the block diagram of the gastrin control. The block configuration and parameters for the gastrin control are:

Blockname	Type	Input1	Input2	Input3	par1	par2	par3
[G]	INT	G			0.41	1	
G	DIV	IGSUM	Vb				
Vb	CON				5		
IGSUM	SUM	eg	igv	ig	-1	1	1
eg	MUL	k	[G]				
k	CON				10^{-2}		
igv	DSP	Vs			0	0.5	1.8
igm	LIM	[H+]			0	0.42	70
igmax	CON				0.42		
ig	SUB	igmax	igm				

Note that par1 of IGSUM is equal to -1 because input1 is eg, which is an *outflow* of gastrin from the blood.

The input Vs in the DSP block igv is the output from the volume control block diagram. The input [H+] in the LIM block igm is the output of the acidity control block diagram.

The relation between gastrin secretion and stomach volume as shown in Figure 24.2A can be simulated with a DSP block. Therefore, the parameters of the igv block (a DSP type block) are: par1=0, par2=V_0=0.5 L., and par3=K_d=1.8. The input of this block is the stomach volume block Vs (see Fig. 24.4A). The DSP block can be used because no negative stomach volumes appear in the model.

The relation between gastrin secretion and H^+ concentration in the stomach, $[H^+]$ (Figure 24.2B) can be simulated with a LIM block and a SUB block. The output of a LIM block with par1=0, par2=igmax=0.42 and par3=Kp=70 represents ig when subtracted from igmax. The input of the LIM block is $[H^+]$. By doing this the blocks igm, igmax, and ig form the relationship between the gastrin secretion and the H^+ concentration in the stomach.

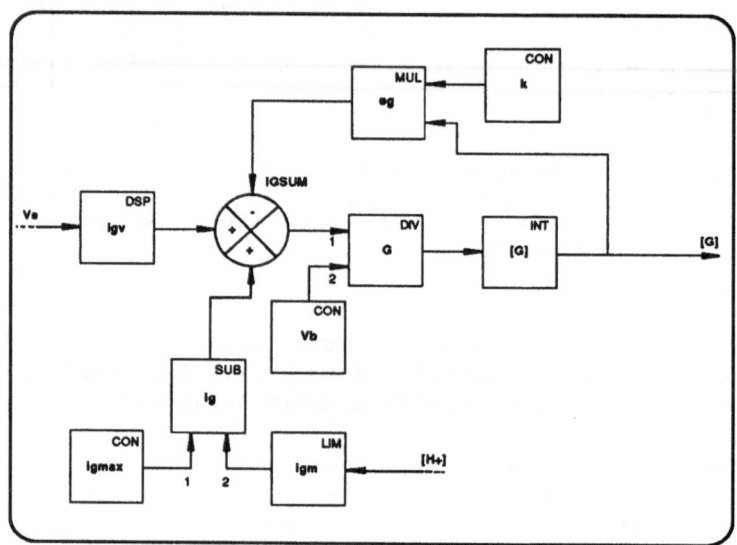

Figure 24.4B. BIOPSI structure diagram of the gastrin secretion control. Block type is indicated in the upper right corner of each block; the block name is indicated in the middle of each block. Circles indicate a SUM block; names of the sum blocks are next to it.

• *Acidity Control*

To design the structure diagram of the acidity control we require the equations (24.4), (24.5), (24.6), (24.7) and (24.15). Figure 24.4C shows the block diagram of the acidity control. The block configuration and parameters for the gastric acidity control are:

Blockname	Type	Input1	Input2	Input3	par1	par2	par3
HTOT	INT	HSUM		$1.9.10^{-4}$	1		
[H+]	DIV	HTOT	Vs				
pH	LOG	[H+]		-1	1	2	
VsH	MUL	[H+]	Vs				
eh	DIV	VsH	R				
R	CON			3000			
ihr	LIM	[G]		0	$1.1.10^{-5}$	$1.5.10^{-7}$	
FOOD	MUL	Q	[H+]f				
[H+]f	CON			10^{-7}			
SALIVA	MUL	Qs	[H+]s				
[H+]s	CON			10^{-7}			
Qs	CON			10^{-5}			
ihsum	SUM	FOOD	SALIVA		1	1	
HSUM	SUM	eh	ihr	ihsum	-1	1	1

Note that par1 of the SUM block named HSUM is equal to -1 because the input1 of this block is eh, which is an *outflow* of H^+ ions from the stomach.

The relation between the H^+ flow into the stomach, ihr, and the gastrin concentration in the blood as shown in Figure 24.3 can be simulated with a LIM block. The parameters of this block are: par1=0, par2=ihrmax=1.1×10^{-5} mol.s^{-1} and par3=Kg=1.5×10^{-7} mol.L.ng^{-1}.s^{-1}. The input [G] in the LIM block ihr is the output of the gastrin control block scheme. The input Q in the MUL block FOOD is the output from the meal taking behavior block scheme. The CON blocks R and Qs are also present in the volume control block scheme.

• Behavior of Meal Intake

To simulate meal intake behavior we use steps in the food and fluid flow, Q, into the stomach. These are steps from 0 to 3.10^{-4} L.s^{-1} with varying pulse lengths depending on the type of meal taken. The pulse duration of breakfast was chosen to be 2200 seconds, that of lunch 2800 seconds and that of dinner 4600 seconds.

To achieve this behavior in Q we use PLS blocks. For one meal we need one PLS block. For three meals a day we need three PLS blocks whose outputs are summed. Figure 24.4D shows the block scheme of meal intake behavior. The block configuration and parameters for meal intake are:

Blockname	Type	input1	input2	input3	par1	par2	par3
BREAKF	PLS				3600	5800	3.10^{-4}
LUNCH	PLS				18000	20800	3.10^{-4}
DINER	PLS				36000	40600	3.10^{-4}
Q	SUM	BREAKF	LUNCH	DINER	1	1	1

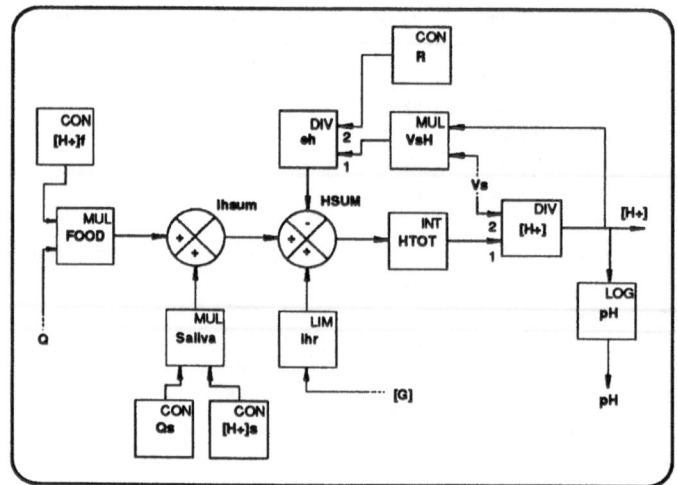

Figure 24.4C. BIOPSI structure diagram of the H^+ secretion control. Block type is indicated in the upper right corner of each block; the block name is indicated in the middle of each block. Circles indicate a SUM block; names of the sum blocks are next to it.

In a 24-hour cycle starting at 8:00 am (TIME=0), breakfast starts at 9:00 am (t_{up}=3600 s., t_{down}=5800 s.), lunch starts at 1:00 pm (t_{up}=18000 s., t_{down}=20800 s.) and dinner starts at 6:00 pm (t_{up}=36000 s., and t_{down}=40600 s.). The cycle ends when TIME=86400 seconds.

24.4.3 Possibilities and Limitations

With the use of the described model we can study the regulation of the gastric acid secretion both during rest (no food in the stomach) and during or after eating (food present in the stomach). We are able to analyze the influence of parts of the regulation system (e.g. volume and H^+ controlled gastrin secretion) in the eating and non-food conditions. From this analysis we can gain a better insight into the regulation of gastric acid secretion and an understanding of the relative importance of each part of the system under various conditions. Many diseases concerning gastric acid secretion are known, and they can be easily simulated using changed parameters in the presented model. The influence of drugs used to prevent or treat these diseases can also be simulated.

In the next section (Exercises) we analyze the system under various conditions. Nevertheless, it should be noted that conclusions derived from these experiments are limited by the assumptions and reductions made in the design of the model (see sections 24.3 and 24.4.1).

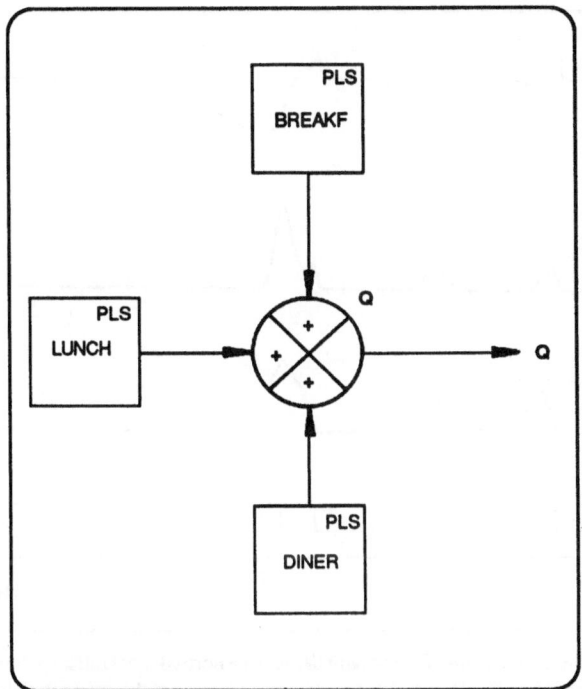

Figure 24.4D. BIOPSI structure diagram of the food inflow control. Block type is indicated
in the upper right corner of each block; the block name is indicated in the
middle of each block. Circles indicate a SUM block; names of the sum blocks
are next to it.

24.4.4 Exercises

24.4.4.1 The Default Demo

The default demo (Figure 24.5) shows the simulation of the complete system during
a 24-hour cycle with the regular consumption pattern of taking three meals. The time
courses of pH in the stomach, of the gastrin concentration in the blood, and of the
stomach volume and the flow of food and fluid into the stomach are illustrated. The
demo shows that the steady-state pH without food present in the stomach is about
2.2.

The intake of food causes an increase in the stomach pH due to dilution of the
stomach content. This increase in stomach pH is followed by an increase in the blood
gastrin concentration (due to less inhibition of the gastrin secreting cells). This
larger gastrin concentration causes increased H^+ secretion, which decreases the
stomach pH. In addition, stomach volumes exceeding 0.5 L cause an extra increase

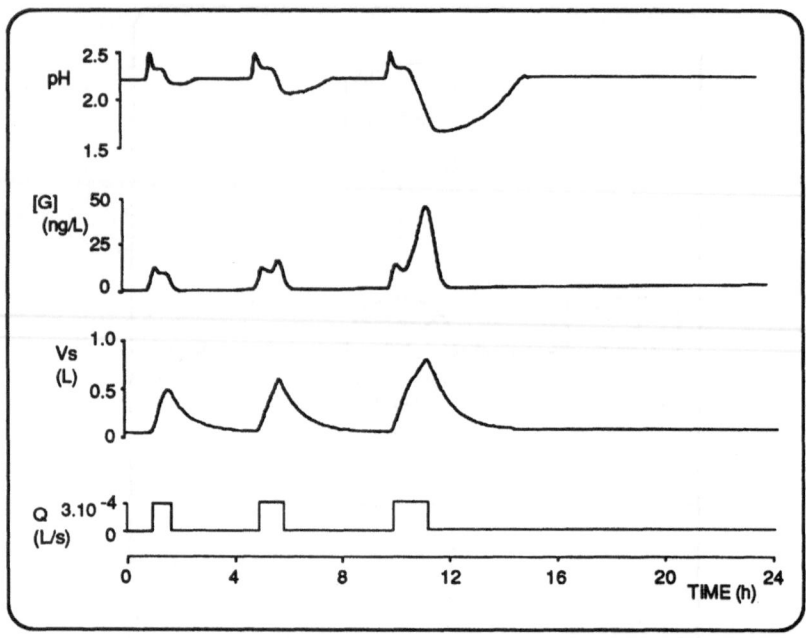

Figure 24.5. The 24-hour run of the default demo for a normal meal intake pattern, showing the stomach pH, the blood gastrin concentration, the stomach volume and the food flow Q into the stomach.

in the blood gastrin concentration leading to additional H^+ secretion, resulting in an even larger decrease in stomach pH. This effect is not present in the first meal pulse (breakfast) and is most clearly visible during and after dinner. After the food inflow has stopped, the pH still decreases (clearly visible after breakfast) due to the large gastrin concentration still present in combination with a decreasing stomach volume (i.e. dilution has stopped, but H^+ secretion will still be large).
In the exercises we shall analyze the system behavior in more detail to find out which system components are responsible for the different phenomena presented in the default demo of the complete system.

24.4.4.2 Stomach Volume Behavior

The behavior of the stomach volume can be studied apart from the gastrin and H^+ secretion control components in the default demo model by changing the value of parameter 3 of the SUM block VSUM from 1 to 0.
The response of the stomach volume upon different flows, Q, into the stomach and different "pyloric time constants", R, can then be simulated. Use this model from 0

to 18000 seconds, so that only one food inflow pulse is present in each simulation. For the output of the model only Vs and Q have to be displayed. The value of Q can be changed by changing parameter 3 of the PLS block BREAKF. The food inflow time can be altered by changing parameter 2 of the PLS block BREAKF. The "pyloric time constant" R can be changed by changing the value of parameter 1 of the CON block named R.

Use the RX command to display simulations with different parameter values on the screen in the same figure.

• **Study the difference between the intake of fluid and solid food.**
• **Simulate fast and slow eating of a meal by changing Q and food intake time simultaneously.**

24.4.4.3 pH-Dependent Acid Secretion Regulation

In the default demo two mechanisms control the gastrin secretion. One of these mechanisms is the pH-dependent gastrin secretion. In the default demo we can study the system behavior with only the pH-dependent gastrin secretion present in the model. To exclude the volume dependent gastrin secretion from the complete model (so that only pH-dependent gastrin secretion is present) parameter 2 of the SUM block IGSUM has been changed from 1 to 0 (see 24.4.2). The value of the food inflow pulse Q is 6.10^{-4} L.s^{-1}.

Run the changed model. To compare the changed model with the complete model (i.e. the default demo model), change parameter 2 of the SUM block IGSUM from 0 to 1 and use the RX command to run the simulation.

To study the behavior of the system with only pH-dependent gastrin secretion for different food inflow pulses, run the changed model from 0 to 18000 seconds. The flow of food into the stomach, Q, can be changed as described in 24.4.4.2. If necessary, change the scaling of the output display.

• **For what values of the food inflow Q (i.e for what times of day) is the regulation of the gastric acidity still sufficient (i.e pH less than 2.2) with only the pH dependent gastrin secretion present?**

24.4.4.4 Stomach Volume-Dependent Acid Secretion Regulation

In addition to the pH-dependent gastrin secretion (24.4.4.3), stomach volume-dependent gastrin secretion is present in our model. To study the contribution of the stomach volume gastrin secretion to the gastric acid secretion control, we exclude the pH-dependent gastrin secretion from the default demo model (so that only stomach volume-dependent gastrin secretion is present). Therefore, parameter 3 of the SUM block IGSUM has been changed from 1 to 0. The pH scaling has been changed in 1 to 7.5 instead of 1 to 2.5. The value of the food inflow pulse Q is 4.10^{-4} $L.s^{-1}$.

To study the behavior of the system with only stomach volume-dependent gastrin secretion present upon different food flow pulses (Q) into the stomach, run the simulation from 0 to 18000 seconds. Q can be changed as described in 24.4.4.2. It is necessary to change the scaling of the output display.

You can compare this to the default demo model by changing parameter 3 of the SUM block IGSUM from 0 to 1 (see 24.4.2) and using the RX command to display simulations with both models simultaneously.

- For what time of day (i.e. what food inflow conditions) is the volume-dependent gastrin secretion sufficient to achieve a well regulated gastric acidity?
- What can be concluded about the necessity of the two gastrin secretion regulating mechanisms in the regulation of gastric acidity?

24.4.4.5   Exercises to Do on Your Own

Starting with the default demo model, several diseases can be easily simulated by changing parameter values as indicated in the text below. The simulation of the application of stomach balloons requires only parameter changes as well. Sensitivity analysis requires more radical changes in the default demo model. These are all described in the text.

Diseases

Several diseases of the digestive tract are known to be due to misregulation of the H^+ secretion in the stomach. In our model we are able to simulate various diseases by changing parameter values in the default demo model.

Achlorhydria

Patients with a chronic gastritis may gradually become atrophic in gastric acid secretion until no oxyntic cell activity remains. A complete loss of gastric secretions results in achlorhydria [Guyton, 1986]. Simulate the development of achlorhydria by decreasing the value of Kg and ihrmax (parameter 3 and 2 respectively of the LIM block ihr). Run the simulation with different reductions of Kg and ihrmax (e.g. 50%, 90%, 99% and 99.9% reduction).

- What can be concluded from these simulations about the H^+ secretion capacity in healthy persons?
- What changes in the gastrin concentration behavior appear when achlorhydria is present?
- What is the influence of eating food with an extremely low pH on the blood gastrin concentration? (change parameter 1 of the CON block named $[H^+]f$ from 10^{-7} into 10^{-2})

Gastrin Producing Tumor

The Zollinger-Ellison syndrome originates from gastrin secreting tumors in the pancreas, duodenum or stomach [Ganong, 1987]. In our model we can simulate this syndrome by the introducing of a constant gastrin flow into the blood. We can achieve this by increasing the value of parameter 1 of the CON block igmax. Run the simulation with igmax=2.42 ng.s^{-1} (instead of 0.42 ng.s^{-1}).

- What will be the symptoms present in patients suffering from the Zollinger-Ellison syndrome (which can be clearly seen in the simulation)?

Peptic Ulcer

A duodenal ulcer is an excoriated (removal of the lining of the stomach) area of the mucosa caused by digestive actions of gastric juice. Defects in the gastric secretion regulation may play a role in generating this disease.

In duodenal ulcer patients a larger gastrin secretion is present due to changes in Kp and igmax. In patients Kp=42 ng.L.mol^{-1}.s^{-1}, and igmax= 0.46 ng.s^{-1}. In addition, gastric acid secretion is abnormally sensitive to gastrin in duodenal ulcer patients; a Kg of about 1.9 x 10^{-7} mol.L.ng^{-1}. s^{-1} with a ihrmax of 1.4 x 10^{-5} mol.s^{-1} is found. [See Lam et al., 1980; Mayer et al., 1974; Walsh et al., 1975.]

To simulate the gastric acidity regulation in duodenal ulcer patients, we change parameter 1 of the CON block igmax to 0.46 and the parameters 2 and 3 of the LIM block igm to 0.46 and 42 respectively. Furthermore, we change parameters 2 and 3 of the LIM block ihr to 1.4x10^{-5} and 1.9x10^{-7} respectively.

Run the default model, change the parameter values as described above, and run the changed model with the RX command.

- What are the most striking differences in gastric acidity regulation under normal conditions and under duodenal ulcer conditions?
- During what time of day will duodenal ulcer patients suffer the most from their disease?
- Can a change in eating behavior decrease the symptoms of duodenal ulcer patients (i.e. what happens when many very small meals are eaten instead of three large meals)?

Treatment with Cimetidine and Ranitidine

Cimetidine and ranitidine, both H2 receptor antagonists, are used in the treatment of duodenal ulcers. Both of these drugs inhibit gastrin stimulated acid secretion [see Richardson, 1978]. To simulate the treatment with cimetidine or ranitidine, we change parameter values in the model used to simulate gastric acidity regulation in duodenal ulcer patients.

Application of cimetidine reduces the acid secretion by 50%. To simulate the effect of cimetidine we reduce Kg from 1.9x10^{-7} to 1.0x10^{-7} and ihrmax from 1.1 x 10^{-5} to 0.55 x 10^{-5} (parameter 2 and 3 of the LIM block ihr are changed to 0.55 x 10^{-5} and 1.0x10^{-7} respectively).

Ranitidine reduces the acid secretion by about 70%. This can be achieved by reduct-

ing Kg from 1.9×10^{-7} to 0.6×10^{-7} and ihrmax from 1.1×10^{-5} to 0.33×10^{-5} (parameters 2 and 3 of the LIM block ihr are changed to 0.33×10^{-5} and 0.6×10^{-7} respectively).
First simulate the gastric acidity regulation in duodenal ulcer patients as described. Then change the parameter values and simulate the model, using the RX command, for treatment with cimetidine or ranitidine.

- Under which conditions is treatment with cimetidine or ranitidine most effective?
- What happens to the blood gastrin concentration during the application of cimetidine or ranitidine?

Stomach Balloon

Intragastric balloons are sometimes used for treating obesity. These balloons introduce a permanent volume in the stomach. A continuous over-stretching of the stomach may lead to misregulation of gastric acidity.
To simulate the presence of a stomach balloon, we must introduce a constant stomach volume that does not contribute to dilution of the stomach content (i.e this volume only contributes to the stomach volume acting on the gastrin secreting cells). Instead of adding an extra volume to the stomach volume, we change the value of the threshold stomach volume V_0. In our model we simulate the introduction of a certain balloon volume by subtracting this volume from V_0.
Simulate the presence of a stomach balloon with volumes of 0.4 and 0.5 L. To achieve this change parameter 2 of the DSP block igv to 0.1 and 0 respectively.

- Which undesired effects may appear when intragastric balloons are used?
- Can these effects be eliminated when cimetidine or ranitidine is used?

Sensitivity Analysis

When a model has been designed, the sensitivity of the output variables to variations in the parameters must be analyzed. It is important to know which parameter variations the model is sensitive to. From a sensitivity analysis it becomes clear whether the model is still able to regulate in a proper way when parameters vary within the physiological range. Sensitivity analysis is also very important in the analysis of diseases, because disregulation will appear easier for pathological changes in sensitive parameters as compared to insensitive parameters. The sensitivity, $s(t)$, of the stomach pH can be defined as:

$$s(t) = \frac{pH(t) - pH_0(t)}{pH_0(t)} \cdot \frac{P_0}{dP} \qquad (24.16)$$

in which $pH_0(t)$ is the pH pertaining to the reference parameter value P_0, and $pH(t)$ is the pH pertaining to the parameter value P+dP. dP is the value of the variation in parameter P. When s(t)=1, this means that the percentage variation in the parameter is equal to the percentage variation in the output variable. When s(t)=0, the output is not sensitive to parameter variation.

The sensitivity of the stomach pH to different parameter variations (e.g. k, Kd, Kp, Kg, etc.) can be analyzed by computer simulations.

To perform this, the complete model has to be copied with different block names (e.g. HTOTs and pHs instead of HTOT and pH). The name of the parameter to be analyzed must also be changed (e.g. Kp varied with 30%). The default model and the model with the changed parameter values have to be run simultaneously. The default model gives the values of $pH_0(t)$ and P_0 and the changed model gives the values of $pH(t)$ and P_0+dP. Some additional blocks have to be made to calculate the sensitivity from these outputs and parameters according to equation 24.16.

Try to find out how sensitive the model is for some parameters and for which parameters the model is most sensitive. Under physiological conditions, the given parameter values do not show variations larger than $\pm 30\%$. Is the sensitivity dependent on the magnitude of the parameter variation used in the simulations? The sensitivity of the model also depends on the value of the input Q; therefore, the sensitivity analysis should be done for different values of this variable.

24.5 Conclusions

From the simulation experiments it is clear that the model correctly simulates the regulation of the gastric acidity if the model assumptions are taken into consideration. For appropriate regulation in the full and empty stomach, the presence of both H^+-dependent and stomach volume-dependent gastrin secretion is required. Several diseases can be simulated with the use of abnormal parameter values in the model (see 24.4.4).

Acknowledgement

The authors wish to thank Prof. A.A. Verveen (Department of Physiology and Physiological Physics, Leiden) and Prof. T.M. Konijn (Cellbiology and Genetics Unit, Leiden) for advice and critical reading of the manuscript.

References

Ganong, W.F.: *Review of Medical Physiology.* 13th edition, Appleton and Lange, Norwalk, Connecticut/San Mateo, California, 1987.

Guyton, A.C.: *Textbook of Medical Physiology.* 7th edition, W.B. Saunders company, Philadelphia, 1986.

Johnson, L.R.: *Physiology of the Gastrointestinal Tract.* 2nd edition, Raven Press, New York, 1987.

Lam, S.K., J.I.Isenberg, M.I. Grossman, W.H. Lane and J.H. Walsh: "Gastric acid secretion is abnormally sensitive to endogenous gastrin released after peptone testmeals in D.U. patients".
J. Clin. Invest., Vol. 65, 1980, pp. 555-562.

Mayer, G., R. Arnold, G. Feurle, K. Fuchs, H. Ketterer, N.S. Track, and W. Creutzfeldt: "Influence of feeding and sham feeding upon serum gastrin and gastric acid secretion, in Control subjects and duodenal ulcer patients". *Scand. J. Gastroenterology*, Vol. 9, 1974, pp. 703-710.

Nicholl, C.G., J.M. Polak and S.R. Bloom: "The hormonal regulation of food intake, digestion, and absorption". *Annu. Rev. Nutr. Vol.* 5, 1985, pp. 213-239.

Richardson, C.T.: "Effect of H_2-receptor antagonists on gastric acid secretion and serum gastrin concentration: a review". *Gastroenterology* Vol. 74, 1978, pp. 366-370.

Sanders, M.J. and A.H. Soll.: "Characterization of receptors regulating secretory function in the fundic mucosa". *Annu. Rev. Physiol.* Vol. 48, 1986, pp. 89-101.

Schiller, L.R., J.H. Walsh and M. Feldman, "Distention-induced gastrin release: Effects of luminal acidification and intravenous atropine". *Gastroenterology* Vol. 78, 1980, pp. 912-917.

Tache, Y.: "CNS peptides and regulation of gastric acid secretion". *Annu. Rev. Physiol.* Vol. 50, 1988, pp. 19-39.

Vander, A.J., J.H. Sherman and D.S. Luciano: *Human Physiology, the mechanism of body function.* 3rd edition, McGraw-Hill Book Company, New York, 1986.

Van Duyn, B., D.L. Ypey, J. de Goede, A.A. Verveen and W. Hekkens: "A model study of the regulation of gastric acid secretion".
Am. J. Physiol., Vol. 257, (Gastrointest. Liver Physiol. 20), 1989, G157-G168.

Walsh, J.H., C.T. Richardson and J.S. Fordtran: "pH dependence of acid secretion and gastrin release in normal and ulcer subjects". *J. Clin. Invest.* Vol. 55, 1975, pp. 462-468.

Walsh, J.H., J.I. Isenberg, J. Ansfield and V. Maxwell: "Clearance and acid-stimulating action of human big and little gastrins in duodenal ulcer subjects". *J. Clin. Invest.* Vol. 57, 1976, pp. 645-657.

Walsh, J.H.: 'Peptides as regulators of gastric acid secretion". *Annu. Rev. Physiol.* Vol. 50, 1988, pp. 41-63.

25
Thermoregulation

Rogier P. van Wijk van Brievingh, Mark J. de Leeuw van Weenen, and Dirk L. Ypey

25.1 Aim of Instruction

By using this model, simulations can be performed to gain insight into the basic processes underlying thermoregulation, including shivering, sweating, and heat loss through the skin. External effects such as clothing, air velocity and humidity are taken into account. The options offered in the model may be used in many combinations to simulate different situations, including physical exercise. The case of a patient who cannot sweat or shiver due to neuronal defects dramatically shows how adequately our thermoregulatory system copes with variations in environmental conditions.

25.2 Theory and References to Standard Textbooks

Thermoregulation is a standard subject in physiology [Ganong, 1987; Guyton, 1986]. Body temperature is determined by a balance between heat production and heat loss to the environment with clothes as shielding, if any. Heat is produced by all metabolic activities in muscles and other organs and is lost by convection, radiation, and evaporation of water in the respiration and from the skin. Normal functioning of the body is only possible within narrow temperature limits of the body core. Thermoregulation includes many neuronal and hormonal control mechanisms.

Neuronal circuits in the hypothalamus (the part of the brain that controls many unconscious functions) process information from temperature receptors located superficially in the skin as well as deep in the core of the body and in the hypothalamus itself. They can detect temperature changes under or above a certain body temperature. The controller affects all body mechanisms involved in heat production and loss.

The human body temperature changes only about 1 degree centigrade for every 27 degrees change in ambient temperature, which means that the feedback gain of thermoregulation is about 27. This is rather high for a biological control system.

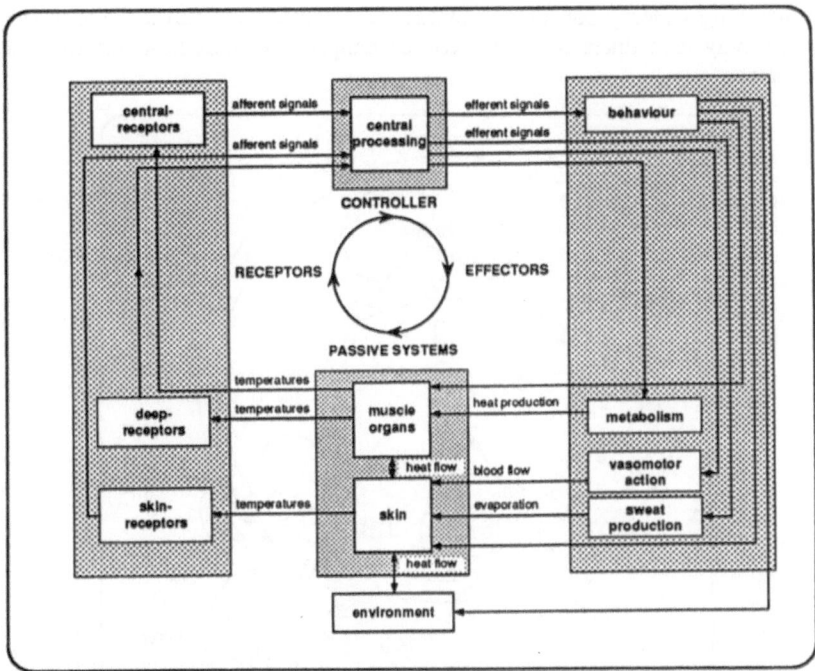

Figure 25.1. Simplified scheme of thermoregulation (Werner, 1983, by permission)

Besides by unconscious thermostatic processes, body temperature is also influenced by behavioral control, which is even more potent. . Signals from the thermostatic brain areas give the personsensations of discomfort, after which the individual will make environmental adjustments to re-establish comfort. Changing clothes, moving to a different environment, and controlling heaters and air conditioners are some examples.

There exists no single level for "normal" body temperature. At rest it lies between 36.5 and 37.5 degrees centigrade. In the morning, at low metabolic level, it may be 36 degrees, while moderate exercise or emotions may raise it to 38 degrees and heavy exercise to 40 degrees.

Temperature is not distributed evenly in the body. Extremities always have lower temperatures than the core. Furthermore, some parts of the body are shielded by clothing while others are not.

25.3 Aspects to be Considered

The model used is even more simplified than shown in Fig.25.1. Spatial aspects are neglected; the body is supposed to be of spherical shape, consisting of a core with homogeneous composition and temperature and an insulating shell (skin) surrounding it. Heat producing processes are distributed uniformly in the core. The model only uses linear relations between temperature, heat flow and thermal resistance.

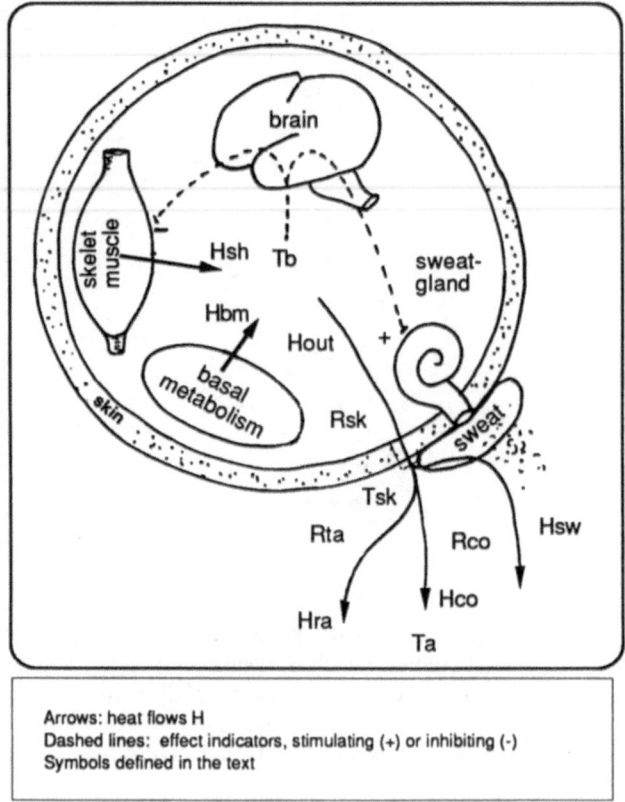

Figure 25.2. Process diagram of the model

25.3.1 Limitations of the Model

As is clear from Fig. 25.2, geometric aspects and non-uniformities are not taken into account in our model. In addition, we do not take into acount time-dependencies at a smaller time-scale than is compatible with heat diffusion processes. Thus, relatively fast changes due to circulation are not considered. Furthermore, no attempt has been made to model behavioral adaptations to or of the environment.

25.3.2   Outlook

Developments in modeling thermoregulation have been described (Werner, 1983) as follows:
- All variables (e.g. temperature, heat-flow, etc.) have to be regarded as functions of time and of three-dimensional local coordinates within the human body;
- All parameters (e.g. density, conductivity index, etc.) have to be considered as locally distributed parameters;
- Geometry and anatomy of the body must be adequately represented, e.g. by photogrammatic analysis;
- The heat transport mechanisms − conduction, convection, and radiation − have to be taken into account separately;
- The locally-dependent effector mechanisms − heat production by metabolism, vasomotor control and heat loss via sweat production and respiration − have to be considered;
- The disturbances to the control process − ambient temperature, humidity, air velocity, radiation and work load − must be incorporated;
- The local definition of the actually controlled variable (i.e. the temperature to be held as constant as possible) has to be adaptable to future results;
- Further developments are characterized by solving a complex system of partial differential equations, which take into account the dependencies on time and on the local three-dimensional coordinates.

25.4   Model Description

25.4.1 Physical Foundations and Assumptions

The model is based on the elementary equation for heat flow:

$$H= dQ/dt = (T_2 - T_1)/R + c.m.dT/dt \qquad (25.1a)$$

with:

Q	amount of heat	[J]	analog:	Electrical charge
H	heat flow	[J/s]		Electrical current
T	temperature	[K]		Electrical potential
R	heat flow resistance	[K.s/J]		Electrical resistance
C	heat capacity	[J/K]		Electrical capacity
m	body mass	[kg]		
c	heat capacity per unit mass	[J/K/kg]		
1,2	indices referring to the compartments considered			

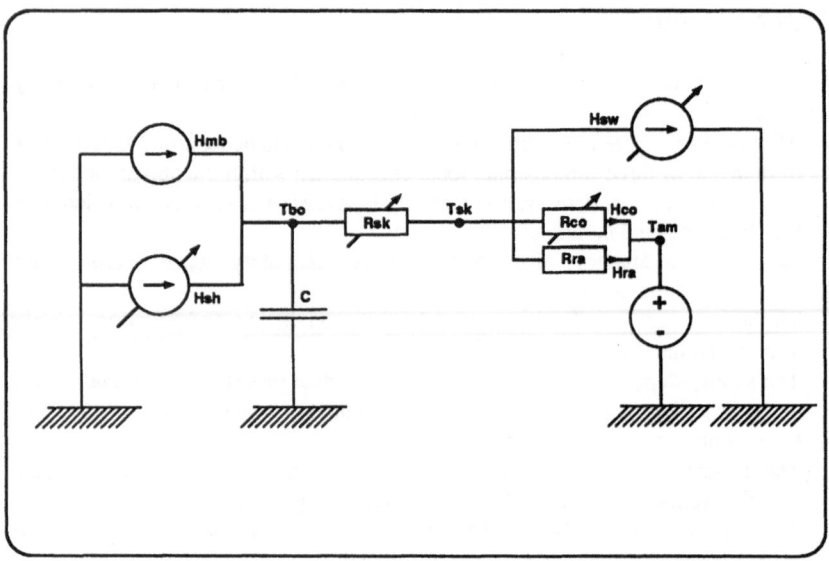

Figure 25.3.   Electrical circuit analogon

25.4.2 Structure

The model is most easily explained with the use of an electrical analog:

with:			
	Tbo	body temperature	dependent variable
	Tsk	skin temperature	dependent variable
	Tam	ambient temperature	independent variable
	Tsp	threshold temperature	Tsp = Tsp(Tsk)
	Rsk	skin resistance	Rsk = Rsk(Tam)
	Rra	heat resistance to radiation	Rra = Rra(clothing)
	Rco	heat resistance to convection	Rco = Rco(v,clothing)
	Hmb	metabolic heat production	Hmb = Hmb(Tbo,exercise)
	Hsh	heat production by shivering	Hsh = Hsh(Tbo,Tsp)
	Hra	heat loss by radiation	Hra = Hra(Rra)
	Hco	heat loss by convection	Hco = Hc(Rco)
	Hsw	heat loss by sweating	Hsw = Hsw(Tbo,Tsp,Rh)
	Hsk	heat flow through skin	Hsk = Hsk(Rsk,Tam)

The Differential Equations for Tbo and Tsk

These follow directly from equation 25.1a and Fig. 25.3:

$$dTb/dt = Htot/(c \cdot m) \qquad (25.1b)$$

with $\qquad Htot = Hmb + Hsh - (Hsw + Hra + Hco) \qquad (25.1c)$

$$Tsk = Tbo - (Hsw + Hra + Hco) \cdot Rsk \qquad (25.1d)$$

To solve these equations, Runge-Kutta integration is used for (25.1b). and whereas Hsw, Hra and Hco depend on Tsk, an iterative scheme is applied to solve (25.1d). The time step of iteration was chosen on grounds of numerical stability, regarding the fact that the equations for Hsw and Hsh have discontinuous derivatives.

25.4.3   Equations and Parameters

25.4.3.1 Metabolism

The speed of chemical reactions depends on temperature. Thus, heat production by metabolic processes is temperature-dependent:

$$Hmb = 72 \cdot 2^{((T_{bo} -37)/5.6)} \qquad (25.2)$$

The constants for the basic metabolic state have been derived from [Spain, 1982] with corrections according to data from [Guyton, 1986]. According to [Talbot, 1973], physical exercise can increase the metabolic heat production ten to twenty times the basic level.
Measurements of oxygen consumption during physical load indicate that in the stationary state the body has an overall efficiency of 25%, meaning that the power delivered mechanically is one-third of metabolic heat production.

25.4.3.2 Shivering

The structure of this part of the model corresponds with [Spain, 1982], who uses a fixed threshold temperature of shivering (Tspsh).
From Fig.25.4., however, it is clear that Tspsh is not constant in the present model:

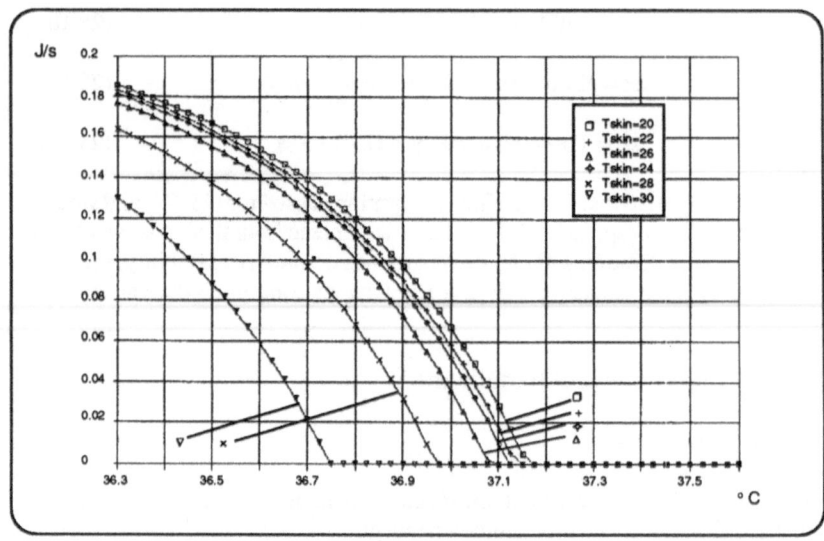

Figure 25.4.   Heat production due to shivering as a function of internal head temperature. The set-point temperature where shivering begins depends on skin temperature (derived from [Guyton, 1986], Fig.72-8, according to [Benzinger, 1969]

The relationships between heat production by shivering (Hsh) and body temperature (Tbo) have been approximated by a third-order equation with a least-squares method. The temperature for which the amount of heat production due to shivering is half the maximum (here 0.74 °C) has been derived from the first derivatives of the curves in the same way. [Guyton, 1986] states that Hsh has a maximum of four to five times the normal basic metabolism.
The equations become:

$$Hsh = 4.5 \cdot Hmb \cdot (Tbo - Tspsh)/(Tbo - (Tspsh + 0.74)) >= 0$$
$$(Tbo < Tspsh) \qquad (25.3)$$
$$Tspsh = 48.3 - 1.45Ts + 0.063Ts^2 + 0.001Ts^3$$
$$(36.40 < Tspsh < 37.16) \qquad (25.4)$$

25.4.3.3 Sweating

Again, the model structure of [Spain, 1982] is used with constants derived from [Guyton, 1986].
The latter states that sweating is maximally 700 ml/h, which leads to a maximum Hsh of 0.488 J/s, whereas the minimum is given as 0.014 to 0.019 J/s. The maximum is corrected for the relative humidity of air, Rh.

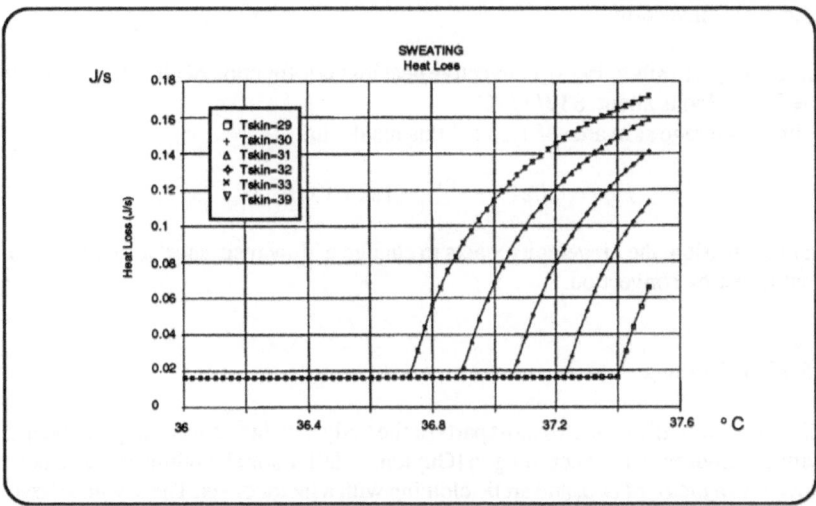

Figure 25.5. Heat loss due to sweating as a function of internal head temperature. The threshold temperature where sweating begins (Tspsw) depends on skin temperature (derived from [Guyton, 1986], Fig.72-7 according to [Benzinger, 1969])

The relationships between heat loss by sweating (Hsw) and body temperature (Tbo) have been approximated by a first-order equation with a least-squares method. The temperature for which the amount of heat loss due to sweating is half the maximum (here 0.34 °C) has been derived from the first derivatives of the curves in the same way. The equations become:

Hsw   = 0.488 • (1-Rh) • (Tbo - Tspsw)/(Tb - (Tspsw - 0.34)) >= 0.016

(Tbo > Tspsw)   (25.5)

Tspsw = Tspsw(Tsk) = 42.3 - 0.17Tsk   (36.5 < Tspsw < 37.7)   (25.6)

25.4.3.4 Radiation

Although radiation heat loss is actually a function of the fourth power of temperature, a valid approximation according to [Spain, 1982] and [Talbot, 1973] is:

Hra = 0.0116 • (Tsk - Tam)   (25.7)

25.4.3.5 Convection

According to [Talbot, 1973], convective heat loss is a function of air velocity v; for
v = 3 m^2, Hco is about 83 J/h / °C.
With an average skin area of 1.83 m^2 this results in:

$$Hco = 0.0245 \cdot v^{0.6} \cdot (Tsk - Tam) \qquad (25.8.)$$

In this equation, the air velocity v acts to change a 'flow resistance' through which
heat is lost by convection.

25.4.3.6 Clothing

Clothing thermally insulates most parts of the body and thus reduces heat exchange
with the environment. According to [Guyton, 1986], normal clothing reduces heat
flow with a factor of two, and arctic clothing with a factor of six. These values have
been implemented in the model for convection as well as for radiation.

25.4.3.7 Skin Conduction

Skin conduction is dependent on the reactions of cutaneous blood vessels on
temperature: heat causes dilation and cold leads to constriction. Blood flow
transports heat from the core of the body to the skin. Thus, the skin acts as an efficient
cooler of the body at Tam < Tbo. From [Guyton, 1986], Fig.72-2, a fourth-order
approximation was obtained by least-squares method for Rsk = Rsk(Tam):

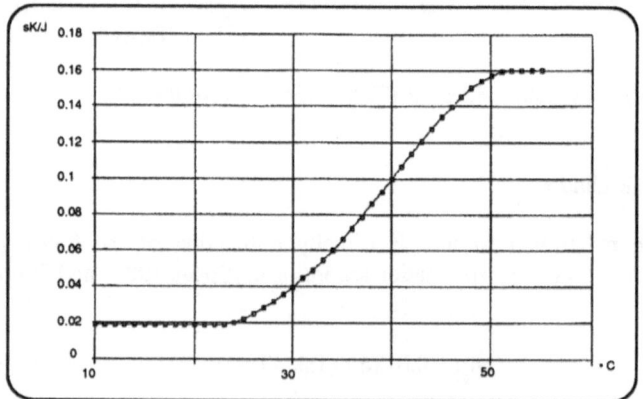

Figure 25.6. Skin conductance as a function of ambient temperature

The heat-flux through skin conduction was derived from [Ganong, 1987], Fig. 14-21, a graph of skin conductance as a function of head temperature. This figure shows that above a sharp threshold of 36.9 °C the cutaneous blood flow causes linear increase of skin conductivity with internal temperature. (The same threshold governs the onset of sweating).

With help of Rsk the skin temperature Tsk is obtained, which plays a pivot role in thermoregulation.

25.4.3.8 Fever

Fever is simulated by increasing the thresholds in the model, thus causing the system to operate at higher temperature levels (Cf. [Ganong, 1987]).

25.4.4 Possibilities and Limitations

The model allows us to gain insight into the basic role of the various heat flows and in body temperature regulation with the use of parameters derived from physiological literature. The effects of the onset of shivering and sweating are included. Fever is included, as well as the effects of clothing and air velocity. Geometrical (local) aspects of heat flow have not been included.

25. 5   Experiments

When starting the model for simulation, a screen is presented with an explanation of the environment chosen and instructions on how to perform a simulation and how to input data. One will also find some hints for exploring the characteristics of the model in the current environment.

After selecting or entering data, the simulation programme will compute all heat fluxes and temperatures for a period of 2.5 hours. It also produces a graph of the body and skin temperatures as well as a graph depicting the extent of heat lost or gained by shivering and sweating.

25.5.1 The Default Demo

The demonstration computes simulations of three different environments in the same way as the exercises with "normal" parameters.

At the top of the screen a short message indicates the relevant condition. At the right-hand side of the screen the data used are displayed. No pre-computation is used; the results are actually derived. The demo introduces the simulation program.

However, you need to do simulations on your own in the following exercises with variations of parameters in order to explore the behavior of the model.

25.5.2 Exercises

The model allows us to choose ambient temperature, relative air humidity and wind velocity as continuous variables within physical limits, as well as three states of clothing. Thus, many combinations are possible to test the regulation of body temperature.

25.5.2.1 Normal Thermoregulation

The model in this environment includes normal thermoregulation, so you will see that body temperatures will remain fairly constant under normal conditions over a period of time. The thermoregulation system seems to be able to cope with a large range of conditions! Explore the limits of thermoregulation from tropical to arctic environments.

25.5.2.2 Patient without Thermoregulation

In this case, neuronal defects (spinal cord transection high in the neck region) cause the inability to shiver. Just a basal amount of sweating has been included in the model. This results in the patient's inability to maintain a constant body temperature in the physiological range. Thus, even under not such extreme circumstances the patient will die when the fatal temperature limit of 42 °C is passed. Find out about these limits by choosing combinations of ambient conditions.

25.5.2.3 Fever

The model in this environment includes normal thermoregulation, so it will produce the same results as in the "normal" case. If you raise the 'threshold temperatures' Tspsh and Tspsw, however, you will see that the thermoregulating system tries to reach this higher temperature. Find out which threshold temperatures the model considers to be fatal.

25.5.3 Conclusions

Experimenting with this simple thermoregulation model shows that results can be obtained that correspond well with daily experience and with experimental observations. Insight into the thermoregulation can be gained by performing simulations with a great variety of parameter combinations.

References

Benzinger, J.T.: "Heat regulation: Homeostasis of central temperature in man". *Physiol. Rev.* , 49:671, 1969.

Ganong, W.F.: *Review of Medical Physiology*. 13th ed, Appleton & Lange, Norwalk, Connecticut/San Mateo, California, 1987, pp. 204-208.

Guyton, A.C.: *Textbook of Medical Physiology*. 7th ed, W.B. Saunders Company, 1986, Ch. 72.

Spain, J.D.: *Basic Microcomputer Models in Biology*. Addison-Wesley Publishing Company, 1982, pp 209-214.

Talbot, S.A. and U. Gessner: *Systems Physiology*. John Wiley & Sons, 1973, Ch. 19.

Werner, J.: "Mathematical Models of the Thermoregulatory System". In: *Modelling and Data Analysis in Biotechnology and Medical Engineering*, Eds.: Vansteenkiste, G.C. and P.C. Young, North-Holland Publishing Company, 1983, pp. 83-91.

Werner, J.: "Mathematical Simulation of the Human Thermal System". In: *Advanced Simulation in Biomedicine*, Ed.: D.P.F. Möller, Series Advances in Simulation, Vol. 3, 1990, pp.141-171.

26
Muscle Control

Rogier P. van Wijk van Brievingh

26.1 Aim of Instruction

This chapter is based on the model descriptions given as signal-flow graphs in Talbot and Gessner [1973]. There are many different kinds of muscle, that behave in a variety of ways, however, fundamental properties e.g. the force exerted divided by the maximum force can be studied by means of models in which normalized values are presented. The same goes for other physiological variables such as the firing frequency of neurons.

Thus, these simulations give insight into the basic processes of muscle and its control, without referring to specific preparations.

26.2 Theory and Reference to Standard Textbooks

Fundamental to an understanding of skeletal muscle contraction in vertebrates is the *sliding filament theory* [Wilkie, 1979]. This rests strongly on electronmicrographic evidence, while the force-producing mechanisms are explained by the formation of crossbridges between specialized sites on actin filaments and on myosine rodlets. The muscle directly converts chemical energy into mechanical and thermal energy, and as such is an energy transducer [Zierler, 1974]. It fulfills the function of mechanical effector for the organism, and is controlled by the central nervous system by means of action potentials. After depolarization of the muscle cell membrane, the *activation-contraction coupling* via Ca^{++}-ions governs the energy conversion process. The core of muscle function is the dynamic behavior of the *active contractile part* of the muscle fiber, constituted by the actin-myosine apparatus in the sarcomeres. The static performance of the sarcomeres is described by their isometric tension-length relationship, which appears to be the same for different types of muscle.

* Thanks are due to R. Jager, research assistant, for programming the models.

Results of dynamic measurements during isometric twitches can also be explained in part by the sliding filament theory. Resting muscle is elastic, but it does not follow Hooke's law in that it becomes more and more inextensible the further it is stretched. The resting elasticity results largely from the meshwork of connective tissue within the muscle. Thus, the tension-length relationship of the active muscle depends strongly on the initial length of the muscle at the moment of stimulation.

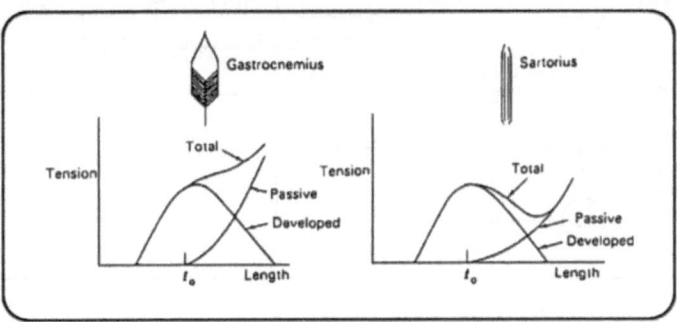

Figure 26.1

Dynamic measurements on fibers loaded isotonically and stimulated tetanically indicate a hyperbolic force-velocity relationship [Hill, 1964], which is widely applicable in normalized form [Talbot and Gessner, 1973]. It appears to express the fact that the rate of chemical reactions in the muscle is somehow linked to the force on it.

$$(f/f_m + 0.25).(v/v_m + 0.25) = 0.3125 \qquad (26.1)$$

One practical consequence of this shape is that the power output of the muscle, f.v , has a maximum of 0.1 $f_m.v_m$ at about one third of f_m and v_m. For the efficient performance of mechanical work, the load presented to the muscle must be chosen accordingly, e.g. by the use of a gear.

In every type of contraction, a sequence of changes in both length and tension takes place. Measurements, however, are mostly made *isometrically* or *isotonically*, so the length (at a chosen initial value) or the force developed can be kept constant. The change in the contractile component need not be very different in either case.

Muscle and its load are regulated by two feedback pathways, one signaling length and velocity through spindle receptors and the other signaling muscular force through tendon organs. The CNS initiates movement and modifies feedback by sending control signals to various neurons in the spinal cord (α-, and γ-motor neurons and interneurons in tendon organ pathways).

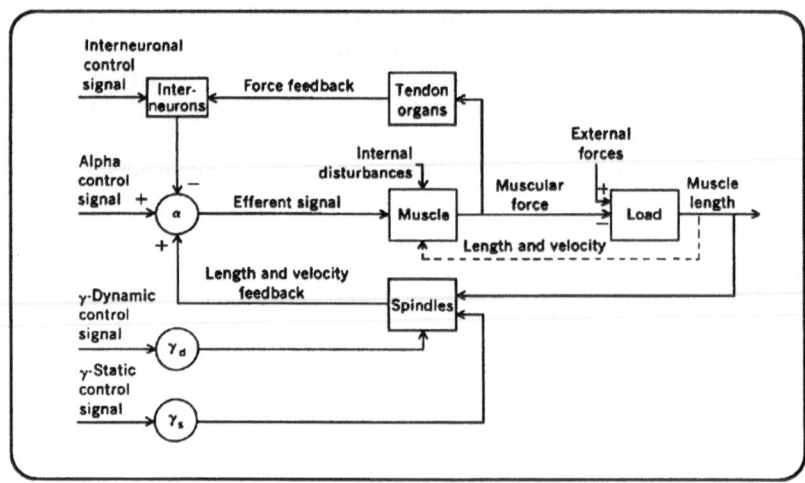

Figure 26.2

The output of the stress transducer within the spindle is carried by way of group Ia afferent fibers to the dorsal root of the spinal cord; then it directly excites the motoneurons belonging to the pool innervating the same muscle, thus closing the feedback loop. This feedback is negative in that increased activity at the annulospiral endings, due to augmented stretch, causes the muscle to shorten. This in turn releases the stress on the spindle, assuming that the τ efferent excitation stays at a constant level. The afferent signal from the muscle spindles has the opposite effect on the antagonistic muscle. This is apparently achieved by way of the internuncial cells connecting to a motoneuron of the antagonist's side.

26.3   Aspects to Be Considered

26.3.1   Limitations of the Model

In our model for an intact muscle system, the relation between anatomical structures and the model elements will not be considered. Furthermore, the model will be used in a *small-signal linear approximation* at a certain rest-length, so that nonlinearities in the springs are neglected and frictional forces are assumed to be proportional to velocity (thus choosing a working-point at the Hill hyperbola).

Since our simulation program BIOPSI precludes the presentation of three-dimensional graphics, the interesting point of force-length-velocity characteristics in the large-signal behavior had to be excluded, although these may be constructed from the two-dimensional results.

Time delays in neuronal pathways are neglected, since only a few synaptic transmissions are engaged. The gain at the motoneuron level may not be constant and the integration of supraspinal input to the α-motoneurons may be nonlinear in contrast to the assumptions in our model.

26.3.2 Outlook

To obtain the surface of the contractile compartment, including its velocity-limiting elements, in principle, it is possible to subtract from the tetanic force-length-velocity surface the forces exerted by the elastic elements and the viscous forces seen in dynamic tests of passive muscle.

Also, the simple model of Fig. 26.3 may be modified by introducing the nonlinearities of the elastic elements, the dependence of fc on distension, and the nonlinearity of the dashpot during the twitch [Bahler et al., 1967]. During an isometric twitch, the observed force equals the difference between f_c and the force on the viscous element due to "internal" velocity. The length of the contractile part is obviously the total muscle length minus the length of the series spring. From the model of the agonist-antagonist pair, a hand tracking task model can be constructed with the feedback loop closed by the visual system [Talbot and Gessner, 1973].

26.4 Model Description

26.4.1 Physical Foundations and Assumptions

26.4.1.1 The Spring Model

The basic model combines in a mechanical system the actin-myosine apparatus as a *contractile component CC* with the connective tissue elasticity modeled as a *parallel elastic component PEC*. The other elastic structures, partly in the tendons, constitute a *series elastic component SEC*. The viscous properties of the material are represented by a *linear dashpot R*.

26.4.1.2 The Neuromuscular Spindle

The neuromuscular spindle essentially performs as a muscle length sensor. The properties of the noncontractile regions containing the nucleated bags and free nerve endings within the spindle are represented by a series spring. The intrafusal fibers of the spindles are lumped into a parallel arrangement of a spring, a dashpot and a tractor element.

26.4.1.3 The Muscle-Spindle Feedback Loop

The muscle-spindle feedback loop model is built from the models of the spring and the neuromuscular spindle acting together as a length servo. The entire system has two inputs, and its output is either the change in muscle length or the force. In voluntary contraction both α and τ inputs may so increase that upon shortening of the muscle, the spindle is also actively shortened, thus increasing (or maintaining) the firing frequency so as not to counteract the purposeful movement.

26.4.1.4 The Agonist-Antagonist Pair

An agonist-antagonist pair can be modeled by two muscle-spindle feedback loops interacting such that the inhibitory action of the spindle afferent of one muscle is fed into the motoneuron pool of the other and vice versa.

26.4.2 Structure

26.4.2.1 The Spring Model

The model used to describe the short-term behavior and fast responses of muscle has the form of a simple mechanical system [Levin and Wyman, 1927]:

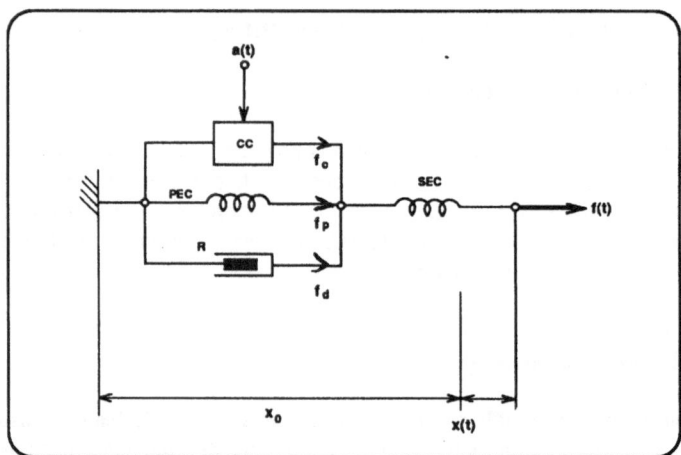

Figure 26.3

The length of CC, which is about the tendon length, x_c, is added to with the elastic length, x_e, to give the muscle rest-length:

$$x_o = x_c + x_e \qquad (26.2)$$

In the small-signal analysis, the total muscle length is composed of the rest-length x_o and the length change $x(t)$:

$$x = x_o + x(t) \qquad (26.3)$$

The force f_c of the contractile "motor" is proportional to the concentration of Ca^{++}-ions released in the sarcoplasmatic reticulum, caused by the net activity of the α-motoneuron pool, $a(t)$:

$$f_c = A \cdot a(t) \qquad (26.4)$$

The force of CC and SEC is the same because of the series configuration. The muscle is said to be in the 'active state' as long as fc exists.

The force f_p of the PEC for the small-signal approximation is proportional to the length change $x(t)$

$$f_p = B \cdot x(t) \qquad (26.5)$$

The force f_d of the damping element R in the small-signal analysis is proportional to the velocity $dx(t)/dt$ according to the slope of the Hill hyperbola at the actual working-point.

$$f_d = C \cdot dx(t)/dt \qquad (26.6)$$

The parallel configuration yields the total force:

$$f = f_p + f_c + f_d \qquad (26.7)$$

From these equations, $f(t)$ and $x(t)$ are calculated as a function of $a(t)$ via the Laplace domain. The model is composed of three parts.

Ca^{++}-flow Submodel

This submodel calculates the Ca^{++}-flow as an exponentially decaying function of time. A limiter is used to keep the signal within prescribed values.

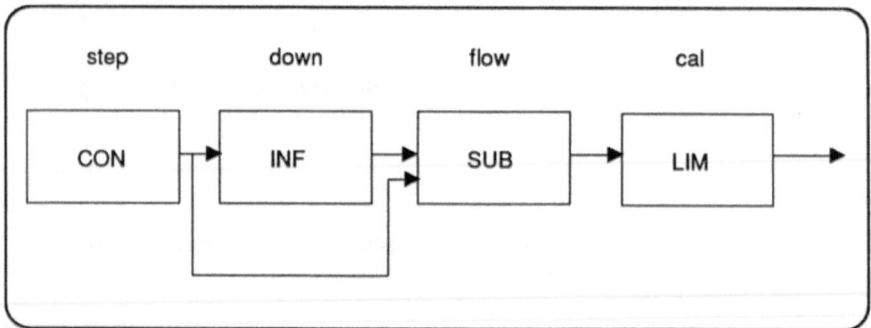

Figure 26.4

Ca^{++}-series Submodel

This submodel generates a series of at most five Ca^{++}-flow pulses according to previous section. A relay switch selects between one or five pulses upon command of the parameter More.

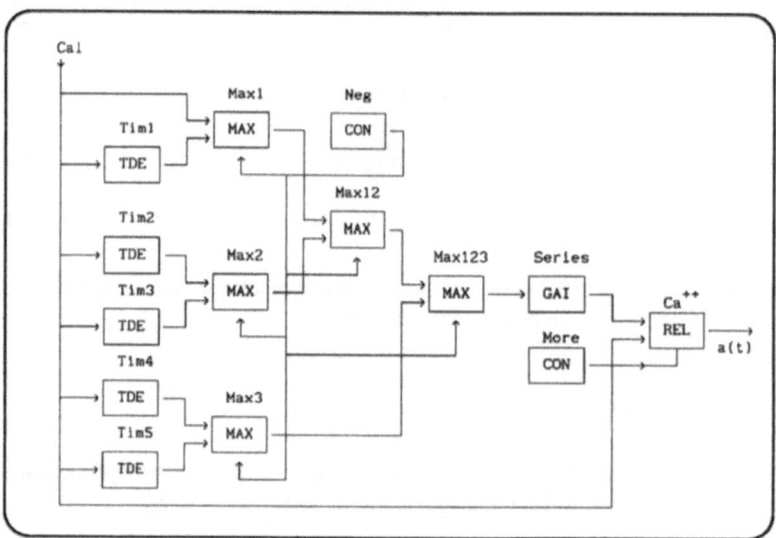

Figure 26.5

Force and Length Submodel

This submodel constitutes the spring model of Fig. 26.3 with the Ca++-flow as input.

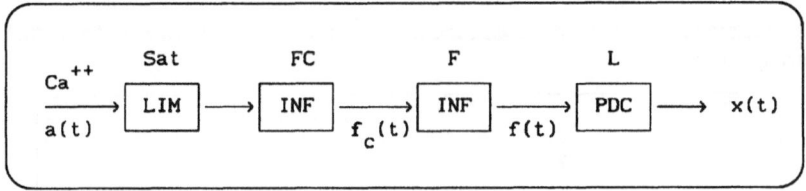

Figure 26.6

26.4.2.2 The Neuromuscular Spindle

Basically the same model as shown in Fig. 26.3 is used to represent the action of neuromuscular spindles.

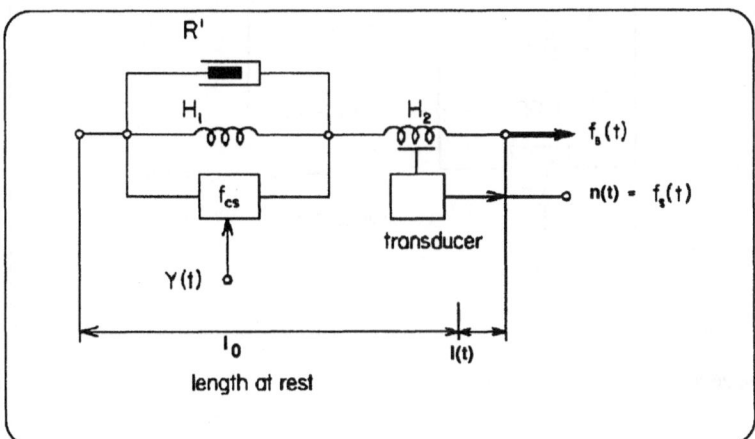

Figure 26.7

The tractor element is linearly innervated by the γ-motoneurons:

$$f_{cs} = G.\gamma(t) \tag{26.8}$$

The annulospiral transducer converts the stress in the series elastic spring linearly into the afferent discharge rate n:

$$n(t) = D \cdot f_s (t) \qquad\qquad (26.9)$$

26.4.2.3 The Muscle-Spindle Feedback Loop

The muscle-spindle feedback loop model is a combination of the models from 26.3 through 26.7.

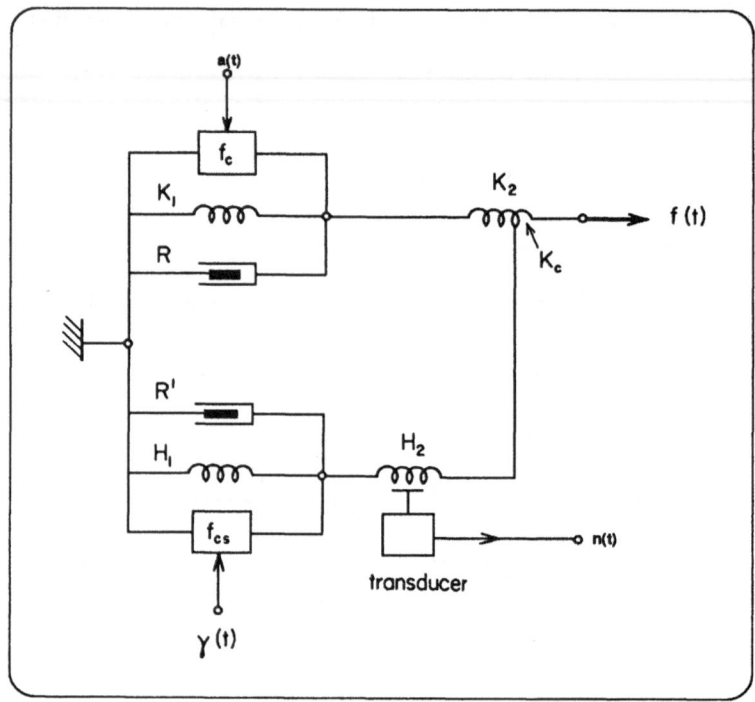

Figure 26.8

The spindles are assumed to be essentially in parallel with the muscle fibers according to physiological evidence. Part of the SEE is also in series with the spindle. The system acts essentially as a length servo.

26.4.3 Parameters

STRUCTURE AND PARAMETERS OF MODEL MUSCLE.PS6

Block	Type	Input1	Input2	Input3	Par1	Par2	Par3
CAL	LIM	FLOW			.0000	1.000	1.000
DOWN	INF	STEP			.0000	1.000	1.0000E-02
F	INF	FC			.0000	1.000	1.5000E-02
FC	INF	SAT			.0000	1.000	1.0000E-03
FLOW	SUB	STEP	DOWN				
L	PDC	F			.0000	1.000	1.0000E-02
SAT	LIM	CAL			.0000	4.000	1.0000E+06
STEP	CON				1.0000E-05		

STRUCTURE AND PARAMETERS OF MODEL MUSCLE2.PS6

Block	Type	Input1	Input2	Input3	Par1	Par2	Par3
CA++	REL	MORE	SERIES	CAL			
CAL	LIM	FLOW			.0000	1.000	1.000
DOWN	INF	STEP			.0000	1.000	1.0000E-02
F	INF	FC			.0000	1.000	1.5000E-02
FC	INF	SAT			.0000	1.000	1.0000E-03
FLOW	SUB	STEP	DOWN				
L	PDC	F			.0000	1.000	1.0000E-02
MAX1	MAX	NEG	CAL	TIM1	.0000		
MAX12	MAX	NEG	MAX1	MAX2	.0000		
MAX123	MAX	NEG	MAX12	MAX3	.0000		
MAX2	MAX	NEG	TIM2	TIM3	.0000		
MAX3	MAX	NEG	TIM4	TIM5	.0000		
MORE	CON					-1.000	
NEG	CON					-1.000	
SAT	LIM	CA++			.0000	4.000	1.0000E+06
SERIES	GAI	MAX123			1.000		
STEP	CON				1.0000E-05		
TIM1	TDE	CAL			.0000	1.3000E-02	.0000
TIM2	TDE	CAL			.0000	2.6000E-02	.0000
TIM3	TDE	CAL			.0000	3.9000E-02	.0000
TIM4	TDE	CAL			.0000	5.2000E-02	.0000
TIM5	TDE	CAL			.0000	6.5000E-02	.0000

STRUCTURE AND PARAMETERS OF MODEL SPINDLE.PS6

Block	Type	Input1	Input2	Input3	Par1	Par2	Par3
BASIS	CON				1.000		
CALCIUM	SUB	STEP	FIRST				
END	TDE	FASTEP			.0000	.6000	.0000
FALL	SUB	FALL1	FALL2				
FALL1	TDE	RAISE			.0000	.1000	.0000
FALL2	TDE	RAISE2			.0000	.4000	.0000
FAST	SUB	START	END				
FASTEP	GAI	STEP			.2000		
FC	INF	SATUR			.0000	1.000	1.0000E-02
FIRST	INF	STEP			.0000	1.000	2.0000E-02
FR	LIM	FREQ			-1.000	10.00	1.000
FR/B	ADD	BASIS	FR				
FREQ	PDC	LIN/B-1			.0000	.4000	.1500
LIN/B-1	SUM	FAST	SLOW	TWITCH	5.000	.0000	.0000
RAISE	INT	STEP			.0000	1.000	
RAISE2	INT	STEP			.0000	2.000	
SATUR	LIM	CALCIUM			.0000	.3000	1.000
SHOCK	INF	FC			.0000	-4.600	1.5000E-02
SLOW	LIM	FALL			.0000	.2000	1.000
START	TDE	FASTEP			.0000	.2000	.0000
STEP	CON				1.000		
TWITCH	TDE	SHOCK			.0000	.4000	.0000

26.4.4   Possibilities and Limitations

Isotonic and isometric twitches appear to have the same shape if they are normalized to f/f_m.

Creep and stress relaxation are not accounted for in this model, as they are slow phenomena and would have to be modeled by more complex mechanical systems.

26.5   Experiments

26.5.1   The Default Demo

Demonstrations are simulation runs with default values of the parameters and are not adjustable by the user.

26.5.2   Exercises

The exercises allow interactive choice of one parameter each within minimum and maximum values, checked by the menu program.

26.5.2.1 Muscle

Twitch and Tetany

A twitch is the reaction of the muscle fiber on one Ca^{++}-pulse; the resulting F is shown.
Contents of the corresponding AUTOEX.PSI file are:

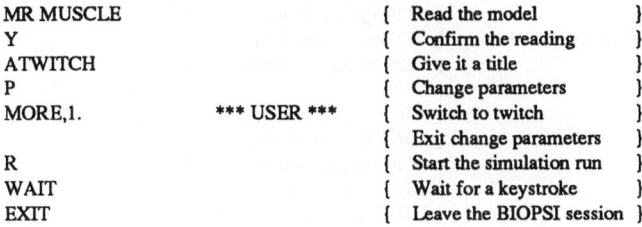

```
MR MUSCLE                          {  Read the model            }
Y                                  {  Confirm the reading       }
ATWITCH                            {  Give it a title           }
P                                  {  Change parameters         }
MORE,1.          *** USER ***      {  Switch to twitch          }
                                   {  Exit change parameters    }
R                                  {  Start the simulation run  }
WAIT                               {  Wait for a keystroke      }
EXIT                               {  Leave the BIOPSI session  }
```

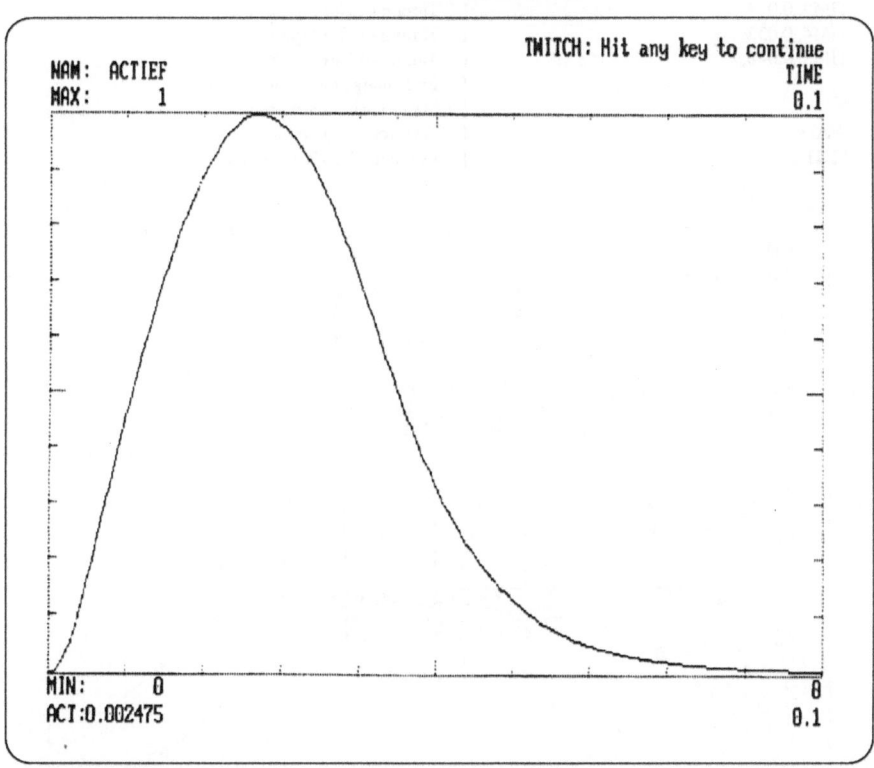

Figure 26.9

Ca^{++} and Tetany

Contents of the corresponding AUTOEX.PSI file are:

MR MUSCLE2	{ Read the model	}
Y	{ Confirm the reading	}
A TETANIC CONTRACTION	{ Title	}
R	{ Start the simulation run	}
WAIT	{ Wait for a keystroke	}
P	{ Change parameters	}
MORE,-1 *** USER ***	{ Choose tetanic stimuli	}
	{ Exit change parameters	}
ASMOOTH TETANY	{ Title	}
R	{ Start the simulation run	}
WAIT	{ Wait for a keystroke	}
P	{ Change parameters	}
TIM1,,0.008,	{ Timing of Ca++-pulse # 1	}
TIM2,,0.016,	{ Timing of Ca++-pulse # 2	}
TIM3,,0.024,	{ Timing of Ca++-pulse # 3	}
TIM4,,0.032,	{ Timing of Ca++-pulse # 4	}
TIM5,,0.040,	{ Timing of Ca++-pulse # 5	}
	{ Exit change parameters	}
R	{ Start the simulation run	}
WAIT	{ Wait for a keystroke	}
EXIT	{ Leave the BIOPSI session	}

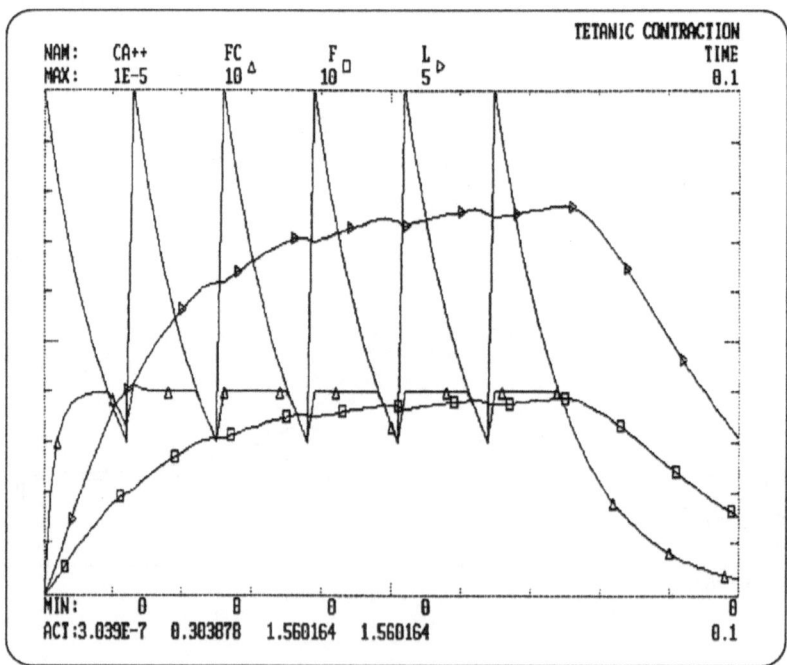

Figure 26.10 a

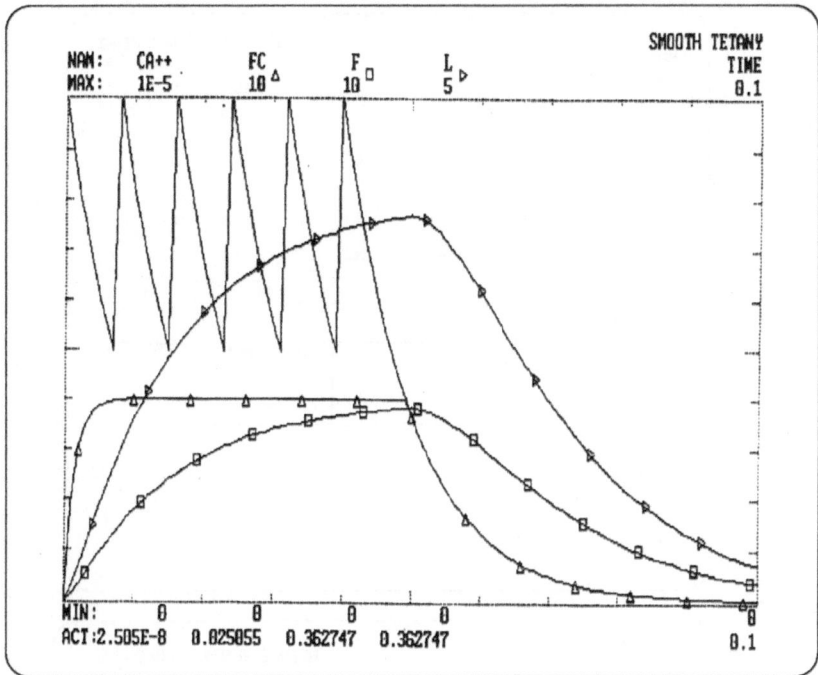

Figure 26.10 b

26.5.2.2 Spindle

Contents of the corresponding AUTOEX.PSI file are:

```
MR SPINDLE                                  {  Read the model                    }
Y                                           {  Confirm the reading               }
DSA -1,4                                     {  Scale the presentation            }
ARESPONSE ON FAST LENGTH CHANGE {  Give it a title                            }
R                                           {  Start the simulation run          }
WAIT                                        {  Wait for a keystroke              }
P                                           {  Change parameters                 }
LIN/B-1,0,5,0          *** USER ***         {  Fire freq. - base freq.          }
                                            {  Exit change parameters           }
ARESPONSE ON SLOW LENGTH CHANGE    {  Give it a title                          }
R                                           {  Start the simulation run          }
WAIT                                        {  Wait for a keystroke              }
P                                           {  Change parameters                 }
LIN/B-1,0,0,1          *** USER ***         {  Fire freq. - base freq.          }
                                            {  Exit change parameters           }
ARESPONSE OF FIRE-FREQUENCY ON TWITCH       {  Title                            }
R                                           {  Start the simulation run          }
WAIT                                        {  Wait for a keystroke              }
EXIT                                        {  Leave the BIOPSI session          }
```

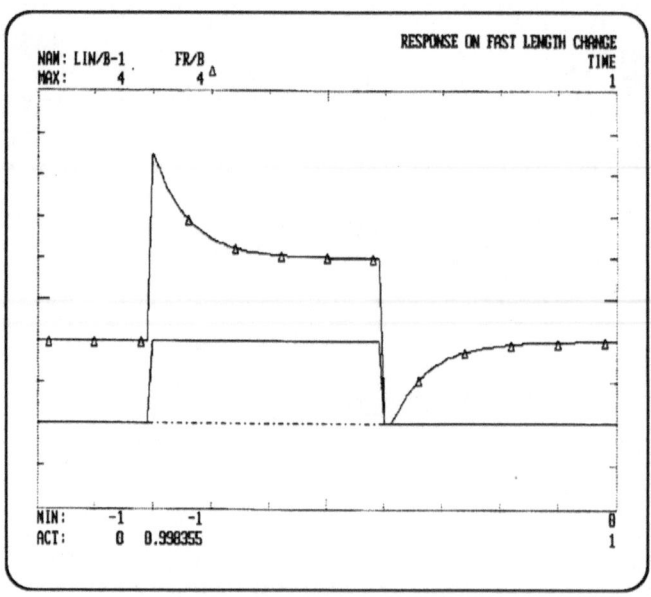

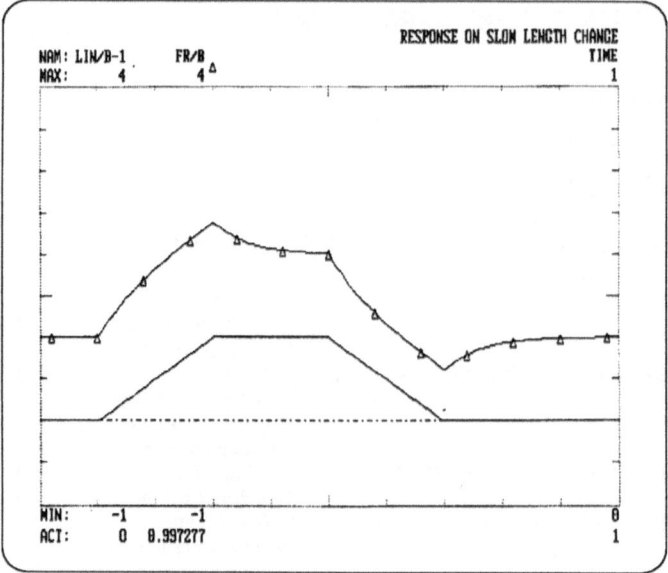

Figure 26.11 a, b

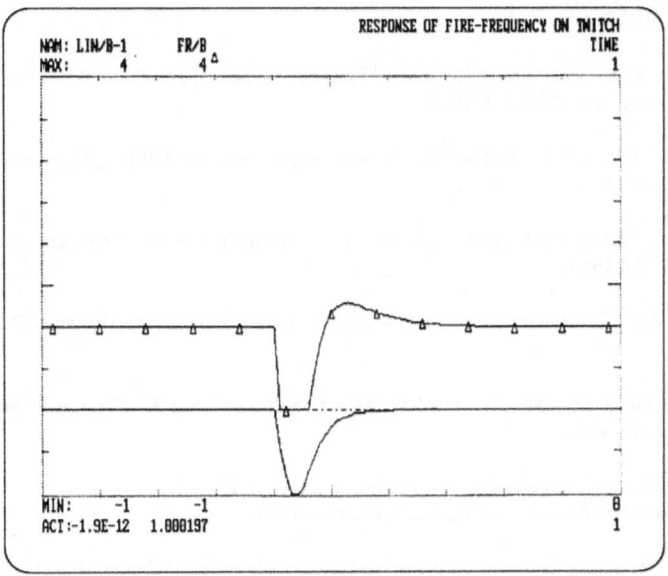

Figure 26.11 c

References

Bahler, A.S., J.T. Fales and K.L. Zierler: "The active state of mammalian skeletal muscle". J. Gen. Physiol. 50:2239-2253, 1967.

Carlson, F.D. and D.R. Wilkie: Muscle Physiology. Prentice-Hall Inc., Englewood Cliffs, New Jersey, 1974.

Hill, A.V.: "The effect of load on the heat of shortening of muscle". Proc. Roy. Soc. [Biol] 159: 297-318,1964.

Huxley, A.F.: "Muscle structure and theories of contraction". Progr. Biophys. 7: 255-318, 1957.

Levin, A. and J. Wyman: "The viscous elastic properties of muscle". Proc. Roy. Soc. [Biol] 101: 218-243, 1927.

Linkens, D.A.: Biological systems, modeling and control.
The Institution of Electrical Engineers, London, 1979.

Talbot, S.A. and U. Gessner: Systems Physiology. Wiley-Interscience, New York, 1973.

Wilkie, D.R.: Muscle. second edition, Study in Biology no 11,
The Camelot Press, Southampton, 1979.

Winter, D.A.: Biomechanics of Human Movement. Wiley, New York, 1979.

Zierler, K.L.: "Mechanisms of muscle contraction and its energetics". In: Mountcastle, V.B.: Medical Physiology, Vol I,
The C.V. Mosby Company, Saint Louis, 1974, pp 77-120.

27
How to Use the Student Programs

Rogier P. van Wijk van Brievingh and Ignacio A. García Alves

27.1 What to Do First

27.1.1 Equipment Check

The programs supplied with this book require:
- An IBM XT, IBM AT, IBM PS/2 or compatible machine with minimal 512 kB RAM and 3 MB free space on the hard disk;
- A video adapter and monitor capable of high resolution graphics (CGA, EGA, VGA, or Hercules);
- A version of operating system (DOS) of 2.1 or later.

The installation program checks these requirements and reports violation of limitations before aborting.

Options:
- We recommend a printer capable of making screen dumps via the PrtScr facility (the utility GRAPHICS.COM from your DOS diskette has to be loaded first, preferably by including it in your AUTOEXEC.BAT file; refer to your DOS manual for this); EPSON^{TM} FX-80 or IBM Pro Printer or compatible.
- A mouse with a Microsoft^{TM} or Mouse Systems^{TM} compatible driver will be detected automatically at program startup.
  The right mouse button executes <ENTER>, the left one <ESC>, the middle one, if present, executes <F1>.

27.1.2 Registration

The book package contains a user registration form/order card. In order to provide you with the latest technical information, announcements of future updates and up-to-the-minute information on new products, it is essential that you return this card directly after breaking the seal of the package. Otherwise, no warranty is given, as stated on the card.

27.1.3 Software Backup

In order to safeguard the programs contained in the diskettes, you **must immediately** make a set of backups using DISKCOPY or another disk-copying program.

Note: The distribution diskettes are not copy-protected, but the model programs are contained in many subdirectories.

Moreover, the diskettes have volume labels used by the installation program. So, do not use COPY or XCOPY or your backups will not work properly. Put away your distribution diskettes in a safe place.

27.1.4 Last-Minute Notice

The diskette labeled "Installation" contains a file (in ASCII-format) named *READ.ME!* This contains last-minute information on the programs. Make a hard copy with the DOS PRINT facility for further reference.

27.2 Files on the Distribution Diskettes

The distribution diskettes have the 5 1/4" (360 kB capacity, DoubleSide)format. Three kinds of diskettes are distinguished by their volume name:
- *INSTALL disk*, containing the installation program INSTALL.EXE configuration files, the READ.ME! file and others;
- *PROGRAM disk*, containing the main program MENU.EXE and the simulation program BIOPSI.EXE as well as utility programs.
- *MODEL disks*, containing the simulation models and stand-alone programs to be called from the menu.

To get a full survey of the directories' tree structure and what will be present on your hard disk after running the installation procedure, refer to the (ASCII-format) file *MENCONF.TXT* on the INSTALL disk. This file is to be regarded as the ultimate list of contents of the installed programs, as well as of the menu configuration.

27.3 Installation on your Computer

Start up your computer, preferably without installing any resident utility programs, printer-spoolers or RAM-disks.

Change the current drive to A: and insert your *working* copy of the INSTALL disk (without a write-protect sticker).

Type INSTALL <ENTER> and follow the instructions on the screen.

After system-check and choice of printer number you are asked for the drive letter of the hard disk to be used and the name of the directory under which the BMTperPC programs will be installed. Hitting <ENTER> here will install from the root directory. (Refer to your DOS manual if necessary).

Make a note of the path shown: drive:\<..>\BMT.

The installation program then offers four options:

- Full install (first time only);
- Add model files, which is possible any time (NB: corresponding files on the hard disk will be overwritten by newer files present on the MODELS diskette);
- Uninstall (removes all files and directories down to the parent directory and saves the menu configuration on the INSTALL disk for future use);
- None of the above (possible any time); a new configuration file is created, containing possible different hardware items such as video adapter or printer.

After a successful first-time installation, include the path noted above in your DOS path, since the program will check for it at start-up. A corresponding error-message will be shown and the program will abort if the environment does not contain this path!

We recommend making an AUTOEXEC.BAT file containing the PATH statement as well as a call to GRAPHICS.COM.

If you use a serial printer, include the relevant MODE command to configure it as LPT1: (see your DOS manual for details).

A CONFIG.SYS file containing "files=15" and "buffers=20" as well as "device=ANSI.SYS" is needed for some of the program used, to install your mouse, "device = MOUSE.SYS" is appropriate.

27.4 Setting up the Student System

After a successful installation, make the "parent" directory current, and see that a subdirectory "BMT" has been created on top of it. Type CD BMT <ENTER> to make this directory the current one, and type MENU <ENTER>. On the opening screen presented to you now, you have the possibility of setting the current date and time using the appropriate editing keys, or to agree with the values offered by typing <ESC> two times.

Next, an identification screen asks for the *name* and *course number* of the student, and a choice between *"credit"* and *"free"* mode is offered. A password is needed in the credit mode, for which the student has to insert a diskette prepared for him by the teacher.

If this has not been satisfactorily entered, the program continues in the "free" mode. In the "credit" case, the student's activities will be logged on his diskette and can be used by the teacher to determine his progress.

The menu is presented in the *inactivated* state now; to activate, press <F10>. Use the cursor keys or mouse to navigate along the menu bars, and <ENTER> to choose a menu, according to the status bars at the bottom of the screen. <ESC> returns you to the preceding menu level. The identification window shows the student name and

number as well as the menu selection made and the current time.
In the "credit" mode, the name appears in red and blinking.

The student menu contains two pages with four themes of five items with up to six exercises each. With <F1>, context-sensitive help is available at all levels of the menu system. The indication bar gives the keys to be used at all levels.

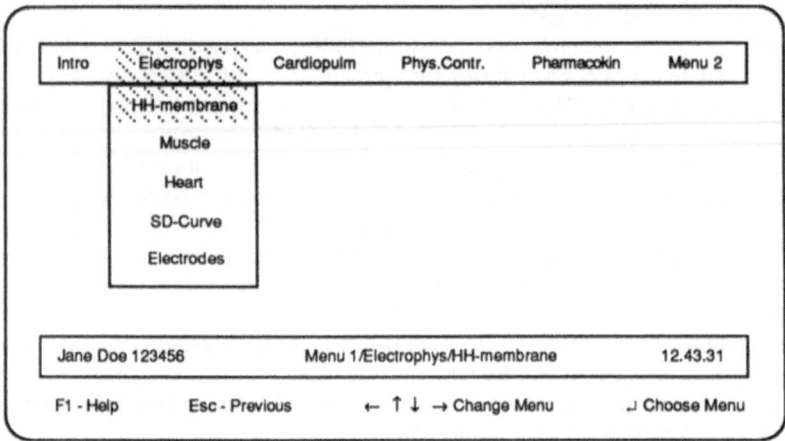

Figure 27.1. Menu screen (page 1)

After finishing a session by entering <ESC> repeatedly, the program returns to the opening screen. If in the credit mode, the student log-file is saved, indicated by the window "Booking your Credit …".
The program does not allow the student to exit to DOS, since the call <CTRL><BREAK> has been disabled. (See next section for the appropriate method.)

27.5   Use of BMTPER PC on Local Area Networks

The program have been tested in a Novel™ network environment only. The installation procedure must be performed by the system operator or someone with sufficient authority to add files on the hard disk and create a path, which in the case of a LAN also must contain A:. It is assumed that a diskette is present in a drive A: at student login and must stay there during the session.
The BMTperPC program is run with "N" as a command-line parameter, so use MENU N <ENTER> if it is to be used in a network environment.
The user returns to the network menu by pressing <ESC> repeatedly.

27.6   What if ...?

The MENU program as well as the programs it accesses called by it contain an extensive error-detection system. DOS errors as well as program errors are captured and signalled in a special error-window.

If you try to use an item or exercise that has not been configured, the message "This menu is not implemented yet" is shown; if <F1> is typed at the corresponding level, an error message stating the name of the lacking HELP-file appears. Some of the programs called from the menu require a first-time installation of their own for a color monitor. They use a configuration file created automatically at first set-up. Other programs require setting of parameters the first time an exercise is chosen. The next exercises will share these parameters, until the item is selected again.

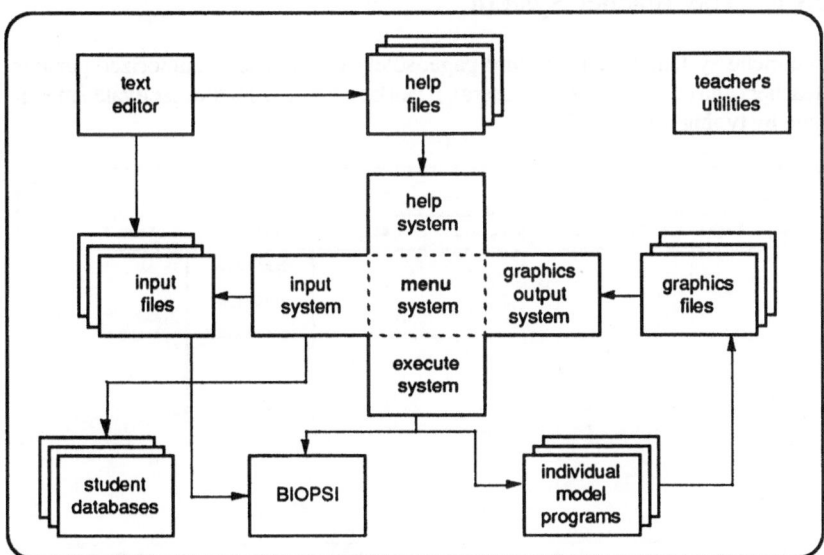

Figure 27.2. Structure of the menu systems

28
How to Use the Teacher's Facilities

Rogier P. van Wijk van Brievingh, Ignacio A. García Alves,
and Gerard L. van Eendenburg

28.1   The Teacher System

The menu system contains a third page, solely for the use of authorized persons (teachers and other experienced users). It can be entered from the inactivated menu-state by typing <ALT><F10>.

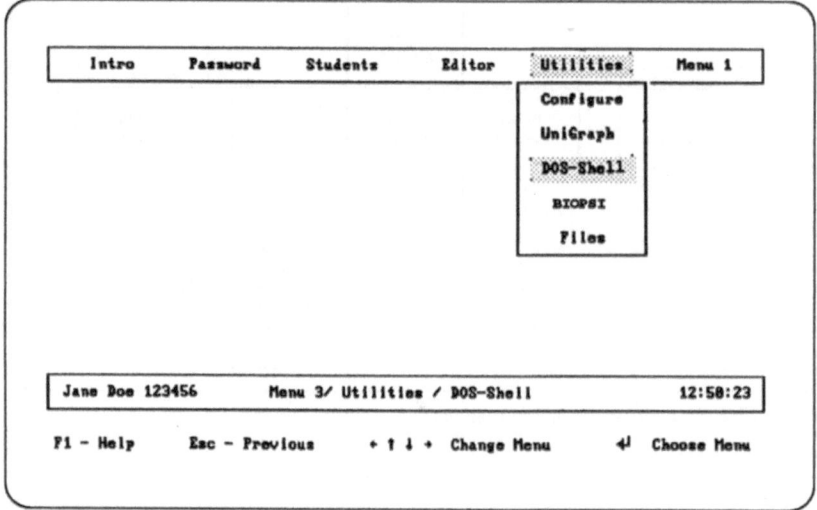

Figure 28.1

28.1.1   Passwords

Insert a formatted diskette in A: on which your password will be recorded in a hidden password file. After typing <ALT><F10> a window is presented, asking for a password.

The default password is TEACHER, valid only once after startup.
After entering this the first time:

- select "Teacherdisk" from the password submenu, and enter a new password of your choice;
- select "Number" to check that the number of passwords = 1;
- you may enter more passwords (for your colleagues) if you like;
- keep (secret) notes of the passwords in use.

Students who apply for credit are given their personal password from the database system, using their own diskette. The student obtains his password from the teacher, so that it can be used for credit evaluation (see section 29.6.5).

The only way of quitting the MENU program to DOS is to choose "Exit" from the password submenu (<CTR>C does not work).

If any changes have been made in the menu configuration (see 28.1.4.1) and these have not been saved, or when a 'credit student' ends his session, a reminder window comes up, asking "Do you want to save changes?". If so, a ASCII text file named "MENCONF.TXT" is generated automatically in the parent directory for future reference.

28.1.2 Students

This submenu is used to document students' requirements and progress. The program keeps a database of student data and allows surveys to be made by the teacher, sorted on various criteria. Data are kept on the teacher's diskette as well as on each student's diskette.

The database system is described in detail in chapter 29.

28.1.3 Editor

This submenu allows you to create ASCII textfiles in the format required by the MENU program. If one of the submenu selections is made, the menu gets into the "editmode". You can travel freely around the menu as usual, as well as use the menu programs. The program automatically switches to the subdirectories corresponding with the menu items. By pressing <F4> the editor is activated to create or edit a file that will reside in the proper subdirectory. See appendix 1 for the editing commands. The files also may be printed directly from the editor.

With <ALT><F4> you can leave the "editmode" without editing and return to the teacher's menu. The type of file that will be edited is determined by the submenu selection.

28.1.3.1 Help Files (HELP.TXT)

The help files are contained in the corresponding subdirectory, encoded into the name of the help file to edit. This name has the structure HELPnnn.TXT for an item, and HELPnnnn.TXT for an exercise, with "n" a number corresponding to the (sub)directory: drive:<parent>\BMT\n\n\n. The filename is generated automatically by the program.

28.1.3.2 BIOPSI Files (AUTO.PSI and AUTO.AUT)

The AUTO.PSI editor facility only operates at the menu exercise level. With this editor you can generate an AUTOnnnn.PSI file, which will automatically be renamed to AUTOEX.PSI by the menu system. It is read by the BIOPSI program during execution if AUTOnnnn.PSI was entered as the execution file name during configuration. In that case, the same numbers are to be used as mentioned for the HELPnnnn.TXT file in the configuration window. Refer to Chapter 30 for the autoexecution facility of BIOPSI.

   The AUTO.AUT facility offers the extra opportunity to create files that interactively change ONE of the parameters in the AUTOEX.PSI file. If you want to use this facility, proceed as follows:

-    if an AUTOnnnn.PSI file is present, this will be loaded to serve as the basis for
     the AUTOnnnn.AUT to be edited,   with nnnn as specified in previous
     section; otherwise an empty file of the proper name will be created by the editor;
-    edit this file such that ONE parameter value <par> to be interactively changed
     is replaced by:
                    <min>,<par>,<max>
     where <min> and <max> are the minimum and maximum values   between
     which <par> must be kept. Asterisks and commas are essential here; <par> is a
     REAL value.

If the menu program encounters a *.AUT file that has been specified during configuration, a temporary file AUTOEX.PSI is created from it with the user-entered value of <par> substituted in it. This file is subsequently run by BIOPSI.

28.1.3.3 Graph Form

This facility operates at either the item level or the exercise level. The editor generates a *.AUX and a *.INP file, with * the same filename as the *.DAT file specified during configuration. The name *.DAT must be the first commandline item after the *.COM or *.EXE program that generates it. In this manner, the data generated by the executable program will be automatically presented by the utility

UNIGRAPH (see section 28.1.4.2). The format used in the graphical presentation is determined by the *.INP and *.AUX files (see appendix 2).
This facility can also be used to copy, rename or delete *.DAT, *.AUX, or *.INP files.

*.DAT File (ASCII-format)

This type of file is generated either by the user-written programs or by the user himself. The user may choose between the special editor and the "free" editor. A *.DAT file constitutes the data to be presented by UniGraph as curves. The format is specified in appendix 2.

*.INP File (ASCII-format)

This type of file may be constructed with the special editor, and it contains only integer values, each on a separate line (see appendix 2). It specifies the numbers of screens and curves to be presented. If a line is left empty, the corresponding default value is taken. If the file *.INP is absent, values are taken from UDEFAULT.INP at execution time.

*.AUX File (ASCII-format)

This type of file may be constructed with the special editor, and it contains only integer values, each on a separate line (see appendix 2). It specifies the axes, grids and colors. If a line is left empty, the corresponding default value is taken. If the file *.AUX is absent, values are taken from UDEFAULT.AUX at execution time.

28.1.3.4 Free Editor

This editor allows you to create textfiles with path/filename to be freely chosen by the user. It may be used to edit and print files residing in known subdirectories or as a stand-alone facility.
The maximum file size is about 64 kB. The editor generates no backup files. See appendix 1 for the editing commands.

28.1.3.5 Graph Editor

This editor allows you to interactively create *.DAT, *.INP and *.AUX files for use by the utility program UNIGRAPH. Besides editing, this program can be used to copy, rename or delete *.DAT, *.AUX and *.INP files in the directory selected.

28.1.4 Utilities

The submenu "Utilities" offers the teacher several facilities:

28.1.4.1 Configuration

Running external programs from within the menu system requires a configuration procedure. If this item has been chosen, the system comes in the "configuration mode". You can travel freely around the menu as usual, and run programs as well. The menu program automatically switches to the subdirectories corresponding to the menu items. By pressing <F2> the configuration window appears. The current subdirectory is indicated by the help file name, see section 28.1.3.1. First, the name to appear in the menu window can be edited. Second, the number of menu selections is to be selected -at least one-. Next, the name of an executable file (*.BAT, *.COM, or *.EXE) -together with possible command line parameters including a numerical output filename, (*.DAT) or the local BIOPSI command file name (*.PSI or *.AUT)- can be entered. At the item level, these filenames indicate the programs to be executed as a preparation for the exercises as soon as the first exercise is run. In this manner, a parameter file may be generated which is to be used by all subsequent exercises. The default command line name is DUMMY.COM, yielding the message "This menu is not implemented yet" when selected. Finally, confirmation of the entries is requested, so that possible typing errors may be corrected. After confirmation, the new menu selection can be activated at once. The configuration changes are saved with <ALT><F2>; if this is neglected the user is reminded before leaving the program. The teacher menu cannot be configured by the user.

28.1.4.2 UniGraph

This menu item offers you the opportunity to directly run files with the UNIGRAPH program. These files can be *.DAT with corresponding *.INP and *.AUX files, or *.PIC files created by previous runs. The options "s(tudent)" for viewing curves from data files or "t(eacher)" for viewing recorded pictures as well as options path\filename are entered interactively. The program contains several levels; by using <ESC>, the next lower level is selected throughout. The following options are offered:

Help

Context-sensitive help can be activated throughout the program with <F1> and quit with <ESC>.

Swap

This menu item allows you to swap between two screens if present. The options for one or two screens and the distribution of the curves are set by the *.INP file. "Swap" can be activated by pressing <F3>.

Change

This menu item allows you to change several items of the picture shown on the screen. By pressing <F4> the "change" menu appears. It is made up of three parts:

Change Curve.
By pressing <F3> you can change the symbols of the data points and/or the line style of a curve. If there is more than one curve you will be asked to select one. There are eight different symbols for the data points (numbered 1-8). A value of 0 indicates that no data points will be drawn. There are 4 different line styles. A value of 1 indicates that the data points will not be connected by a curve on the screen.

Change Axes.
By pressing <F4> you can make changes on the axes.
The "Change/Axes" menu has three options:
Option 1 allows you to choose if and where the axes and scale division are drawn on the screen.
Options 2 and 3 select the X-axis and Y-axis respectively for changing the layout of the axes chosen. The screen shows the current value of each item, which can be changed by pressing the number of that item.
The items that can be changed in the "Change/Axes/Standard" menu are:

a. Scale: The scale of the current axis can be linear or logarithmic. It is possible to scale the axes differently. If any of the coordinates of the data points for this axis is smaller than 1E-37, then you cannot select a logarithmic scale.
b. Intersection with the other axis: You can change the intersection of the current axis with the other axis. The point of intersection does not have to be within the range of the current axis.
c. Number of decimals: The representation of values at the main marks depends on the "number of decimals", which can have the value -1, 0, 1, 2 or 3:
    -1   = no decimal point and no decimals are drawn
    0   = only a decimal point is drawn
    1..3 = a decimal point and 1..3 decimals are drawn
d. Range: The range of the current axis may be any interval within [-1E37,1E37] for a linear axis and [1-E37,1E37] for a logarithmic axis.

e.   Grid: The grid can be 'on' or 'off'. If grid is 'on' then a grid is drawn through every main mark on the current axis.
f.   Position of label: The label can have three positions: no label, parallel to the axis or at the end of the axis.
g.   Label: The name of the label can be changed (maximum length = 20 characters).
h.   Distance between main marks: This item can only be changed by the user if the scale of the axis is linear. A value is printed at each main mark if there is enough space.i.
      Number of marks between main marks: With this item the axis between two main marks can be subdivided into small   intervals.

You may have noticed that the number of items to change on   the current axis is not always the same. If an axis has a logarithmic scale, the user is not allowed to change the distance between the main marks or the number of marks between the main marks. If the label position has the value 'no label', the item 'change label' is not shown.

General Changes.
After pressing <F5>, a menu with 4 options is presented:

a.   Interpolation: Interpolation can be 'on' or 'off'.
      If interpolation is 'on' then the curve points are interpolated. When at least one of the curves is a relation rather than a function, then neither of these curves can be interpolated. Although the first and last data point of an interpolated curve will be drawn, the program will only draw the curves between the second and next-to-last data point.
b.   Clipping: Clipping can be 'on' or 'off'.
      When it is 'on', your curve will be clipped at the edge of the screen. Only the parts of the curves inside the ranges of the visible screen will be drawn. When clipping is 'off', the parts of the curves that should be outside the visible screen are also drawn on the screen. This effect is called 'wraparound'.
c.   Grid: Grid can be 'on', 'off' or 'on/off'.
      Grid is 'on' or 'off' if the grid of both axes is 'on' or 'off' respectively. If the grids of the x-axis and y-axis do not have the same value, the grid is 'on/off'.
d.   Title: The title of the graphic can be changed (maximum length = 78 characters).

Zoom

This menu item allows you to zoom in on the picture shown on the screen. By pressing <F5>, a rectangle appears that represents the outlines of the new picture. All details inside the rectangle will be shown in the new picture.

The position of the rectangle can be chosen using the cursor keys. The size of the rectangle can be changed by pressing <SHIFT> and a cursor key together. Press <ENTER> to have the new picture drawn.

If <ESC> is pressed, the rectangle disappears and the previous menu is shown. Press <F10> for returning to the main menu without affecting the picture shown on the screen. <F6> and <F7> are keys for restoring the previous and original picture respectively.

Print

This menu item allows you to make a hard copy using a printer that has graphic capabilities. The DOS program GRAPHICS.COM must be installed before running UNIGRAPH. The printer must be switched on and it must contain enough paper.

Press <F3> to create a vertical print "portrait". Compared with the screen this print is rotated counter-clockwise 90 degrees on the printout page. The screen's upper right corner appears on the paper's upper left corner.

Press <F4> to create a horizontal print "landscape". The horizontal print is smaller than the vertical print.

Save

This menu item allows you to save the screen. Press <F8> to save the screen. It is stored on disk in a file with the same name as the *.DAT file but with the extension *.PIC. Only students can save a screen and only teachers can read and draw a saved screen.

Note that for each *.DAT file only one screen can be stored.

New Curve

This menu item allows you to select a new curve. After pressing <F9> the user is asked to enter the name of a (new) *.DAT file (no extension allowed). It is possible to select a file from another drive and/or (sub)-directory. The corresponding *.INP and *.AUX files are taken from that same directory. However, if these files are not found the UDEFAULT files from the current directory are used.

Escape

Throughout the program, the control key <ESC> is used to quit, to return to the preceding menu or to terminate user-input without changing the default value presented by the program.

Commandline Parameters

UNIGRAPH can operate with various command-line parameters:

1. Standard use of UNIGRAPH
    Syntax: UNIGRAPH M/C N/S Path_to_help S/T [filename]

M/C selects mono or color monitor. N/S selects network or stand alone system. The third command-line parameter is required. It sets the path to the directory containing the HELP files. This path may be different from the path in the filename. The last command-line parameter represents the filename of the *.DAT file to be used. Only the filename must be entered and not the extension .DAT. The filename may include a path. The second command-line parameter represents the options: 's' for student or 't' for teacher.

2. Installing a new configuration file
    Syntax: UNIGRAPH i

This single command-line parameter results in a new configuration file BMTPERPC.CNF. After installing a new configuration file, UNIGRAPH continues operating interactively.

3. Running UNIGRAPH with a different graphics adapter
    Syntax: UNIGRAPH [number]

This option does not use the configuration file BMTPERPC.CNF, but uses the graphics adapter selected by the user.
[number] may have the following values:

0	Autodetect	
1	Color Graphics Adapter	(CGA)
2	Monochrome Color Graphics Adapter	(MCGA)
3	Enhanced Graphics Adapter	(EGA)
4	Enhanced Graphics Adapter 64 lines	(EGA64)
5	Monochrome Enhanced Graphics Adapter	(EGAMono)
7	Hercules	(HercMono)
8	ATT & Olivetti Graphics Adapter	(ATT400)
9	Video Graphics Array	(VGA)
10	IBM PC3270	

After the filename and the option have been entered interactively, UNIGRAPH checks if the correct graphics adapter has been selected.

Error Messages

Most of these messages are preceded by "Error:" and followed by "Press any key" or "Press ESC to quit or any key to try again". All messages are self-explanatory:

Both default files needed but not found or damaged
Curve .. is not a function
Curves .. are not functions
Data-file corrupt: errors detected
Disk full
Drive .. not ready
File Udefault.aux needed but not found or damaged
File Udefault.inp needed but not found or damaged
Illegal filename and/or path
Illegal option
Incorrect range -> Linear x-scale inserted
Incorrect range -> Linear x- and y-scale inserted
Incorrect range -> Linear y-scale inserted
Invalid directory
Less curves found (.DAT-FILE) than requested (.INP-FILE)
No help available
No printer-port detected
Not enough memory to read all curve-data
Not enough memory to zoom in
Printer not ready
Printer out of paper
Range is too small. Press F6, F7 or ESC
Unable to write to disk
Use UNIGRAPH <(path to unigraph) filename> <option>
Your input exceeds maximum value
Your input exceeds minimum value
Your input is not an integer
Your input is not a value
Your screen has not been saved

28.1.4.3 DOS Shell

This menu item gives access to the operating system without having to leave MENU. It loads a second command processor with which all standard DOS operations can be performed. This especially facilitates the copying of files into the subdirectory chosen or checking free disk space.

The cursor is arrow-shaped and contains the current directory.

For viewing the contents of subdirectories, refer to section 28.1.4.5.

To leave, type EXIT. The menu program automatically returns to the previously current directory.

28.1.4.4 BIOPSI

This menu item allows you to use the program BIOPSI interactively. For details refer to the Chapter 30. Preferably, the model files created or used should be placed on a diskette before introducing them in a subdirectory.

28.1.4.5 Files

This menu item puts the system in "view-mode", as indicated by the status bar. You can travel freely through the menu with mouse or cursor keys and view the subdirectory contents by pressing <F3> at any level chosen. Leave this mode with <ALT><F3>.
The directorywindow is presented automatically in case the AUTOnnnn.PSI file specified during configuration is not found by the program during an attempt to execute it.

28.2 Including Your Own Programs

The menu program comes equipped with many simulation programs. The items and exercises still open can be filled with programs supplied by you:

-    Choose Utilities\Files from the teacher's menu
-    Navigate through the menu and select an exercise
-    Press <F3> to enter the files screen
-    Inspect the files present in order to prevent duplicates
-    You may delete existing files with <CTRL><D> after selection or copy these to diskette with <CTRL><F>
-    Insert the diskette containing your programs in drive A:
-    Press <CTRL><A> to present the directory contents of the diskette
-    Select the file(s) to be copied and copy these with <CTRL><F> to the subdirectory chosen (no name change allowed)
-    End the session with <ESC>
-    Configure the menu item and exercise according to section 28.1.4.1. Do not forget to include possible command line items to be used by your program
-    Produce the corresponding help files with the Help editor according to section 28.1.3.1.

28.3 Designing Your Own Menu

The menu structure is contained in three files: MENUDATn.MEN with n = 1..3, residing in drive:<parent>\BMT. The menu pages 1 and 2 (page 3, the teacher menu, is fixed) may be altered at will by the configuration procedure as described in section 1.4.1. You can also start afresh by deleting or (preferably) renaming these structure files. In that case, when you start the menu program again, the menu will be filled with default names containing the numbers explained in section 28.1.3.1. Then you can set up your own menu. Take care in deleting existing program files and keep the subdirectory structure intact!

Appendix

1. Editor Commands

The commands for the "WYSIWYG" text editors are the well-known WordStar™ and Turbo PascalTM editor commands:

^ = <CTRL>   <RET> = new line   ^K^S = save file   ^K^Q = quit without saving

Character Left:	^S	or	LtArrow	Character Right:	^D	or	RtArrow
Word Left:	^A	or	^LtArrow	Word Right:	^F	or	^RtArrow
New Line:	^M			Insert Line:	^N		
Line Up:	^E	or	UpArrow	Line Down:	^X	or	DnArrow
Scroll Up:	^W			Scroll Down:	^Z		
Page Up:	^R	or	PgUp	Page Down:	^C	or	PgDn
Beginning of File:	^Q^R	or	^PgUp	End of File:	^Q^C	or	^PgDn
Beginning of Line:	^Q^S	or	Home	End of Line:	^Q^D	or	End
Top of Screen:	^Q^E	or	^Home	Bottom of Screen:	^Q^X	or	^End
Delete Curr. Char:	^G	or	Del	Delete Char Left:	^H	or	<-
Delete Word:	^T			Delete Line:	^Y		
Delete to EOL:			^Q^Y	Tab:	^G	or	Tab
Previous Cursor Position:			^Q^P	Print File:	*	^P	
Find:			^Q^F	Find Next:	^L		
Find and Replace:			^Q^A	Restore Line:	^Q^L		
Begin Block:	^K^B	or	F7	End Block:	^K^K	or	F8
Top of Block:	^Q^B			Bottom of Block:	^Q^K		
Copy Block:	^K^C			Move Block:	^K^Y		
Delete Block:	^K^Y			Hide Block:	^K^H		
Read Block from File:			^K^R	Write Block to File:			^K^W
Mark Current Word as Block:			^K^T	Print Block:	^K^P		
Toggle Insert Mode:	^V	or	Ins	Toggle Autoindent Mode:			^O^I
Set Marker n (0..3)			^O n	Jump to Marker n (0..3)			^K n

* NB No printer check is performed !

2. File Structure for UniGraph Files:

The *.DAT File Structure

Items 1 through 7 have to be output from the program generating the data or may be entered with one of the editors. The file has the following structure:

Line #	Description	Conditions
1.	Title	Between single quotes ('), <= 70 characters
		if " , no title will be entered
2.	Label x-axis	Between single quotes, <= 20 characters
		(if longer, will be truncated) if " ,no label will be entered
3.	Label y-axis	As for label x-axis
4.	R1,R2,R3,R4	Real values, separated by commas
or	spaces	
		R1: abscissa of intersection of axes
		R2: distance between x-axis mainmarks
		R3: ordinate of intersection of axes
		R4: distance between y-axis mainmarks
		If line 4 is empty, the programme will determine "reasonable values";
5.	Mindomain,	Real values
	Maxdomain,	Same conditions as for line #4
	MinRange,	
	Maxrange	
6.	Npoints	Integer value 0<= Npoints<=1024,
		number of data points, must be on a separate line.
7.	Xi,Yi, ...	Reals, separated by commas or spaces, or EOL more
		coordinate pairs on one line are allowed.

Pairs of 6 and 7 may be repeated at will.

The *.INP File Structure

The *.INP file contains only integers. In all cases, if a line is empty or contains a non-integer, a default-value from DEFAULT.INP is substituted.

Line #	Description	Conditions
1.	Nscreens	= 1 or 2, number of screens
2.	Ncurves	integer, Ncurves >= 0, number of curves per screen
3.	Symbol	integer, 0<=Symbol<=8, symbol identifying curve
4.	Scolor	integer, 1<=Scolor<=6, symbol color
5.	Linestyle	integer, 1<=Linestyle<=5,
		1: no line connecting datapoints
		2: drawn line
		3: dotted line
		4: short stripes
		5: stripes
6.	Linecolor	integer, 1<=Linecolor<=6.

The *.AUX File Structure

This file only contains integers; if a line is left empty, the corresponding value from UDEFAULT.AUX is taken.

Line #   Description   Conditions
1.   Nxmarks   -1<=Nxmarks<=9, number of X axis marks between main values:
                               -1 :   logarithmic
                               0..9:   linear

2.   Nymarks   same for Y axis
3.   Nxdecs   -1<=Nxdecs<=3, number of decimals of values at the X axis mainmarks:
no value
4.   Nydecs   same for Y axis
5.   Labelx   0<=Labelx<=2, label position for X axis:
                               0:          no label
                               1:          label perpendicular to axis
                               2:          label parallel to axis
6.   Labely   same for Y axis
7.   Scale   0<=Scale<=2, appearance of axis scales:
                               0:          no scale divisions
                               1:          standard divisions
                               2:          box
8.   Axes   0:   absent
                               1:          present
9.   Gridx   grid lines perpendicular to X axis
                               0:          absent
                               1:          present
10.   Gridy   same for Y axis
11.   Interpol   interpolation-flag:
                               0:          no interpolation
                               1:          interpolation
12.   Wrapping   clipping-flag
                               0:          wrap-around
                               1:          clipping.

Registered trademarks:

Turbo Pascal Borland International, Inc.
IBM   International Business Machines, Inc.
WordStar   MicroPro International Corporation.

29
The Student Database

Rogier P. van Wijk van Brievingh and Patrick Min

29.1 Introduction

This section of the manual describes the use of TEACHER.EXE, the student database management program. As its name suggests, the program is intended for teacher's use.

Some of its features are:

- Normal database management with options like "Append", "Edit", "Search", "Browse" and "Pack"
- Moving data on the exercises a student is required to make to a "studentdisk" -a personalized disk containing name, password and exercise status-. The reverse is also possible, after which a student's exercise status is automatically updated.
- Generation of reports on one student, all students, all students doing a certain course or all students of a certain group.

29.2 Screen Layout

Before actually starting to work with the program, you should be familiar with the different parts of the screen and their functions, several keys and their functions, and the fields of a student record. Sections 29.2, 29.3 and 29.5 can be used as a quick reference. Some of the terms that may seem unfamiliar will be explained:

The Status Line
The status line, at the top line of the screen, displays the current mode, the current database in use and the number of records in this database.
The Menu Window
The menu window, right below the status line, displays a list of possible options. One of the options is inverted. Use the cursor left and right keys to move the inverted bar. By pressing <ENTER>, the option that is shown inverted is activated.

The Main Window
The main window, the large space at the center of the screen, is used for displaying directories and reports, and for displaying and editing student records.

The Status Window
The status window, at the bottom of the screen, is used for program messages and short communications with you, such as questions or line input. The first line of the status window informs you of what is happening or what has happened; the second shows possible errors. Questions and error messages are preceded by a short beep.

The Key Bar
The key bar, at the bottom line of the screen, displays which keys you can press and the functions they perform. The arrow symbols indicate the cursor keys. The only key that is never displayed on the key bar is the "F9, Toggle Password" key (see next section).

29.3 Keys

Standard keys throughout the program are:

<F1>	-	Help
<ESC>	-	End, Exit, No, or any equivalent meaning

Standard editing keys for all string input are:

Cursor Keys	-	move one character left or right, or one field up or down (when editing a record)
<BACKSPACE>	-	deletes the character to the left of the cursor
<HOME>	-	moves one word left
<END>	-	moves one word right
<ENTER>	-	the same as <Cursor DOWN> when editing a record, or the standard "accept" when editing a line (e.g. when the name for a new database is entered)
<CTRL><Y>	-	delete line
<F9>	-	toggle password (see section 26.5)
<ESC>	-	end edit

Browse keys:

<PgUp>/<PgDn>	-	move to previous or next student
<F9>	-	toggle password
<ESC>	-	exit "Browse"

Report viewing keys:

<Cursor UP>/		
<Cursor DOWN>	-	move one line up/down
<PgUp>/<PgDn>	-	move ten lines up/down
<HOME>/<END>	-	move to start/end of report
<ESC>	-	exit report viewing

29.4   Startup and Database Selection

29.4.1 Startup

The program can be started from the command line (type TEACHER and <ENTER>) as well as from the teacher page of the BMTperPC program. When started, the program first reads the BMTperPC menu structure from the current directory. If a command line parameter is passed to Teacher, that parameter should contain the parent directory (e.g. TEACHER C:\BMT), which contains the files MENUDATn.MEN.

Next, the program reads the A: directory in search of database files. Most often when a disk error occurs (anywhere in the teacher program), the user is asked to try again or abort the last operation. This allows for the correction of small errors, such as leaving the disk drive door open.

Four files are kept for each database:

-    a DataBase Main file, extension DBM
-    a DataBase Backup file, extension DBB
-    an IndeX Main file, extension IXM
-    an IndeX Backup file, extension IXB

After the directory has been displayed, an existing database can be selected or a new one can be created. The top line of the directory is accompanied by a message "(NB: More than one screen)" if there are more than 77 database files found (very unlikely). The displayed filenames then scroll along with the inverted bar.

If no database files were found, the "Create a Database" option is entered automatically.

Make sure that each of your database disks contains the (hidden) file TEACHER.FIL (containing the encrypted teacher password). The teacher program uses this file for identifying a disk as a database disk!

29.4.2   Selecting a Database

Move the inverted bar over the name of the database you want to use and press <ENTER> to open the corresponding database and index files.

When the main database or index file cannot be read because of, for example, a corrupted sector, the program tries to select the backup. These backups are updated every time you exit a database. If a database file has been successfully opened but both index files were not, the index is rebuilt.

After a database has been successfully selected, its name is displayed on the status line (at the top line of the screen) and the main menu appears. The status line also displays the number of records in the database.

29.4.3   Creating a Database

It is not possible to create a database on a non-teacher disk. This means the file "TEACHER.FIL" should be on the diskette. (Refer to section 28.1.1.)

If there are fewer than 128 kBytes free (about the size of a completely filled database plus index) on the disk in drive A:, it is not possible to create a new database. Between 256 and 128 kB you are given a warning. Whether or not you are going to take this seriously depends on whether you are likely to fill up your database files (one student takes about 0.5 kB, with a maximum of 128 students per database).

If there is sufficient room for a new database, you are asked to enter a name. An existing database with the same name will only be overwritten after your confirmation. The database just created is selected as the current one.

29.4.4   Exiting a Database

After pressing <ESC> in the main menu ("Edit Database / Requirements / Survey"), and confirming your exit, the current database will be closed. When one of the options in the set ["Append", "Edit", "Pack", "Move To", "Move From"] was selected for that database, the program will attempt to backup the database and index files.

29.5   Student Record Fields

A student record contains several special fields. Before you start editing and appending, it is helpful to read over the following list.

surname   -   Since the database is indexed on surname, you should standardi-
                  zethe way names are entered. In practice this means that, for in-
                  stance, names like "DE VRIES" (all record input is capitalized)

should be entered as "VRIES, DE" to make it appear with "V" instead of "D"in an alphabetical list. Also for standardization purposes, leading spaces in a surname will be removed.

initials   -   used to distinguish different students with the same surname (internally in the teacher program this is done by assigning a unique code to each student).

course   -   in the "Survey" menu, reports can be generated on all students doing a certain course, so with this field it is important that "same course" means "same course field". Thus if you want a report on all students doing physics, there should not be physics students in your database with "PH" as well as with "PHYSICS" in the course field. If there are, and you generate a report on all students doing "PHYSICS", the students doing physics with "PH" in their course field will be skipped. In short, standardization is required again.

department - present for your convenience. May be left blank.

group   -   the same as for the course field; reports on all students of a certain group are also possible.

course number,
year   -   convenience fields.

password - student disks are not only identified by a student's surname, but also by a student password. The password is checked when a student wants to do exercises in the BMTperPC program in "credit mode" and when his/her exercise status is updated after some (or all) of the exercises have been made (this is done with the "Move From" option of the Requirements mode). A password should be one word containing only characters in the range A-Z.

For reasons of privacy there is a "Toggle Password" key, <F9>, which toggles the display of a student's password. When this is not displayed, the field name "Password" is invisible, too. The <F9> key is active during editing and browsing only, and it is the only key never displayed on the key bar.

percentage - this field is updated after each "Move From" operation (in the Requirements mode), and reset after each "Move To" operation (of the same mode). It cannot be edited, and displays the percentage of exercises a student has done to date.

mark   -   can be set to any value between 0.0 and 10.0. Non-numbers are changed to 0.0 and values larger than 10.0 are set to 10.0. When this field is not equal to 0.0 and the percentage field is larger than zero, it will be included in reports made in the "Survey" mode

deleted   -   gives an easy way to delete student records: by setting this field to "Y", the whole record is marked to be deleted. At that time it is not yet deleted, and can still be edited ("Y" can be changed to "N" again, in effect undeleting the record).

After choosing the option "Pack" (of the "Edit Database" mode), all records marked "Y" will effectively be deleted.

29.6   Edit Database

29.6.1   Append

An empty record form is presented, which can be filled in. The fields "surname", "course", "group" and "password" are considered to be essential. If you press <ESC> (= End Edit) and one or more of those fields are still empty, you are given the opportunity to fill them in or to abort the edit (in which case the record will not be appended).

The maximum number of students in one database is 128. When this number has been reached, no more students can be appended.

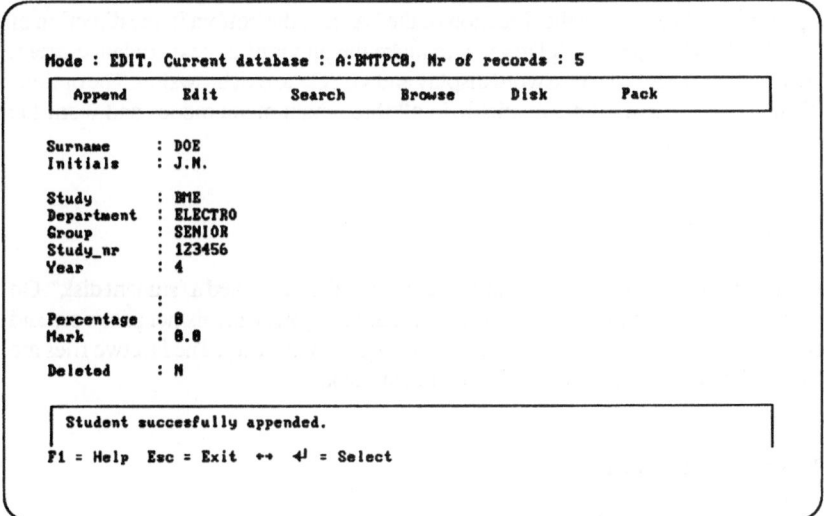

Figure 29.1

29.6.2   Edit

This option allows you to edit the current student record, i.e. the one visible. Press <F9> to toggle the display of the student's password. You can only edit the password when it is visible. When you end editing by pressing <ESC> and the contents of the record have been changed, you will be asked to confirm the changes. Answering "No" to this question restores the original record. As in the "Append" option, the program will not accept records with one or more empty "essential" fields ("surname", "course", "group" and "password"). When you end editing with such fields empty, you will be asked if you want to return to editing or ignore the changes.

29.6.3 Search

Through the use of "Search" and "Browse", a student record can easily be made current. You are asked to enter the surname (or part of it) of the student you are searching for. If you press <ESC> or enter an empty string, the search is aborted. In all other cases the student with a surname equal to, or the first one larger than, the input string is located and made current. The search fails if no such student is found.

29.6.4 Browse

With "Browse" you can move up or down in the database in alphabetical order. The top of the database lies in the direction of the "A" and the bottom in the direction of "Z".Use the <PgUp> or <PgDn> key to make the previous or next student current. Press <F9> to toggle the password display and <ESC> to exit browsing. The options "Search" and "Browse" are available in all three main menu modes and work the same in each mode.

29.6.5 Disk

A student keeps his/her exercise data on a special diskette called a "student disk". On it, three files are kept: one with the student's surname, one with his/her password and one with his/her exercise-status (called the requirements file). The last two files are hidden. There are four operations for student disks.

29.6.5.1 Test Student disk

First you are asked to insert a student disk. After that:

- the surname file is read and the surname found is displayed
- the same occurs for the password file, plus a test to check if it is hidden or not
- the requirements file is read and tested to check if it is hidden

Then you must re-insert the teacher disk. When the surname and password file are successfully read, the corresponding student record is searched for and made current in the current database. The password found on the student disk and the one found in the database are compared. If there are more of the same surnames in the database, they are all tested for a password match.

29.6.5.2 Correct Surname

Creates or overwrites the surname file for the current student on a student disk, a useful feature when the original is corrupted. You are asked to confirm this, so make sure you insert the right student disk before answering "Yes".

29.6.5.3 Correct Password

Password files are treated like surname files, as described above.

29.6.5.4 Create Student disk

Performs both actions, i.e. writes the surname file as well as the corresponding password file. With the option "Move To" from the "Requirements" mode, a requirements file can be added.

29.6.6 "Pack"

This option deletes, after your confirmation, all student records with a "Y" in the "Delete" field.
Thus before you do a "Pack" you can undelete any student record by changing the "Y" in the "Delete" field back into an "N".

29.7 Requirements

29.7.1 Move To

After this option is selected, you are asked to insert a student disk. Its data is read, and after you have re-inserted the teacher disk, the current database is searched for the corresponding student record. When this has been found, the most recently saved requirements (in the BMTperPC program) are moved from the parent directory to the database as well as to the student disk. The percentage field of this student is then reset to 0%.
If a student's surname is found in the database (one or more times) but there is no password match, the password found on the student disk is displayed. This is to check if either:

-    the wrong database was selected
-    you have changed the student's password after the student disk was created,or
-    the student has changed the password file

In the last two cases you can use the "Correct Password" sub-option of the "Disk" option (in the "Edit Database" mode) to correct the fault.
If the student disk is damaged in any way, you should also use the "Disk" option to repair it.

29.7.2 Move From

The disk in drive A: is identified in the same way as the option "Move To".
After successful identification, the requirements of the current student in the database are updated and the remaining requirements (the exercises that still have to be made) are made the current ones in the BMTperPC program (i.e. copied to the parent directory). This is so you can add or remove exercises to those requirements and move them to the student disk again. Note that after each "Move To", the percentage field of a student record is reset to 0%.

29.8 Survey

Several different reports can be generated.

29.8.1 Report on One

A detailed report is presented of all data on the current student, together with a list of the exercises that still have to be made.

29.8.2 Report on Group

Three sub-options may be used:

-    "All students": a list of all student names, the year, the number out of the required number they have made, their percentage score and mark
-    "By course": after you have entered a course, the same is done for all students doing that course.-
     "By group": the same for groups.

Please note that in "Report on Group", "group" means "a collection of students", and in "By group", "group" means the field with the same name in a student record.
The percentage and mark average and SD (Standard Deviation) are shown below the list. A mark is included in the survey if it is not equal to 0.0 and the percentage for that student is bigger than 0. When there are marks included in the survey, a histogram of the marks is printed below the student list (all marks rounded to the nearest integer).

29.8.3 Report Viewing and Printing

You can view the whole report with the aid of the <Cursor UP> and <Cursor DOWN> keys (move up/down one line), the <PgUp> and <PgDn> keys (move up/down ten lines), and the <HOME> and <END> key (move to start or end of report).

After pressing <ESC>, you are asked if you want to print the report. If you do, the top three lines are printed first, then the list, and then the bottom three lines (as they appear in the main window). While printing, you can press <ESC> anytime to pause and then decide whether to continue or abort printing. Each unaborted report print is ended with a form feed.

Appendix

1. Error Messages

Apart from all standard Dos error messages (which are generated in the Teacher program when they occur) there are several specific error messages. Some speak for themselves, some require further explanation. Below you will find a reference list.

Database full. Max no. of students is ...
You are trying to put more students in a database than is allowed.
Disk in drive A: is non-teacher disk
reported by "Create a Database". It is not possible to create database files on a non-teacher disk!
Disk read failed
Reported by "Test Student disk" when the surname, password or requirements file could not be read.
Error: the password file should be hidden, but it isn't
Reported by "Test Student disk".
Error: the requirements file should be hidden, but it isn't
Reported by "Test Student disk".
Error occurred while packing
"Pack" failed due to Dos error.
Failed, Retry (R) or Abort (A)?
Reported after almost every failed disk action. Press "R" to try again, "A" to abort.
Index lost
Reported when both the main and the backup index file could not be opened. Rebuilding will start automatically.
Index rebuild failed
Rebuild failed due to Dos error.
Insufficient disk space
Reported by "Create Database" when there is not enough disk space for a new database.
Insufficient disk space for more students
Unlikely to occur. Probably because you created too many small databases on one disk and filled them all up afterwards.
Matching surname(s) but no matching password found
Password found on student disk is ...
Reported by "Move To" and "Move From". You have either selected the wrong database, or you have changed the student's password after the student disk was created, or the student has been tampering with the file. Use the Disk option (of the "Edit Database" mode) to correct faults.
Password mismatch: the one found in the database is ...
Reported by "Test Student disk". See also the error "Matching surname(s) but no matching password found".
Password move failed
Reported by "Correct Password".
Requirements move failed
Reported by "Move To".
Requirements update failed
Reported by "Move From".
Search failed
There is no student in the current database with a surname that is equal to or larger than the one you have entered.
Severe error: Teacher disk not identified
Reported by "Move To" and "Move From". First try again to find the right teacher disk, and if you cannot, exit the program and look for the correct disk.
Student data incomplete
One or more of the "essential" fields (surname, course, group and password) is missing. Reported after ending "Append" or "Edit".
Student data read failed
Either the surname file, the password file or the requirements file could not be read. Check the disk with

the "Test Student disk" sub-option of the "Disk" option ("Edit Database" mode).
Student data update failed
Reported after "Edit" failed to update changes you made in a student record (due to Dos error).
Student disk creation failed
Reported by "Create Student disk".
Surname move failed
Reported by "Correct Surname".
System error: Check Search Proc. in Req Unit
Should never occur : warn System Supervisor or Author.
System error: Check Update Proc. in Dbase Unit
Should never occur: warn Systrem Supervisor or Author..
System error: Check store Proc. in Screen Unit
Should never occur: warn System Supervisor or Author..
System error: Report store capacity inadequate
Should never occur: warn System Supervisor or Author..
System error: Too many Restore calls
Should never occur: warn System Supervisor or Author.
This data is not available:
The BMTperPC menu structure could not be read
Reported by "Report on One", when the menu structure read at the startup of the program failed.
This database is empty
Reported by the "Requirements" and "Survey" mode. You cannot manage the exercise data of empty databases or generate reports them.
This student was not found in the current database
Reported by "Move To", "Move From" or "Test Student disk": the student whose disk is in the A: drive is not present in the current database.
Unable to open ...
Reported after selecting a database when either both the database files or both the index files could not be opened. In the first case the database is lost altogether, in the second the index is rebuilt.

2. *Menu Structure*

[Select a Database]
[Create a Database]

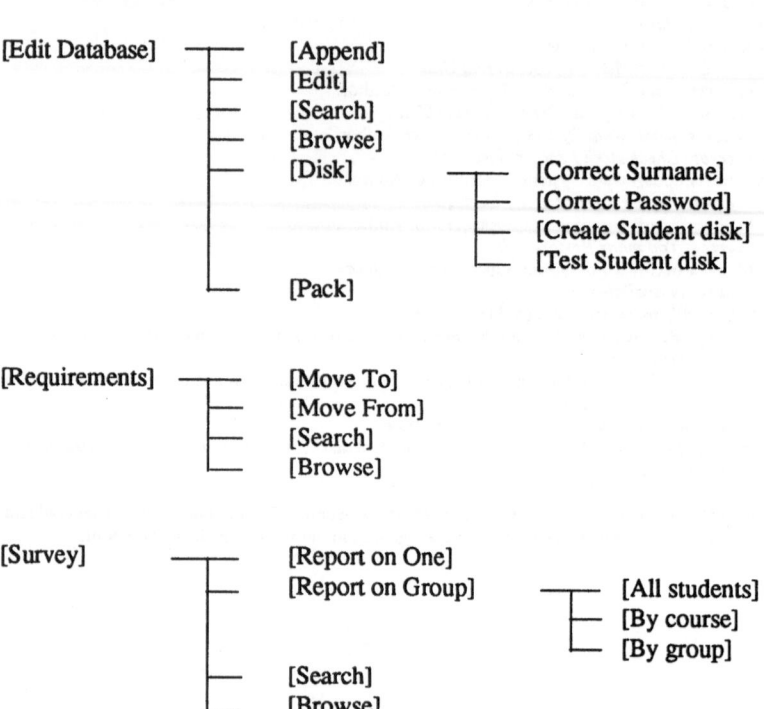

[Edit Database] ──┬── [Append]
                  ├── [Edit]
                  ├── [Search]
                  ├── [Browse]
                  ├── [Disk] ──────┬── [Correct Surname]
                  │   ├── [Correct Password]
                  │   ├── [Create Student disk]
                  │   └── [Test Student disk]
                  └── [Pack]

[Requirements] ──┬── [Move To]
                 ├── [Move From]
                 ├── [Search]
                 └── [Browse]

[Survey] ──┬── [Report on One]
           ├── [Report on Group] ──┬── [All students]
           │   ├── [By course]
           │   └── [By group]
           ├── [Search]
           └── [Browse]

30
Interactive Simulation Program BIOPSI

Paul P.J. van den Bosch

Abstract

BIOPSI is an interactive simulation program for studying the behavior of dynamical systems. An interesting application area concerns the analysis of biomedical and physiological phenomena described by means of non-linear differential equations. This chapter introduces the use of BIOPSI with the block diagram as a way of describing dynamical models; the most-used blocks and commands are defined and illustrated. Moreover, it focusses on some of BIOPS's dedicated facilities.

30.1   Introduction

BIOPSI, as used for illustrative purposes in this book, has been derived from PSI. PSI is an interactive block-oriented simulation program for studying the behavior of dynamic continuous and discrete systems. This program was developed at the Laboratory for Control Engineering of the Delft University of Technology in order to have a portable, computer-independent simulation program.

This simulation program is used in thousands of installations in The Netherlands and abroad. Owing to its highly interactive communication with the user, clear graphical presentations and powerful set of simulation elements and commands, applications of PSI can be found in many different areas such as:

-   Modeling and identification of physical, chemical and biomedical systems,
-   Analysis and design of control systems,
-   Analysis of econometric models.

The block diagram is used to describe the topological structure of the simulation model of the system to be simulated. About 60 different block types are available, both dynamic continuous and discrete linear and non-linear types as well as some

logic, Boolean block types. Therefore, PSI can be used to solve sets of differential, difference, logical or algebraic linear and non-linear equations, or any mixture of these. Each block has a maximum of three inputs and three parameters.
Some facilities offered by PSI are:

- A command language to realize the interaction between user and program. About 100 different commands are available.
- About 60 powerful block types are available, including which several types of integrators (limited, resettable, mode-controlled), delays, function generators (one or two inputs with linear or quadratic interpolation), controllers, pulse-width modulation, etc.
- Memories to store signals during a simulation run. The signals stored in these memories can be studied after the simulation run, used as inputs in future runs or saved on disk.
- Symbolic block names.
- Facilities to solve algebraic loops iteratively.
- Five numerical integration algorithms. Four algorithms have a fixed step (Euler, Adams-Bashfort 2, Runge Kutta 2, Runge Kutta 4) and one algorithm has a variable integration step (Runge Kutta 4).
- Optimization facilities. Up to eight parameters of the simulation model can be arbitrarily defined to minimize any user-defined cost function. Moreover, any parameter can be optimized by using constraints and can be scaled to improve convergence. This facility allows the design of optimal controllers for linear or non-linear systems. Optimization can also be used for identification by minimizing the difference between a measured response and the response of an adjustable model.
- Initial, dynamic and terminal facilities.
- Many multi-run facilities.
- Output that can be directed to a printer or graphical display.
- Format-free input.
- Extensive tests on the validity of all user-supplied information, which can result in about 75 different error messages.
- External data files can be stored into function generators in order to realize parameter estimation.

PSI is written in FORTRAN 77 with a limited number of system dependent program parts. These parts concern the interfaces with

- Operating system (error messages of operating system, etc.)
- File I/O system (keyboard interrupts, disk I/O, etc.) and
- Graphics system.

These parts have been isolated in PSI and must be written for each new computer system. Consequently, it is quite easy to adapt PSI for different computer systems, both hardware and software. Moreover, provisions are made inside PSI to allow all alphanumerical and graphical output to be directed to one screen or two different screens. In practice this implies that PSI can be run on nearly any computer system.

PSI is available as a ready-to-run program for a number of personal computers, namely the IBM-PC/XT/AT and their compatibles. Moreover, a number of MS.DOS computers are supported with their own, non-standard, proprietary graphics interfaces.

If you require even more facilities than offered by these ready-to-run programs of PSI, you can use either all object-versions or the full Fortran sources of PSI. These offer you the following additional facilities:

1. In PSI a number of empty block types are present, namely the XXi-blocks. In a separate file these blocks are represented by means of an empty subroutine:

```
SUBROUTINE XX1 (in1,in2,in3,par1,par2,par3,out,s1,is1)
RETURN
END
```

You can easily add a user-defined function into a subroutine in order to define your own block type. As with all block types in PSI, this function can have up to three inputs (in1,in2,in3), up to three parameters (par1,par2,par3) and only one output. Moreover, each block can have two state variables, namely one integer and one real variable. Such a block type can be used many times in your simulation model. If you have added the required function as Fortran statements in XXi, you have to compile these subroutines and then link PSI, which yields a new, user-defined PSI.

2. You can easily add your own screen layout for presenting the calculated responses. For example, if you are studying ships it is quite convenient to show the horizon, several indicators, the shape of the ship, etc. on the screen as a moving picture instead of using just responses.

This facility is offered to you by means of a software interface between PSI and your own Fortran programs. It transfers the numerical values calculated in your simulation model, via the DRW-blocks, to your Fortran program. This user-written Fortran program can draw the required pictures on the screen. In drawing these pictures, you can utilize the graphics interface of PSI.

3. There is a program called PSICOM for compiling your PSI models into a Fortran program. If this program is linked together with other PSI programs your model will run very rapidly. On an average, calculation speed is reduced by two or three times. All facilities of PSI still exist and can be used. Only the B-command to modify the structure of a model is no longer effective.

4. This previous facility allows you to incorporate FORTRAN statements into your model description. Consequently, portions of a simulation model that are

difficult to model with PSI can be programmed with FORTRAN statements. This facility can be useful to represent, for example, coordinate transformations in using robots, representing vectors and matrices, to implement parameter estimators, etc. In doing so, you have full access to all variables (outputs and parameters) of the other blocks.

These object versions are only available for a limited number of operating systems and compilers, e.g. MS.DOS/PC.DOS with MS-Fortran and VAX/VMS with the DEC-Fortran 77 compiler.
If you use the full Fortran sources, you have direct access to all PSI-codes. You can tailor PSI to meet all your requirements. For example you can very easily change the following maxima:

- the number of blocks available
- the number and size of function generators
- the number and size of the responses to be shown on the screen
- the number of integrators
- the number and size of the time delay blocks

The modifications consist of modifying just one number in only one file, to compiling all programs and linking them, which yields PSI. This goal can be achieved in less then one hour.
You can also modify the names and/or functions of existing block types or increase the number of different block types. Then, a new programming effort is necessary. For all previous operations, documentation has been supplied.

If you want to replace an existing numerical integration method by a new, and better one, you can easily replace the appropriate Fortran statements in the corresponding Fortran file. If you want to replace the optimization algorithm, you have to reprogram the corresponding Fortran file. In fact, you can modify nearly any function or command in PSI. However, various modifications will require considerable effort and programming skills.

30.2 Using BIOPSI as a Simulation Tool

In systems engineering, a simulation program is used to study the behavior of dynamic, continuous systems. These kinds of systems are assumed to be both continuous and parallel. With an analog computer both characteristics offer no additional problems. However, in using a digital computer severe problems may arise: first the continuous integrator has to be calculated with a discrete algorithm and then the parallel structure of the system has to be solved with a digital computer that calculates all elements of the simulation sequentially. These problems can be solved, at the expense of calculation time, with numerical integration methods and by using a proper sorting procedure.

All calculations in PSI are executed with single-precision floating-point numbers. This means an accuracy of about seven digits and a dynamic range of all numbers between 10^{36} and 10^{-36}. So in general no scaling problems can be expected.

30.2.1 BIOPSI Block Schemes

To be able to solve (non-)linear high-order differential equations, these equations have to be reduced to a set of first-order differential equations. Such first-order equations can easily be solved by using an integrator.

$$\underline{\ddot{Y}}(t) + \dot{y}(t) + y(t) = u(t)$$

Suppose:

$$\ddot{y}(t) = u(t) - y(t) - \dot{y}(t)$$

$$\dot{y}(t) = \dot{y}(0) + \int_0^t \ddot{y}(\tau)d\tau$$

$$y(t) = y(0) + \int_0^t \dot{y}(\tau)d\tau$$

This second-order differential equation can be reformulated as two first-order integral equations and one definition equation:

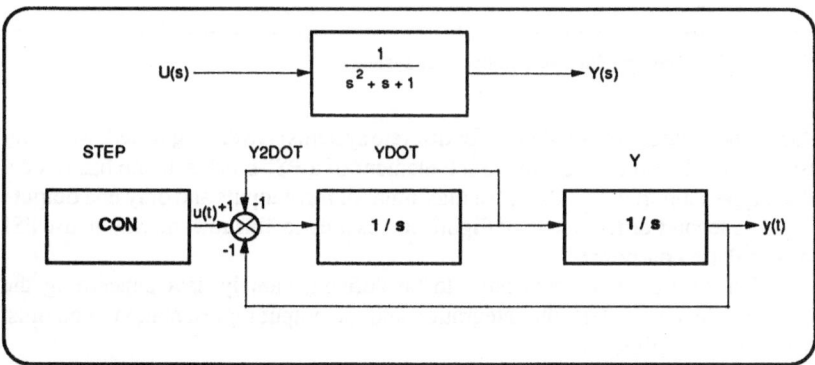

Figure 30.1. A second-order system solved with two integrators

A more general approach for decomposing high-order transfer functions or differential equations to a PSI-block scheme will be given.
Suppose we have a transfer function H(s)

$$H(s) = \frac{A_n s^n + ... + A_1 s + A_0}{s^n + B_{n-1} s^{n-1} + ... + B_0}$$

or, equivalently:

$$y^{(n)}(t) + ... + B_1 \dot{y}(t) + B_0 y(t) = A_n u^{(n)}(t) + ... + A_1 \dot{u}(t) + A_0 u(t)$$

This high-order system can be decomposed by introducing the state variable x(t), as illustrated in Figure 30.2.

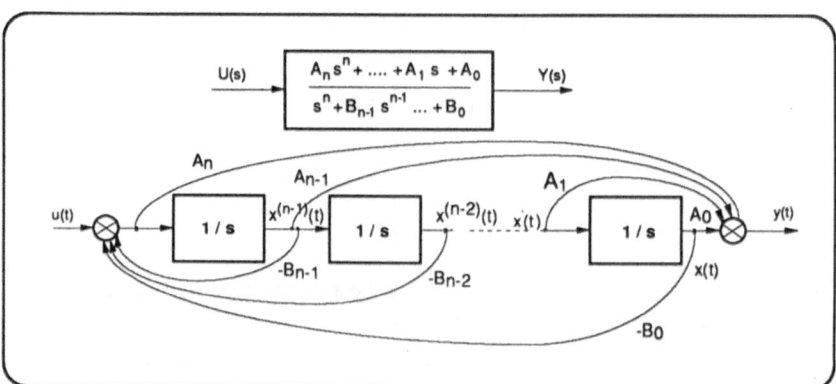

Figure 30.2. Decomposition of a high-order system

Once single integrators (or delays for discrete systems) are distinguished, we easily can derive a PSI-block scheme. Each element of a PSI-block scheme has to be a block type supported by PSI with a maximum of three inputs and only one output.

The second-order system of Figure 30.1 is used to illustrate the way to use PSI for simulation purposes.

First four types of data have to be defined, namely data concerning the structure, the parameters, the integration and the output representation. You must type in the underlined text.

30.2.2 Structure

PSI is a block-oriented simulation program. Therefore, the structure is given by defining the inputs of each block. When the inputs of all blocks have been defined, the structure of the simulation model is known and PSI is able to calculate the required responses. Each block is indicated by a name, which determines both the block and its output, its block type and its inputs. In this example:

$$\text{STEP} \quad = u(t)$$

$$Y \quad = y(t)$$

$$\text{YDOT} \quad = \dot{y}(t)$$

$$\text{Y2DOT} \quad = \ddot{y}(t)$$

With the B command, the block structure can be defined:

```
PSI* B
      Configuration Specification
Block, Type, Input1, Input2, Input3
      B*STEP.CON                        : STEP is a CON block
      B*YDOT.INT.Y2DOT                  : YDOT is an integrator with Y2DOT as input
      B*Y.INT.YDOT                      : Y is an integrator with YDOT as input
      B*Y2DOT.SUM.STEP.Y.YDOT           : Y2DOT is a summer (SUM) to add all inputs
      B*
```

Remark:
In PSI an integrator can have up to three weighted inputs. So to increase calculation speed, the blocks Y2DOT and YDOT can be combined into one integrator. This reduction is not implemented in this example.

30.2.3 Parameters

The parameters of each block can be defined with the P command:

```
PSI* P

      Parameters
Block, Par1, Par2, Par3
      P*STEP.1                          : STEP gets value 1
      P*Y.0.1                           : initial condition=0; input gain=1
      P*YDOT.0.1                        : initial condition=0; input gain=1
      P*Y2DOT.1.-1.-1                   : gains of the corresponding inputs
      P*
```

30.2.4 Timing Data

Three variables define the integration, namely the integration method, the integration interval and the total simulation time. The integration method is defined by using the DI command. The default method is Runge Kutta 2. The other two variables are defined by the T command:

PSI* T

    Integration interval=? 0.1
    Total time=?   10.

With these answers a simulation will be calculated for 10 time units with an integration interval of 0.1 time unit. This time unit is arbitrary and may be a second, an hour or a day, etc.

30.2.5 Output Representation

All blocks are calculated in a simulation, but only a few can be shown on the display. Suppose we want to study the value of the variable y(t). Then we can request by means of the O command that y(t) will be shown on the display screen :

PSI* O

Names of blocks to be shown=? Y

Now all data is defined and PSI is able to calculate the required responses. Moreover, Y = y(t) will be shown on the display. The simulation can be started by the R command.

PSI* R

The response of Y = y(t) will now appear on the screen.

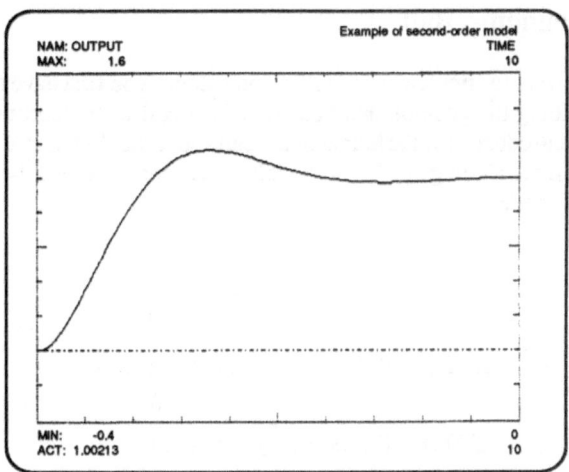

Figure 30.3. Response of y(t) on the screen

PSI allows the construction of more complicated pictures on the screen, as illustrated in Figure 30.4.

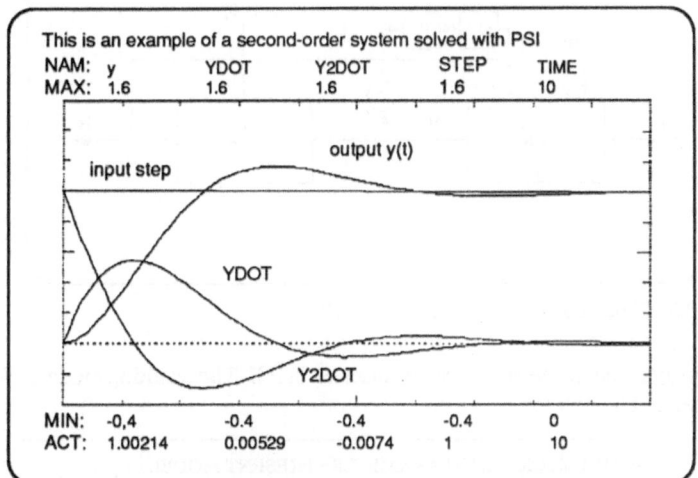

Figure 30.4. Layout of display for four responses

30.3 Illustrative Examples

Three examples will show how to use PSI for solving real problems.

30.3.1   Bouncing Ball

Suppose you want to simulate a ball falling on a floor. You first have to derive the model, describing the position and velocity of the ball with respect to the floor. Suppose you introduce H as the height of the ball above the floor and V as the speed of the ball (positive if going up). Then, the following equations describe a simplified mathematical model:

$$H(t) = H(0) + \int_0^t V(\tau)\, d\tau$$

$$\text{if } H(t) > 0 \text{ then } V(t) = V(0) + \int_0^t g\, d\tau$$

$$\text{if } H(t) = 0 \text{ then } V(t^+) = a\, V(t^-)$$

with "a" as a constant representing the friction in the ball owing to the bouncing process and g the gravity constant.
This simplified model can be inserted in PSI as illustrated in Figure 30.5.

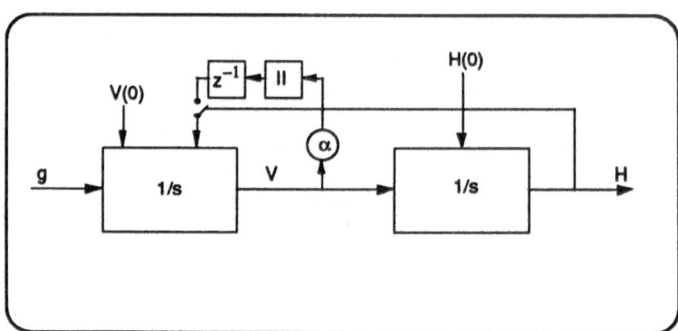

Figure 30.5. Simulation model of a bouncing ball

Based on this model, the blocks can be inserted in PSI. The resulting model in PSI is illustrated in the next text:

		- -STRUCTURE AND PARAMETERS PRESENT MODEL - -					
Block	Type	Input1	Input2	Input3	Par1	Par2	Par3
RESET	BNG	HEIGTH			.0000	1.000	.0000
ACCELAR.	CON				-9.180		
ALPHA	GAI	Z-1			-.7400		
HEIGTH	INT	VELOCITY			10.00	1.000	
VELOCITY	INC	RESET	ACCELAR	ALPHA	.0000	1.000	
Z-1	ZZI	VELOCITY			.000	1.000	

After supplying PSI with timing data, the responses of Figure 30.6 will appear on the screen.

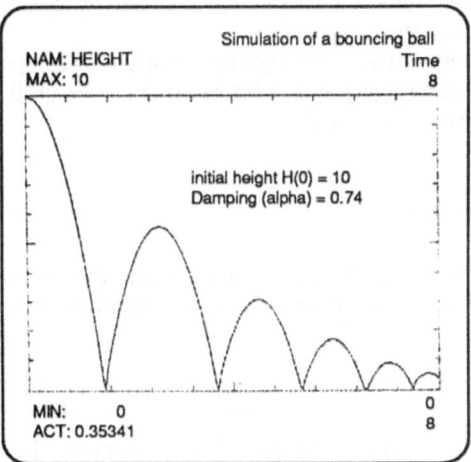

Figure 30.6. Bouncing ball

30.3.2 Parameter Estimation

Suppose you have the measurements of both input and output of a system. These measurements can be inserted in PSI and stored in function generators. Consequently, based on the approach illustrated in Figure 30.7a, the parameters of the unknown process can be estimated by minimizing the error between the output of the adjustable model and the measured output.

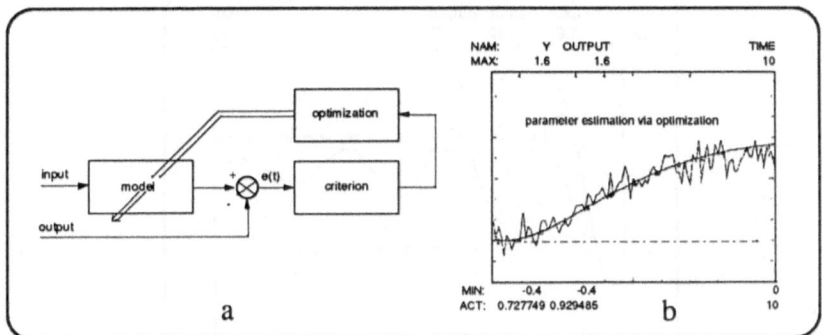

Figure 30.7. Parameter estimation via optimization

An actual example is illustrated in Figure 30.7b. Based on a number of iterations of the optimization process, PSI has adjusted the parameters of the model in such a way that the error is reduced. These parameter values can be used as estimates for the parameters of the unknown process.

This approach is very valuable, although time-consuming compared with the least-squares method, because any apriori information can be used and any model is allowed, even a non-linear model.

30.3.3 Controller Design

If you have a model of a process, then you can utilize PSI for finding an "optimal" controller for it. For that purpose you can insert in PSI a model represented in Figure 30.8:

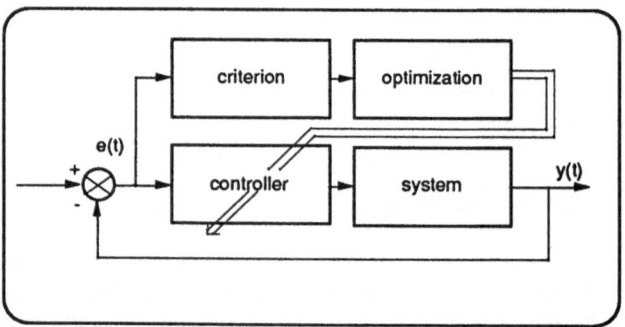

Figure 30.8. Controller design

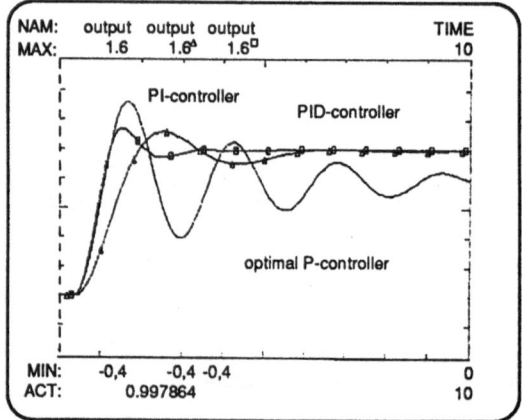

Figure 30.9. Responses of three "optimal" controllers

You, as designer, must select the structure of a controller. Then you can ask PSI to search for "optimal" values of the parameters of this controller in order to minimize a criterion between the reference and actual values of the output based on the error signal. In Figure 30.9, the "optimal" controllers are illustrated for a process, if a P-, a PI- or a PID-controller, was selected.

30.4 Command Language

Communication between you and PSI is realized by means of a command language. PSI asks you to give the next command, and when it is accepted by PSI, it will be executed immediately. Then, PSI will ask for a new command. It is allowable to give more than one command in one command string, but then the commands must be separated by a comma.

When an error occurs in a command string, PSI will indicate the error and ask for a new command string. However, sometimes only a part of the command string is illegal. Then, by means of a question and answer approach, PSI will ask for the missing information.

A short description of some basic commands (about 30) will be given in this section. The PSI manual contains a comprehensive description of all commands (about 100) available in PSI. The commands are divided into eight groups. These groups are:

1. Simulation-data definition statements (B, P, T, O)
2. Simulation-data definition commands (Dx)
3. Parameter show commands (Sx)
4. Signal show commands (Xx)
5. Control commands (C, R, Exit, Empty)
6. Model-save commands (Mx)
7. Function generator commands (Fx)
8. Miscellaneous commands (A, L, Plot, ?)

In starting PSI, or when a previous command has been executed, PSI is ready to receive a new command. The following text is written on the terminal:

PSI*

Now a command or a sequence of commands separated by commas can be typed. Each command string must be terminated by a carriage return.

The use of the sign (*) indicates that the associated numbers may be different in your implementation of PSI.

30.4.1   Simulation-Data Definition Statements

With the B, P, T and O commands you will jump to program parts in which the topological or block structure of the simulation model (B), the parameters of this model (P), the timing data (T) and the output representation (O) can be defined. In these local program sections you can only supply the required information. All other commands and other information is illegal.

B - Configuration statements

The configuration statement describes the topological or block structure of the simulation model. This goal is achieved by defining the block type and the inputs of each block. You are in the B-program section when PSI writes the text B* on your terminal. After this text you can insert a configuration statement. The configuration statement is:

    B*NAM,TYPE,NAM1,NAM2,NAM3

or equivalently

    B*NAM=TYPE(NAM1,NAM2,NAM3)

with:NAM - block name that represents both a block and the output of this block. This name must be unique. It may consist of up to 8 characters. Any character is allowed except for spaces, equal signs and commas.
TYPE - block type of three characters
NAM1 - name of block that is input 1
NAM2 - name of block that is input 2
NAM3 - name of block that is input 3

The number of inputs is defined by the block type. Some block types have no inputs, others have a variable number of inputs. A check is performed to test the number of inputs. If this check is not satisfactory, the configuration statement will not be accepted.

The name TIME can be used as an input. This name is assigned to the independent variable of the simulation, usually the time.

If a block name that has already been used is selected, the oldest block will be deleted and replaced by the new block.

Deleting a block name from the active blocks is achieved by just typing its name:

    B*NAM

An existing block can be modified by retyping only the modified input, while the other inputs are represented by commas:

    *NAM,TYPE,,NAM2,

Then the inputs NAM1 and NAM3 are not modified.

The configuration statements are employed to represent a parallel system. They may be defined in any order. The program uses a sorting procedure to determine a calculating sequence that takes into consideration the sequential calculation of each block in a digital computer. Owing to this calculating sequence, the parallel description of the system is maintained. Certain errors related to the configuration specification can be noted by this sorting procedure. A message will then explain the error, and suggest improvements to the description of the model. This sorting procedure becomes active when you terminate the configuration specification phase.

The configuration specification phase can be terminated by just pressing the carriage return.

    B*

P - Parameter statement

Parameter statements are employed to provide the simulation model with the parameter values associated with the block type, as introduced in the configuration statement. You are in the P program section when PSI writes the text P*. After this text you can insert one parameter statement. The parameter statement is:

    P*NAM,p1,p2,p3

with: NAM -   the name of the block, whose parameters have to be defined
      p1   -   first parameter
      p2   -   second parameter
      p3   -   third parameter

The number of parameters and their meanings depend on the block type of block NAM. A check is performed on the number of parameters and on their actual values. If one of these checks discovers illegal values, an error message will be generated.

With the following statement an arbitrary parameter can be redefined, while the other parameters remain unchanged:

    *NAM,,p2,

Only parameter p2 gets a new value while p1 and p3 remain unchanged. The parameter specification phase can be terminated by just pressing the carriage return.

    P*

T - *Timing data*

In this section the program will ask for the integration interval (T) of the numerical integration method and the total simulation time (TTOT). Both values are expressed in user-defined time units. They can be given by typing the appropriate values just after the texts:

    Integration interval =? ...

    Total time =? ...

Q - *Output representation data*

The print interval is assumed to be TOTAL TIME/200 (*).
    The blocks whose outputs have to be shown on the screen must be defined. The desired block names can be inserted just after the text:

    Names of blocks to be shown =? ... , ... , ... , ...

The outputs of the blocks defined in the previous statement are shown stored in the memories AT THE SAME TIME. The position of the block names in this statement determines in which memory the output will be stored. For example, the statement

    Names of blocks to be shown =? PRESSURE,,FLOW

will show both PRESSURE and FLOW on the screen and at the same time store PRESSURE in memory 1 and FLOW in memory 3. The content of the memories will be saved until they are redefined. So signals from earlier simulation runs can be saved to compare them with signals from the present or future runs.

30.4.2 Simulation Data Definition Commands

You can use a number of commands for directly defining some simulation variables. These commands avoid the use of questions and answers.

DIi
Selects the i-th integration algorithm.

i	name	order	fixed/variable
1	Euler	1	fixed step
2	Adams-Bashfort 2	2	"
3	Runge Kutta 2	2	"
4	Runge Kutta 4	4	"
5	Runge Kutta 4	4	variable step

When a method has been defined, it stays active until it is redefined. In starting PSI Runge Kutta 2 is active.

DSi min,max
Defines the scaling of memory i (i=1-6(*)).
By choosing i=A (DSA), all memories get the defined scaling (minimum, maximum).
Default values: (-4,1.6).

30.4.3 Show Commands

You can use a number of commands to show the actual values of a number of simulation parameters on the terminal.

SB NAM
Shows actual data of the block "NAM", namely block type, inputs, parameters and actual output value.

SF
Shows all function generator tables that are used by the simulation model.

SI
Shows the actual integration data, such as integration interval (T), integration method, total simulation time (TTOT), TIME and, with variable-step integration methods, the error boundaries for adjusting the step size of the integration interval.

SM
Shows the actual simulation model, both structure and parameters. The blocks are shown alphabetically according to block type. Blocks with the same block type are shown alphabetically according to block name.

SS
Shows the actual scaling, the names of the blocks to be stored, the names of the blocks stored in the memories and the color assigned to each memory.

SX

Shows the maxima of this version of PSI, namely the maximum number of blocks, integrators, function generator, tables, memories, dead times and characters for the title (A-command).

30.4.4 Signal Show Commands

These commands enable you to study the signals stored in the memories. An active memory is defined by means of the O-command.

XA

Shows the signals stored in the active memories on the screen with an initialization of the screen shown first.

XTi

Shows a signal stored in memory i on the screen as a function of the time.

XX

Shows all signals stored in the active memories on the screen without initialization of the screen.

A negative number for the Y-axis (i with XT) clears the screen before the required response is shown. With a positive memory number the response is added to the already existing responses. The scaling can be adjusted with the DSi command.

Once a response is shown on the screen, a cursor will appear. This cursor can be moved forward along this response by typing, behind the text X*, a positive number followed by a carriage return and backwards by typing a negative number followed by a carriage return. An extra carriage return repeats the last command. Not only the cursor position, but also the numerical values of the point indicated by the cursor are shown on the screen.

If you use the XT command, only the numerical values of the indicated responses are shown. If you use the XA or the XX command, then the numerical values of the signals stored in all active memories are shown on the screen.

In order to leave this display phase, indicated with X*, give the L(eave) command (so X*L). You can also give any other PSI command. Then this command will be executed immediately.

30.4.5   Control Commands

You can use the following run-control commands to control a simulation run:

C
Continues a simulation where it was stopped previously. Neither the TIME nor the blocks nor the display are initialized. This simulation run continues until TIME=TTOT. If the command is given when TIME=n.TTOT (n=1,2,3...), the simulation continues until (n+1).TTOT. In this last situation the screen is partly cleared.

R
Starts the simulation run. The outputs of all blocks with an initial condition are set to their defined value (p1-parameter). The independent variable of the simulation TIME is reset to zero. The display is initialized. The simulation continues until TTOT, the final response time.

↵ (carriage return)

A carriage return stops any output from PSI to the terminal, the line printer or the screen. A simulation run that has been stopped in this way can be continued by means of the C command.

EXIT
Stops a PSI session and returns control to the operating system.

EMPTY
Clears all simulation data and restarts PSI.

30.4.6   Model Save Commands

You can save all simulation data (thus simulation model, parameters, function-generator tables, scaling, etc.). All these data can be written on disk in order to have them available for later use. The general command is:

Mx NAM
with: x   - determines the action
        NAM   - name of the file

You can choose this filename NAM arbitrarily. However, it must contain only alphanumeric characters, and the first character has to be an alphabetic one. The maximum number of characters is 8. Some operating systems allow you to include a device name in the filename. For example, with MS-DOS you can use the filenames A:BOILER or B:SHIP, where A: and B: define the required device.

MR NAM
Reads the file NAM.

MS
Shows the names of all stored files. This command has no filename associated with it.

MW NAM
Writes the simulation model to file NAM.

30.4.7 Function-Generator Commands

You can fill the function-generator tables with the F command. After receiving the F command, PSI will return with the text F*. After this text you have to type the required table number j, followed by all data points that belong to that table. These data points have to be separated by commas. If these points cannot be typed on one line, just continue on the next line, i.e.:

```
F*j,f1,f2,f3, ....,        fj
F*fj+1, ...........,       fk
F*fk+1, ...........,       fn
```

with: j as table number (11,12,51,52,101,102,103, .. , 106(*))
        fi function value

You can use the function-generator commands to read from or write the contents of the memories to disk in order to read/save these signals for later use. The filename NAME must be a legal one. Consult the explanation of legal filenames in section 30.4.6 dealing with the model-save commands. The memory number is from 1 to 6(*).

FRi NAM
Reads file NAM into memory i.

FWi NAM
Writes memory i to file NAM.

30.4.8 Miscellaneous Commands

Atext
Puts all text typed directly after A above each output (max. 60 (*) characters; the comma is not allowed as a character). This command can be used for documentation purposes.

L
Directs all output to the line printer instead of the terminal.

PLOT
Plots contents of the graphical screen on the printer or plotter.

?B
Shows all block types available in PSI.

?B TYP
Shows data concerning block type TYP.

?
Shows titles of ten sections of the commands implemented in PSI.

?i
Shows all commands of section i (i=1,10).

30.5   Description of Some Facilities

30.5.1 Memories

There are six (*) memories available in PSI. During a simulation run, when the output is directed to the screen of the display, the signals shown on the screen are stored in these memories. These signals are saved until the memories are used again. Each memory can contain a maximum of 201 (*) samples. With the O command you can define which blocks are to be stored in which memory. These memories then become a so-called active memory. The memories can be used for studying the stored signals after a simulation run with the aid of the XT, XX or XA command.

Another application of the memories is to use signals, calculated and stored in a preceding run, as inputs in a future run. Then the 201 (*) points of the memory are interpreted as a table for a function generator. By using a function generator with the appropriate values for its parameters p1, p2 and p3 (e.g. p1=0, p2=TTOT, p3=101, 102, 103 or 104 for indicating memory 1, 2, 3 or 4, respectively) and TIME as input, this function generator will reconstruct the original signal stored in the memory with the aid of linear or quadratic interpolation. In case no interpolation is required, and only the values of the 201 (*) points are of interest, a sample-hold (SPL) block can be used, together with a timer block (TIM) with period TTOT/100 (*).

With the FW command you can write the contents of a selected memory to disk to save the contents for later use. With the FR command you can read the contents of a memory from disk. This file can also be created with another program so you can read an input and output signal of a real system for identification.

30.5.2   Function Generators

Each function generator has three parameters. Parameters p1 and p2 define the interval of the input for which interpolation is applied. Parameter p3 is a pointer to a table with function values. Consequently, the (absolute) value of p3 has to be a legal number of a function generator table. By this approach many function generators can share the same table. The function values stored in the tables are assumed to belong to equi-spaced values of the input. These tables can be defined by the F command or filled via the DF command. In this section we assume there are:

- 2 (*) tables with 11 points
- 2 (*) tables with 51 points
- 6 (*) memories, each with 201 (*) points

These numbers can be changed in your actual implementation of PSI . Consult the actual numbers with the SX command.

The following numbers are used to distinguish among the different function-generator tables.

Table numbers 11 and 12 point to tables with 11 function values.
Table numbers 51 and 52 point to tables with 51 function values.
Table numbers 101, 102, ... 106 point to the memories 1, 2, ... 6 respectively with 201 function values.

The one-dimensional tables use either linear or quadratic interpolation and extrapolation. If the third parameter of a function generator (FNG block) is positive, then linear interpolation is used. If this third parameter is negative then quadratic interpolation and extrapolation is applied. The input is interpolated between p1 and p2. Consequently, if the input u equals p1, then the output of the function generator equals f1; if the input equals p2, then the output equals either f11, f51 or f101.

Quadratic interpolation is defined with three points, namely one preceding and two succeeding points between p1 and p2. Consequently, linear interpolation has to be applied between the function values 10, 11 and 12 (tables 11, 12 and 13) 50, 51 and 52 (tables 51, 52 and 53) and between function values 100 ... 106 (tables 101 ... 106). Figure 30.10 illustrates the interpolation and extrapolation of the function generators for tables with 51 points. The table with 11 and 201 points are treated in the same way.

30.5.3   Automatic Program Execution

In PSI two modes can be recognized, namely the interactive mode and the automatic execution mode. The first one is the one described in this manual and is the normal operating mode. The second mode is only used if the commands to control PSI are given in a file called AUTOEX.PSI and this file is present in the PSI default directory.

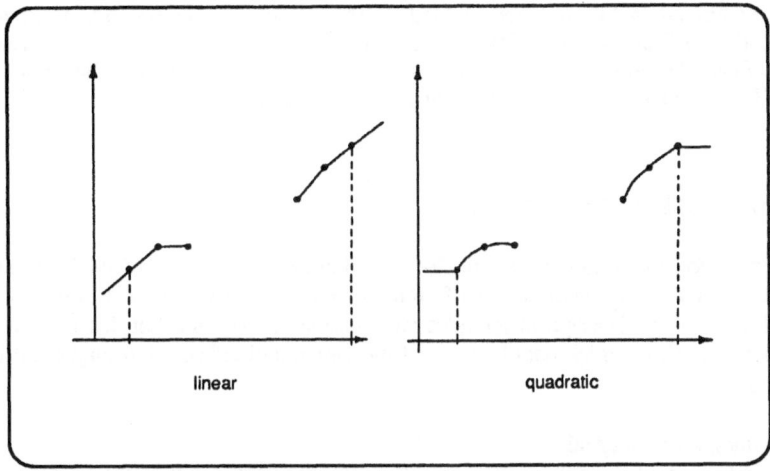

Figure 30.10.   Interpolation and extrapolation for function generators

When you start PSI, it will first search for a file with name AUTOEX.PSI. If this file does not exist, you have access to all commands and facilities of PSI. However, if this file exists, it will take over control of PSI. This file must contain all the commands, as you would have given them, after which PSI will execute them consecutively. If all commands stored in AUTOEX.PSI have been used, you will get control over PSI again.

AUTOEX.PSI is a formatted ASCII file with records of max length=80 characters. So you can pack a maximum of 80 characters in one command line.

Example:
Suppose you want to read the PSI model "TWEEDE.PS6" from disk, change the second parameter of block D to the value -.5 and R(un) it. Then you want to inspect the response with the cursor (XT1), after which you want to see the model (SM). The following sequence of commands, as stored in AUTOEX.PSI will realize these goals:

```
MR TWEEDE
Y              Don't forget this answer!!!!
R
P
D,,-.5,

RX
XT1
SM
WAIT
EXIT
```

The command WAIT has been added. If PSI detects this command, program execution will be stopped. You can continue if you hit any key. The command WAIT can be used as many times as you like. It is not necessary to finish the file with EXIT. If EXIT is not present, you can continue with PSI.

30.6 Block Type Definitions

A description will be given of about 20 most-often used block types available in PSI. A complete description of all 60 different block types is given in the PSI manual. Each block type description includes an equation or diagram that describes the operation performed by that block. The following notation is used throughout this section:

T - integration interval
t - independent variable of the simulation. In general this will be the time.
k - number of integration intervals. Calculations are performed at a finite
      number of calculation points, namely for $t = k \cdot T$, k=0,1,2,3,4, ...
y - value of the output
i_j - value of the j-th input; j=1, 2 or 3
p_j - value of the j-th parameter; j=1, 2 or 3
u - sum of all available inputs, so $u = i1+i2+i3$

Moreover, the number of the inputs, the number of the parameters and the allowable value of each parameter is given.

A block will be represented by a drawing and a mathematical description of the relation between inputs, parameters and output. In the drawing, for example Figure 30.11, inputs above the block represent initial conditions, inputs below the block represent control signals and inputs left of the block represents input signals that directly determine the value of the block output.

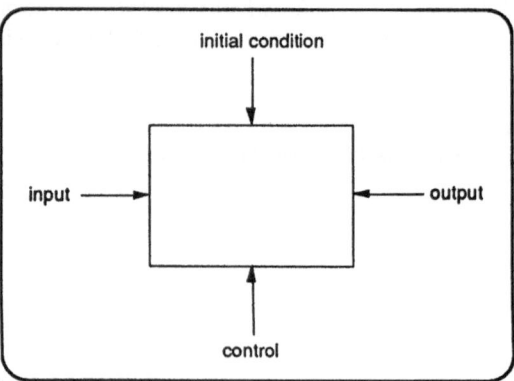

Figure 30.11. General block

The use of the (*) sign indicates that the associated numbers may be different in your implementation of PSI. Consult the SX command.

ABS
Absolute value

> y = p1*ABS(u)

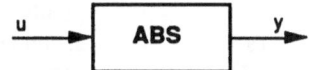

> 1-3   inputs
> 1   parameter

ADD *Adder*

> y = u = i1 + i2 + i3

> 1-3   inputs
> 0   parameters

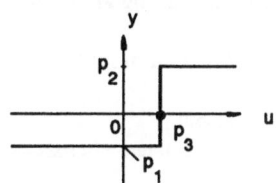

Adder

BNG *Bang - Bang*

> If   u >= p3   y = p2
>   u < p3   y = p1

> 1-3   inputs
> 3   parameters; p1<=p2

Bang-Bang

CON *Constant*

> y = p1

> 0   inputs
> 1   parameter

Constant

DAI - *data input from disk*

> y(o) = p1
> y(k) = value of element k from data file p2 on disk

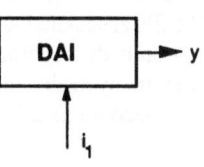

> if i1>0, then a new sample is taken from this data file
> if i1<=0, then the old value of y(k) is maintained

The file name p2 has the following shape:

if p2=1	then filename becomes FIL001.ASC
if p2=100	then filename becomes FIL100.ASC
p2 has to satisfy:	$1<=p2<=499$.

An *.ASC file has (N+2) records and is written as follows (sequential, formatted):

N	number of samples (integer I13)
Ts	sample time (real E13.6)
x(1)	sample 1 (real E13.6)
...	...
x(N)	sample N (real E13.6)

With both the DAI and DAO blocks, the value of the sample time is not very meaningful. Still, it is included because it makes these files compatible with the ASCII-files produced by TRIPSI. Therefore, you can insert its value as parameter p1.

If there are no samples left in the file (reading beyond the end of the file), the last valid data element will be used to define the value of y(k). If you use the same file in two succeeding runs, you have to close it after each use with the DCL command. The C-command in combination with a DAI block will continue at read in the file without starting at the first record.

DAO - Data output to disk

1 input

2 parameters $1<=p2<=499$

$i_2 \rightarrow$ DAO $\uparrow i_1$

If i1>0, then the value of i2 will be written to disk as an element of file p2. Moreover, the total number of elements written to this file and the sample time (p1) are also added to this file. All open data files are closed with the DCL command, which makes them available for future use.

Consult the description of the shape of the filename p2, which depends on the value of p2 as given with the DAI block. If p2=5, the filename will be FIL005.ASC.

Remark:

It is not allowable to use both a DAI and a DAO block with the same file name p2 in one simulation run. Still, you can implement up to 80 DAI and/or DAO blocks in one simulation model as long as they do not use the same filename p2.

The file associated with a DAO block is only closed by using the DCL command. If you omit this command the file will not be created. However, if you leave the program by means of the EXIT command, an intermediate file named FIL5p2.ASC will become available. For example, if you try to create FIL002.ASC, the intermediate file is called FIL502.ASC. Both files differ only by the first entry – the number of samples.

DIV Divider

    y = i1/i2
    if ABS(i2) < 1.E-30 then if(i1=0) y=0
    else y = sign(i1/i2) * 1.E30.

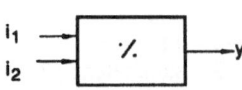

    2   inputs
    0   parameters   Divider

FNG Function generator

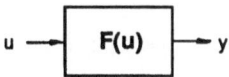

    y = F(u)   Function generator

    1-3   inputs
    3   parameters; p1<p2, p3 = +/- 11-12,51-52,101-106(*)

Function:
A FNG block uses a table to calculate a corresponding value of the output y for each value of the input u. The output is calculated by using interpolation or extrapolation for the input u, depending on the specified interval (p1,p2). The table is specified by the table number p3. The points in a table are assumed to be spaced at equal distances between p1 and p2. The following tables are supported:

    p3=11,12(*)   a table with 11 points
    p3=51,52(*)   a table with 51 points
    p3=101-106(*)   a table with 201(*) points

Linear interpolation is assumed for positive values of p3. With negative values for p3, quadratic interpolation is applied.

GAI Gain

    y = p1*u

                             u ⟶ (P₁)⟶ y

    1-3   inputs
    1   parameter   Gain

INF *Integrator*, pictured as a first order system

$$Y(s) = \frac{K}{s\tau + 1} \cdot U(s)$$

$y(0) = p1$
$K = p2$
$\tau = p3$

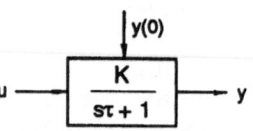

First-order system

1-3   inputs
3   parameters; p3 = 0

INL *Limited integrator*

$$y = p1 + \int u\, dt$$

Function:
The output of the internal state and so
the output is limited between
$p2 \le y \le p3$

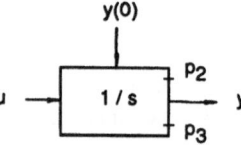

Limited integrator

1-3   inputs
3   parameters; p2 <= p1 <= p3

INT *Integrator*

$$y = p1 + \int (p2 * i1 + p3 * i2 + i3)\, dt$$

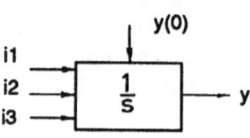

Integrator

1-3   inputs
2-3   parameters

LIM *Limiter*

$y = p3 * u$
$y = p1$ if y<p1
$y = p2$ if y>p2

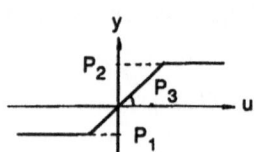

Limiter

1-3   inputs
3   parameters; p1<=p2, p3>

MUL Multiplier

$$y = i1 * i2 * i3$$

2-3 inputs
 0 parameters

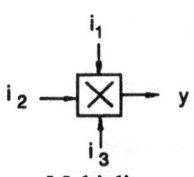

Multiplier

OFS Offset

$$y = p1 * (u + p2)$$

1-3 inputs
 2 parameter

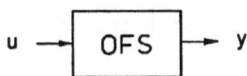

Offset

POL Polynomial

$$y = p1 * u ** 3 + p2 * u ** 2 + p3 * u$$

1-3 inputs
 3 parameters

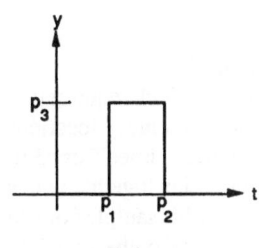

Polynomial

PLS Pulse

$$y = 0 \quad \text{if} \quad t<p1 \text{ or } t>p2$$
$$y = p3 \quad \text{if} \quad p1<=t<=p2$$

 0 inputs
     parameters 0<=p1<=p2

Pulse

SQT Square root

If   u>=0 $y = p1 * \sqrt{u}$
      u<0 $y = p1 * \sqrt{-u}$

1-3 inputs
 1 parameter

Square root

SUB Subtractor

$$y = i1 - i2$$

2   inputs
0   parameters

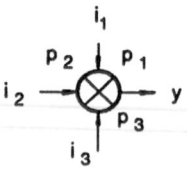

Subtractor

SUM Summer

$$y = p1 * i1 + p2 * i2 + p3 * i3$$

1-3   inputs
1-3   parameters

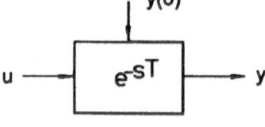

Weighted summer

TDE Dead time/Time delay

$$y(t) = p1 \qquad t < p2$$
$$y(t) = u(t - p2) \qquad t >= p2$$

1-3   inputs
  3   parameters; p2 > 0, T<=p3<=p2

Dead time/Time delay

Remark:
A TDE block requires p2/MAX(p3,T)+1 memory locations. There is a maximum of 250(*) memory locations for up to 10(*) TDE blocks. When more locations are required, either T or p3 has to be increased. This time delay can be considered as a time delay together with a sample and hold with sample time p3 in front of it. With p3=T, this sample hold does not affect the delay action. Then, for each T time unit a new value of the input is taken and stored and after p2 time units it is put at the output of the block. With p3>T a sample is taken each p3 time unit, so that longer but less accurate time delays can be realized. As a consequence of rounding errors the dead time, realized by the TDE block, may be less than p2, because the real dead time DT is calculated as

$$DT = IFIX(p2/p3) * IFIX(p3/T) * T$$

WARNING:
The use of TDE blocks together with variable step-size integration methods will produce erroneous results.

_ZZ1_ *Unit delay*

$$y(k+1) = p2 * i1(k)+p3 * i2(k)+i3(k)$$

$$y(0) = p1$$

1-3   inputs
2-3   parameters

Unit delay

Remark:
The ZZ1 block is only calculated at the calculation points t=k * T.
WARNING: The use of ZZ1 blocks together with variable step-size methods will produce erroneous results.

References

Bosch, P.P.J. van den: "PSI - A Software Tool for Control System Design". *Journal A*, Vol 22, no 2, 1981, pp. 55-62.

Bosch, P.P.J. van den: "Interactive System Analysis and System Design Using Simulation and Optimization". *Proceedings IFAC Symposium on Computer Aided Design of Multivariable Technological Systems 1982*, Lafayette, Pergamon Press London, 1983, pp. 225-232.

Bosch, P.P.J. van den: "Interactive Computer-Aided Control System Analysis and Design". Ed.: M. Jamshidi, C.J. Herget, In: *Advances in Computer-Aided Control Systems Engineering*, North-Holland Publishing Company, 1985, pp. 229-242.

Bosch, P.P.J. van den: "PSI-Interactive Simulation Program".
In: Advances in Computer Aided Control Systems Engineering, Eds.: M. Jamshidi and C.J. Herget, North-Holland Publishing Company, 1985, pp. 362-364.

Bosch, P.P.J. van den and G. Shuling: "Evaluation of Computer Systems for Control System Design". *Proceedings IFAC Symposium on Computer-Aided Design of Control and Engineering Systems*, 1985, Lyngby, Pergamon Press London, pp. 110-115.

Bosch, P.P.J. van den and W.J.M. van Geest: "Computer-Aided Control System Design with mini- and microcomputers". *Special issue on Computer-Aided Design of the Transactions of the Institute of Measurement and Control*, Vol 7, no 2, 1985, pp. 71-77.

Bosch, P.P.J. van den et.al.: "Simulatie". *Cursus 42.20 PBNA*, Arnhem, 1985 (178 pages, in Dutch).

Bosch, P.P.J. van den: *PSI manual.* (98 pages), 1987. Boza Automatisering, POBox 113, NL2640, Pijnacker, the Netherlands,
ISBN 90-71898-01-6.

Linkens, D.A., R.J. Rimmer and P.P.J. van den Bosch: "muPSI, A Microcomputer Simulation Language with Computer Control Capabilities". *Transactions of the Institute of Measurement and Control*, Vol 6, no 2, 1984, pp. 58-66.

Index